Water Resources Assessment, Modelling and Management

PREFACE

Water is one of the most vital resources on Earth and plays a crucial role in sustaining life. With the rapid increase in population and urbanization, the demand for water has been increasing exponentially, leading to an immense pressure on the existing water resources. In this context, the assessment, modelling, and management of water resources have become crucial to ensure sustainable development.

This book, "Water Resources Assessment, Modelling and Management," is a collection of 96 articles that delve into various topics related to water resources. It covers various aspects related to surface water, groundwater, surface water modelling, groundwater modelling, management of water resources, challenges and strategies, advanced techniques in water resources, thrust areas. The book provides a special emphasis on water resources issues and management in India, which is crucial due to the significant water-related challenges faced by the country.

The articles in the book offer a broad overview of the latest techniques, tools, and strategies used in the assessment, modelling, and management of water resources. Each article is concise, with a length of around three pages, and provides a brief yet informative summary of the respective topic. The aim of the book is to provide readers with a quick and accessible overview of each topic, without delving too deeply into technical details or using equations. To keep the content straightforward and easy to understand, the articles do not contain equations. However, readers who wish to explore a topic in more depth are encouraged to consult other relevant books and resources, where they can find more comprehensive information and mathematical formulations.

The book aims to be a valuable resource for water resources professionals, researchers, and students who are interested in understanding the challenges associated with water resources and the strategies for their management. It covers a wide range of topics, including the latest techniques and tools used in water resources management, and the impact of climate change on water resources.

Overall, the book provides a comprehensive overview of the current state of knowledge and practice in the assessment, modelling, and management of water resources. The book is expected to serve as a useful reference for anyone interested in this important and timely topic.

C. P. Kumar
Former Scientist 'G'
National Institute of Hydrology
Roorkee - 247667, India

E-mail: cpkumar@yahoo.com
Web: https://www.angelfire.com/nh/cpkumar/

CONTENTS

1 - Water Resources Management in India: Challenges and Initiatives

Abstract

This article explores the challenges and initiatives in water resources management in India. The uneven distribution of water resources, population growth, and climate change make water management in India a complex and challenging task. The article discusses various policies and programs implemented by the Indian government to manage water resources effectively, including the National Water Policy, Namami Gange, Jal Shakti Abhiyan, National Groundwater Management Improvement Scheme, and Swachh Bharat Abhiyan. The challenges in water resources management in India, such as the lack of awareness, poor infrastructure, policy implementation, groundwater over-exploitation, and climate change, are also discussed. The article concludes that effective water resources management in India requires a multi-pronged approach involving various stakeholders and the adoption of sustainable water management practices.

Introduction

India is a country with a large population and an agrarian economy, which means it has a high demand for water resources. India is also home to many large rivers, such as the Ganges, Brahmaputra, and Indus, which are the lifeblood of the country's agricultural sector. However, despite these abundant water resources, India faces several water management challenges due to the uneven distribution of water resources, population growth, and climate change. In this article, we will discuss the water resources issues and management in India.

Water Resources in India

India has a diverse range of water resources, including rivers, lakes, groundwater, and rainfall. According to the Central Water Commission, India has an annual precipitation of about 4,000 billion cubic meters (BCM), but only 1,869 BCM is available for use due to evaporation, infiltration, and other losses. India has 20 major river basins, and the Indus, Ganga, and Brahmaputra river basins are the largest in the country, accounting for 80% of the surface water resources.

India also has significant groundwater resources, with an estimated annual recharge of 433 BCM. However, over-exploitation of groundwater has led to a decline in groundwater levels in many parts of the country. According to the Central Ground Water Board, 1093 out of 6605 groundwater assessment units in the country are over-exploited, and 253 units are critical.

Water Resources Issues in India

Despite having abundant water resources, India faces several water management challenges, which are discussed below.

Uneven Distribution of Water Resources

India's water resources are unevenly distributed in terms of time and space. The country receives most of its rainfall during the monsoon season, which lasts for only four months. As a result, many parts of the country face water scarcity during the non-monsoon months. Additionally, the water resources are not evenly distributed spatially, with some regions having more water resources than others. For example, the northern region of India has more water resources than the western and southern regions.

Population Growth

India is the second-most populous country in the world, with a population of over 1.3 billion. The population growth rate in India is also high, with an average annual growth rate of 1.2%. The high population growth rate has put significant pressure on water resources, leading to over-exploitation of groundwater and surface water resources.

Climate Change

Climate change is another significant challenge that India faces in water resources management. The country is vulnerable to climate change, with increased variability in precipitation, extreme weather events, and rising sea levels. These changes in the climate can have severe impacts on water resources, leading to droughts, floods, and water scarcity.

Poor Water Quality

Another significant issue in water resources management in India is poor water quality. Many water bodies in the country are polluted, leading to health hazards and environmental degradation. Industrial effluents, untreated sewage, and agricultural runoff are some of the major sources of water pollution in India.

Water Resources Management in India

The Indian government has implemented several policies and programs to manage water resources in the country. Some of the key initiatives are discussed below.

National Water Policy

The National Water Policy of 2012 is the latest water policy of the Indian government. The policy aims to ensure the sustainable development and management of water resources in the country. The policy emphasizes the need for an integrated approach to water resources management, including the participation of all stakeholders.

National River Conservation Plan

The National River Conservation Plan is a flagship program of the Indian government to improve the water quality of rivers in the country. The program aims to reduce pollution in rivers and promote their ecological health. Under the program, several activities are undertaken, such as the construction of sewage treatment plants, the cleaning of riverbanks, and the implementation of effluent standards for industries located along the rivers.

Jal Shakti Abhiyan

Jal Shakti Abhiyan is a water conservation campaign launched by the Indian government in 2019. The program aims to improve the water situation in 256 water-stressed districts in the country. Under the program, several activities are undertaken, such as the construction of water harvesting structures, desilting of water bodies, and renovation of traditional water bodies.

National Groundwater Management Improvement Scheme

The National Groundwater Management Improvement Scheme is a program launched by the Indian government to improve the management of groundwater resources in the country. The program aims to promote sustainable groundwater management practices, including the adoption of water-efficient technologies, the promotion of recharge structures, and the regulation of groundwater extraction.

Swachh Bharat Abhiyan

Swachh Bharat Abhiyan is a sanitation program launched by the Indian government in 2014. The program aims to improve the sanitation facilities in the country, including the provision of toilets and the construction of sewage treatment plants. The program has a significant impact on water resources management, as it helps to reduce the pollution of water bodies.

Challenges in Water Resources Management in India

Despite the initiatives taken by the Indian government, several challenges remain in water resources management in the country. Some of the key challenges are discussed below.

Lack of Awareness

One of the major challenges in water resources management in India is the lack of awareness among the public. Many people in the country are unaware of the importance of water conservation and the need to adopt sustainable water management practices.

Poor Infrastructure

Another significant challenge in water resources management in India is the poor infrastructure. Many parts of the country lack basic water supply and sanitation facilities, leading to water scarcity and poor water quality.

Policy Implementation

The implementation of water policies and programs in India is often inadequate due to various reasons such as bureaucratic hurdles, lack of funds, and corruption. This hampers the effective management of water resources in the country.

Groundwater Over-Exploitation

The over-exploitation of groundwater is a significant challenge in water resources management in India. Many farmers in the country depend on groundwater for irrigation, and the overuse of groundwater leads to declining groundwater levels.

Climate Change

Climate change is another significant challenge in water resources management in India. The changing weather patterns and extreme weather events can lead to droughts, floods, and water scarcity, which can further exacerbate the water management challenges in the country.

Conclusion

Water resources management in India is a complex and challenging task, given the uneven distribution of water resources, population growth, and climate change. However, the Indian government has implemented several policies and programs to manage water resources effectively. The success of these initiatives depends on their proper implementation, effective public participation, and the adoption of sustainable water management practices. The challenges in water resources management in India require a multi-pronged approach involving various stakeholders, including the government, civil society organizations, and the private sector. With the right policies and initiatives, India can effectively manage its water resources and ensure sustainable development.

2 - Rivers of India: A Comprehensive Overview

Abstract

Rivers are an integral part of Indian culture, economy, and geography. The river system of India is vast and complex, with many rivers originating from the Himalayas and flowing through the plains before emptying into the Indian Ocean. In this article, we provide a comprehensive overview of the major rivers of India, their origins, courses, and significance. We also discuss the challenges faced by India's rivers and the need to protect and conserve them for future generations. This article aims to increase awareness and understanding of the importance of India's rivers and their role in the country's development and culture.

Introduction

India is a land of rivers. The country is blessed with an extensive river system that has been instrumental in shaping its geography, economy, and culture. The river system of India comprises many large and small rivers, streams, and tributaries that are spread across the length and breadth of the country. These rivers play a significant role in the lives of millions of people who depend on them for various purposes, such as irrigation, transportation, and power generation.

In this article, we will explore the river system of India in detail, covering its importance, classification, major rivers, and their significance.

Importance of Rivers in India

Rivers are an integral part of Indian civilization and have been revered since ancient times. They are considered sacred and worshipped as deities by many communities. The importance of rivers in India can be understood from the fact that many of the major cities in the country have developed along the banks of rivers.

Rivers play a crucial role in the economy of India. They are a major source of irrigation for agriculture, which is the backbone of the Indian economy. Many industries such as textiles, paper, sugar, and chemicals also depend on rivers for their water needs. Rivers also provide an important mode of transportation for people and goods.

The river system of India is also a major source of power generation. India has a significant hydroelectric power generation capacity, and many of the major dams and hydroelectric power stations in the country are built on rivers.

Classification of Rivers in India

The river system of India can be broadly classified into two categories:

Himalayan Rivers

The Himalayan rivers originate from the glaciers in the Himalayas and flow down to the plains. These rivers are perennial and are fed by snowmelt and rainfall. They are characterized by high velocity, steep gradient, and turbulent flow. The Himalayan rivers are further divided into two categories:

a. Indus River System

The Indus River System is the largest river system in the region and originates from the Tibetan Plateau in China. The river flows through India, Pakistan, and Afghanistan before emptying into the Arabian Sea. The Indus River System comprises many tributaries, including the Jhelum, Chenab, Ravi, Beas, and Sutlej rivers.

b. Ganga-Brahmaputra-Meghna River System

The Ganga-Brahmaputra-Meghna River System is the most extensive river system in India and originates from the Himalayas. The Ganga, which is the longest river in India, originates from the Gangotri Glacier in Uttarakhand and flows through the northern plains before emptying into the Bay of Bengal. The Brahmaputra River originates from Tibet and flows through Assam before emptying into the Bay of Bengal. The Meghna River is a major tributary of the Brahmaputra and flows through Bangladesh.

Peninsular Rivers

The Peninsular Rivers originate from the Western Ghats and the Eastern Ghats and flow towards the east and west coasts. These rivers are non-perennial and are fed by rainfall. The Peninsular Rivers are further divided into two categories:

a. East Flowing Rivers

The East Flowing Rivers originate from the Western Ghats and flow towards the Bay of Bengal. The major rivers in this category include the Godavari, Krishna, Mahanadi, and Cauvery.

b. West Flowing Rivers

The West Flowing Rivers originate from the Western Ghats and flow towards the Arabian Sea. The major rivers in this category include the Narmada, Tapi, and Sabarmati.

Major Rivers in India

India is home to many major rivers, and some of them are listed below.

Ganga River

The Ganga River is the most important river in India and is considered sacred by Hindus. It is also the longest river in India, stretching for 2,525 kilometers. The Ganga originates from the Gangotri Glacier in Uttarakhand and flows through the northern plains of India, passing through the states of Uttar Pradesh, Bihar, and West Bengal, before emptying into the Bay of Bengal. The river is known for its rich biodiversity and is home to many aquatic species, including the endangered Gangetic Dolphin.

The Ganga River is also a major source of irrigation for agriculture, with many large dams and canals built along the river for this purpose. The river is also used for transportation, with many large cities such as Kolkata and Varanasi located along its banks.

Brahmaputra River

The Brahmaputra River is the second-largest river in India and originates from Tibet. The river flows through the northeastern state of Assam before emptying into the Bay of Bengal. The Brahmaputra is a major source of water for irrigation, with many large dams and hydroelectric power stations built along its banks. The river is also home to many aquatic species, including the endangered Gangetic River Dolphin.

Yamuna River

The Yamuna River is a major tributary of the Ganga River and is one of the most polluted rivers in India. The river originates from the Yamunotri Glacier in Uttarakhand and flows through the states of Delhi, Haryana, and Uttar Pradesh before merging with the Ganga River in Allahabad. The river is a major source of water for irrigation and is also used for transportation.

Godavari River

The Godavari River is the second-longest river in India and originates from the Western Ghats in Maharashtra. The river flows through the states of Maharashtra, Telangana, Andhra Pradesh, and Odisha before emptying into the Bay of Bengal. The Godavari River is a major source of irrigation for agriculture and is also used for hydroelectric power generation.

Krishna River

The Krishna River is one of the major rivers in India and originates from the Western Ghats in Maharashtra. The river flows through the states of Maharashtra, Karnataka, and Andhra Pradesh before emptying into the Bay of Bengal. The Krishna River is a major source of water for irrigation and is also used for hydroelectric power generation.

Narmada River

The Narmada River is one of the most important rivers in India and originates from the Amarkantak Plateau in Madhya Pradesh. The river flows through the states of Madhya Pradesh, Gujarat, and Maharashtra before emptying into the Arabian Sea. The Narmada River is a major source of water for irrigation and is also used for hydroelectric power generation.

Tapi River

The Tapi River originates from the Western Ghats in Maharashtra and flows through the states of Maharashtra and Gujarat before emptying into the Arabian Sea. The river is a major source of water for irrigation and is also used for hydroelectric power generation.

Significance of Rivers in Indian Culture

Rivers have been an integral part of Indian culture since ancient times. Many rivers in India are considered sacred, and people worship them as deities. The Ganga River is considered the holiest river in India, and many pilgrims travel long distances to bathe in its waters, which are believed to purify the soul.

Rivers have also played a significant role in Indian literature and art. Many famous Indian epics such as the Ramayana and Mahabharata have references to rivers, and many Indian paintings depict rivers and their associated myths and legends.

Conclusion

The river system of India is a vital component of the country's economy, culture, and geography. The rivers provide a source of water for irrigation, hydroelectric power generation, and transportation. They also support rich biodiversity and are home to many aquatic species.

However, many of India's rivers are facing significant challenges, including pollution, overuse, and damming, which threaten their long-term sustainability. It is important to take steps to protect and conserve India's rivers to ensure their continued use and preservation for future generations.

In recent years, the Indian government has launched several initiatives to clean up and protect the country's rivers, including the Namami Gange program aimed at cleaning up the Ganga River. There is also a need for greater awareness among the public about the importance of protecting and conserving the rivers.

3 - Hydrogeology of India: Aquifer Systems, Groundwater Resources, and Management Challenges

Abstract

India is a country with diverse geological, hydrological and climatic conditions, making it complex for hydrogeologists to study groundwater occurrence, movement, and quality. India's aquifer systems can be classified into unconsolidated and consolidated types, and recharge sources include precipitation, surface water, and lateral flow. India is the largest user of groundwater in the world, with about 25% of the global groundwater extraction, and groundwater resources are used for a variety of purposes. However, the overexploitation of groundwater resources has resulted in challenges related to its use and management, including depletion, deterioration of water quality, climate change, lack of governance, and inefficient irrigation practices.

Introduction

India is a country with diverse geological, hydrological and climatic conditions. The country has a long history of groundwater use for domestic, agricultural and industrial purposes. Hydrogeology is the science of water in the subsurface, which includes the study of groundwater occurrence, movement, and quality. In this article, we will discuss the hydrogeology of India, including its aquifer systems, groundwater resources, and challenges related to groundwater use and management.

Geological Setting

The geological setting of India is complex and varied, with different rock types and geological structures. The country is divided into six major physiographic regions: the Himalayas, the Indo-Gangetic Plain, the Peninsular Plateau, the Western Ghats, the Eastern Ghats, and the Coastal Plains and Islands. The Himalayan region is characterized by young folded mountains with a variety of rock types, including metamorphic, sedimentary, and volcanic rocks. The Indo-Gangetic Plain is a vast alluvial plain that covers a significant portion of northern and eastern India. The Peninsular Plateau is an ancient block of stable continental crust that covers most of southern and central India. The Western and Eastern Ghats are mountain ranges that run parallel to the west and east coasts of India, respectively. The Coastal Plains and Islands are low-lying areas along the coastlines of India and its islands.

Aquifer Systems

The aquifer systems in India can be classified into two broad categories: unconsolidated and consolidated. Unconsolidated aquifers are composed of loose sediments such as sand, gravel, and clay. These aquifers are found in the alluvial deposits of major rivers, coastal plains, and deltaic regions. Consolidated aquifers, on the other hand, are composed of hard rock formations such as granite, basalt, limestone, and sandstone. These aquifers are found in the Peninsular Plateau, the Western and Eastern Ghats, and the Himalayan region.

The aquifer systems in India can also be classified based on their recharge sources. The three main sources of groundwater recharge in India are precipitation, surface water, and lateral flow. Precipitation recharge occurs when rainwater infiltrates the ground and replenishes the aquifers. Surface water recharge occurs when surface water bodies such as rivers, lakes, and ponds recharge the aquifers. Lateral flow recharge occurs when water moves laterally from high rainfall areas to low rainfall areas and recharges the aquifers.

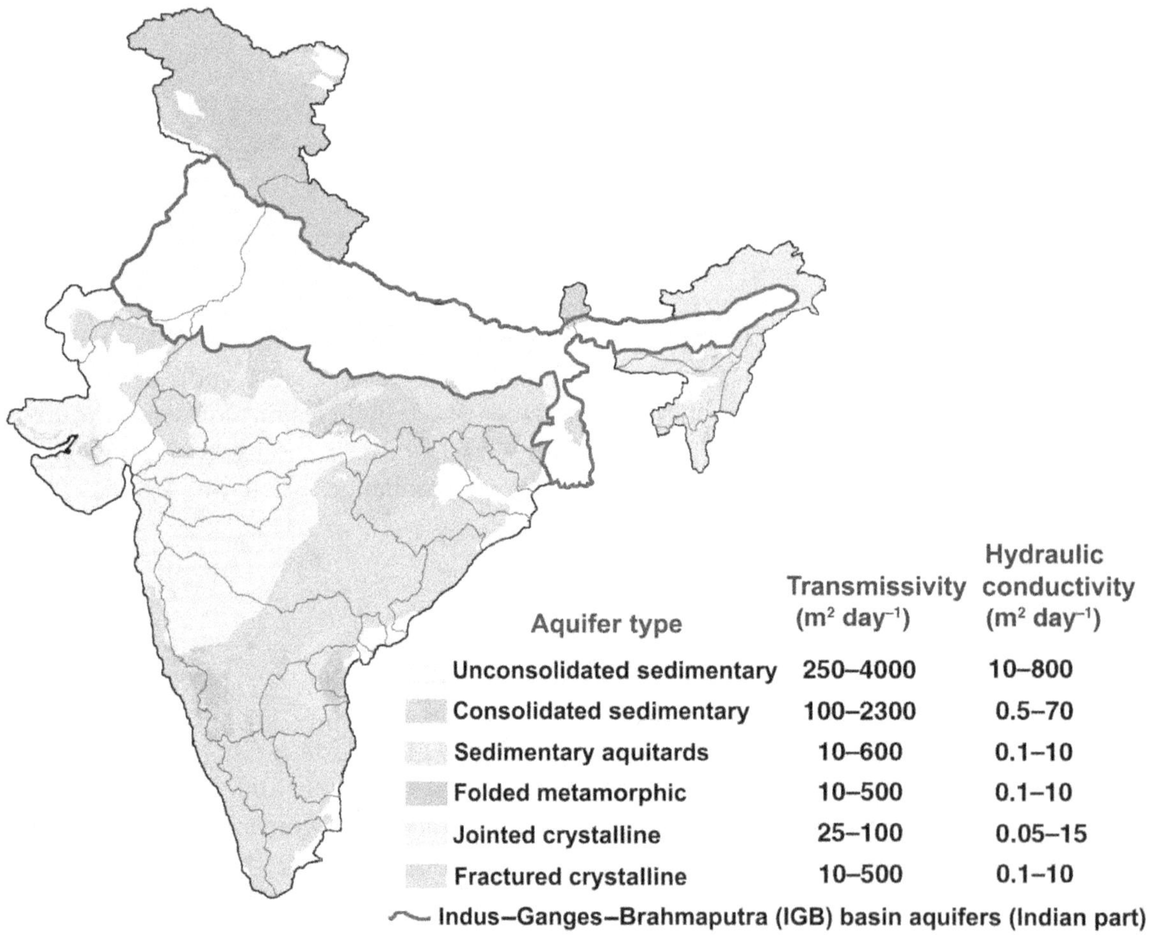

Groundwater Resources

India is the largest user of groundwater in the world, with about 25% of the global groundwater extraction. Groundwater is used for a variety of purposes, including irrigation, domestic and industrial water supply, and mining activities. The groundwater potential of India is estimated to be around 432 billion cubic meters (BCM). However, the utilizable groundwater resources are estimated to be around 395 BCM.

The groundwater availability in India varies greatly from region to region, depending on the rainfall and geological conditions. The major groundwater basins in India are the Indo-Gangetic Plain, the Deccan Plateau, the Coastal Plains, and the Himalayan region. The Indo-Gangetic Plain is the largest and most productive aquifer system in India, providing irrigation water to the agricultural lands in northern and eastern India. The Deccan Plateau is another important aquifer system that provides water to the agricultural lands of central and southern India. The Coastal Plains are also significant sources of groundwater for irrigation and domestic use. The Himalayan region, although less productive, is an important source of drinking water for the people living in the foothills and mountainous regions of northern India.

Challenges of Groundwater Use and Management

The growing population, urbanization, and industrialization have led to increased demand for groundwater in India. However, the overexploitation of groundwater resources has resulted in several challenges related to its use and management. Some of the major challenges are discussed below:

Groundwater depletion: The uncontrolled extraction of groundwater has led to the depletion of aquifers in several parts of India. The depletion of groundwater resources has resulted in a decline in water levels, reduced well yields, and deterioration in water quality.

Water quality deterioration: The indiscriminate use of fertilizers, pesticides, and other chemicals in agriculture has led to the contamination of groundwater with nitrates, phosphates, and other pollutants. The discharge of untreated industrial effluents and sewage into groundwater has also contributed to the deterioration of water quality.

Climate change: Climate change is expected to have a significant impact on groundwater resources in India. The changing rainfall patterns and increasing temperatures are likely to affect the recharge rates of aquifers and increase the demand for groundwater.

Lack of governance: The lack of governance and regulation of groundwater use and management has led to the overexploitation of groundwater resources. The absence of a comprehensive legal framework for groundwater management has resulted in unregulated extraction, especially in the agricultural sector.

Inefficient irrigation practices: The use of inefficient irrigation practices, such as flood irrigation and unlined canals, has resulted in the wastage of water and increased demand for groundwater.

Conclusion

The hydrogeology of India is diverse and complex, with different aquifer systems and groundwater resources. The country is the largest user of groundwater in the world, but the overexploitation of groundwater resources has led to several challenges related to its use and management. To ensure the sustainable use and management of groundwater resources in India, there is a need for a comprehensive legal framework, efficient irrigation practices, and effective governance and regulation of groundwater use. Additionally, there is a need for research and development of technologies for the recharge of aquifers and the conservation of groundwater resources.

4 - Spatial and Temporal Rainfall Pattern in India: Understanding the Variability and Impact

Abstract

India is a diverse country with varying topography, climate, and weather phenomena, resulting in a complex rainfall pattern. The rainfall in India is predominantly influenced by the southwest and northeast monsoons, which contribute to the majority of the precipitation received in the country. The spatial and temporal variability of rainfall in India has a significant impact on various sectors, including agriculture, economy, and water resources. This article provides an overview of the spatial and temporal rainfall pattern in India, including the factors influencing the variability and impact of rainfall on various sectors. Additionally, the article highlights the importance of understanding the rainfall pattern to develop sustainable strategies for development and adaptation to the changing climate.

Introduction

Rainfall is a crucial factor in the growth of agriculture and sustenance of human life. The spatial and temporal variability of rainfall in India has been the subject of many studies due to its impact on the economy and livelihoods of people. In this article, we will explore the patterns of rainfall in India, the factors that influence them, and their impact on various sectors.

Spatial Rainfall Pattern

India is a vast country with varying topography, ranging from the Himalayas in the north to the Indian Ocean in the south. The spatial distribution of rainfall in India is influenced by various factors such as latitude, altitude, distance from the sea, and the direction of the monsoon winds.

The southwest monsoon, which accounts for 75% of the total annual rainfall in India, starts in June and lasts till September. It brings the bulk of the rainfall to the country. The western coast of India, including the states of Kerala, Karnataka, and Maharashtra, receives the highest rainfall due to their proximity to the Arabian Sea. The eastern coast of India, including the states of Orissa, West Bengal, and Andhra Pradesh, receives less rainfall than the western coast due to the influence of the Bay of Bengal.

The Himalayan region receives the highest amount of rainfall in the country due to its high altitude and the presence of the Himalayan mountain range, which acts as a barrier and forces the moist monsoon winds to rise and release their moisture. The northwestern region of India, including the states of Rajasthan and Gujarat, receives very little rainfall due to the presence of the Thar desert.

Temporal Rainfall Pattern

The temporal distribution of rainfall in India is characterized by the presence of two monsoons, the southwest monsoon and the northeast monsoon. The southwest monsoon starts in June and lasts till September and is responsible for the bulk of the rainfall in the country. The northeast monsoon, which occurs during the winter months of October to December, brings rainfall to the southern parts of India.

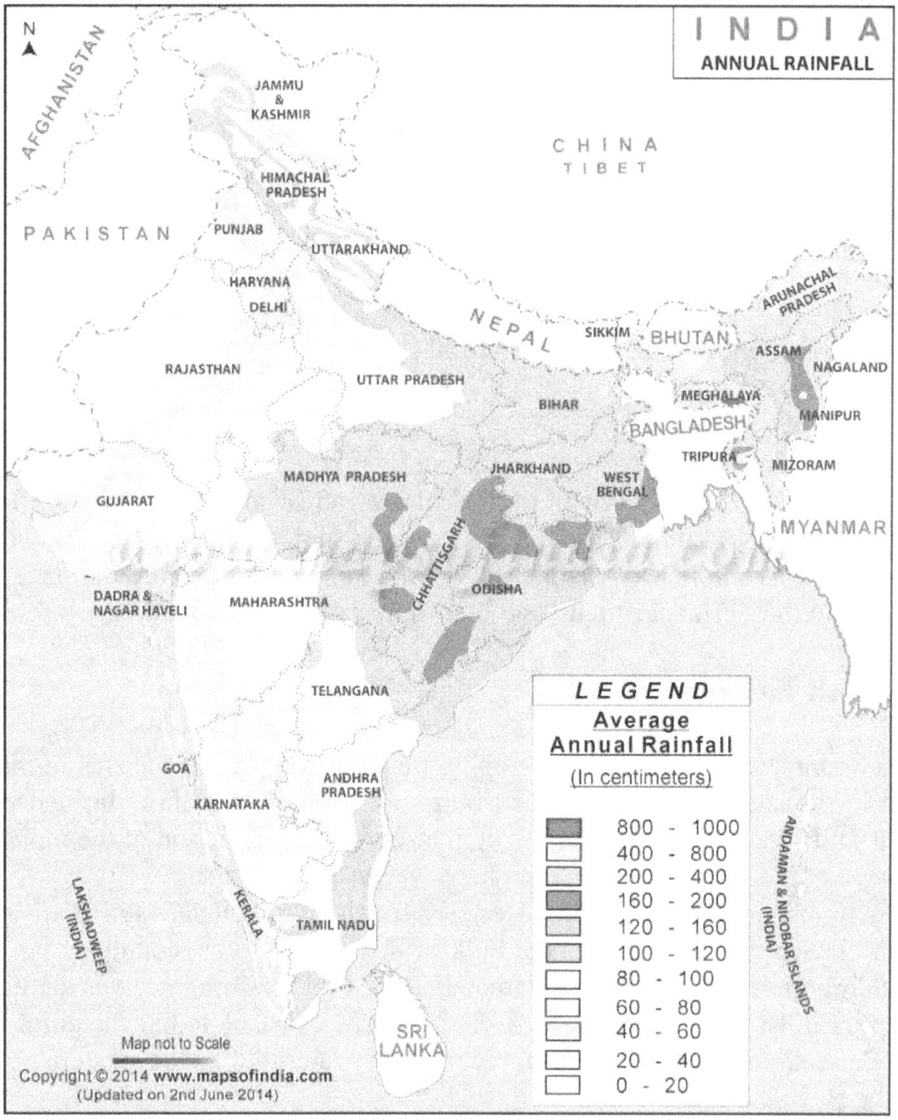

The rainfall pattern in India has been influenced by various factors such as El Nino, La Nina, and Indian Ocean Dipole (IOD). El Nino is a weather phenomenon that occurs due to the warming of the Pacific Ocean, leading to a decrease in rainfall in India. La Nina, on the other hand, is a cooling of the Pacific Ocean, which leads to an increase in rainfall in India. The IOD is a phenomenon where the western and eastern parts of the Indian Ocean experience different temperature variations, leading to an impact on rainfall in India.

Normal Rainfall Ranges

Some typical normal rainfall values for different parts of India are given below.

Southwest Monsoon Season (June to September):

Western Coast: 2500-3500 mm (Kerala, Karnataka, Maharashtra)
Northeast India: 2500-3500 mm (Assam, Meghalaya, Mizoram)
Central India: 1000-1500 mm (Madhya Pradesh, Chhattisgarh)
Eastern India: 1000-1500 mm (Orissa, West Bengal)
Northwest India: 100-500 mm (Rajasthan, Gujarat)

Northeast Monsoon Season (October to December):

Coastal Tamil Nadu: 400-600 mm
Coastal Andhra Pradesh: 200-300 mm
Rayalaseema: 100-200 mm
Telangana: 50-100 mm

Note that these are only typical normal values, and rainfall patterns can vary significantly from year to year due to various factors such as El Nino and La Nina.

Impact of Rainfall Pattern

The rainfall pattern in India has a significant impact on various sectors such as agriculture, economy, and water resources. Agriculture is the backbone of the Indian economy, and the majority of the Indian population is dependent on it for their livelihood. The spatial and temporal variability of rainfall in India has a significant impact on crop production, leading to a fluctuation in food prices and affecting the economy.

The availability of water resources in India is also influenced by rainfall. The majority of the water resources in India come from rainfall, and a change in rainfall pattern leads to a change in the availability of water resources. The spatial variability of rainfall leads to the presence of water-rich and water-scarce regions in the country.

Conclusion

The spatial and temporal variability of rainfall in India is influenced by various factors such as latitude, altitude, distance from the sea, and the direction of the monsoon winds. The rainfall pattern has a significant impact on various sectors such as agriculture, economy, and water resources. It is essential to understand the rainfall pattern in India and its impact to develop strategies for sustainable development and to adapt to the changing climate. Climate change is expected to have a significant impact on the rainfall pattern in India, leading to an increase in extreme weather events such as droughts and floods.

Therefore, it is crucial to implement measures such as water conservation, rainwater harvesting, and crop diversification to mitigate the impact of changing rainfall patterns. The government and various organizations should work together to promote sustainable practices and develop policies to manage the water resources and agricultural practices in the country.

5 - Runoff Coefficients for Various Regions in India: An Overview

Abstract

Runoff coefficients are an essential component in hydrological modelling, as they provide a measure of the amount of rainfall that can be expected to run off a particular area or basin. In India, where water resources are scarce and unevenly distributed, accurate estimation of runoff coefficients is crucial for effective water management. This article provides a comprehensive overview of runoff coefficients for various regions in India. The article highlights the factors that influence runoff coefficients, such as topography, soil types, land use, and vegetation cover. It also presents the methods used for estimating runoff coefficients, including the Soil Conservation Service Curve Number (SCS-CN) method and artificial neural network (ANN) techniques. Finally, the article emphasizes the importance of continually monitoring and updating runoff coefficients to accurately predict and manage water resources in India.

Introduction

Runoff is a critical hydrological parameter that plays a vital role in determining the surface water availability and water balance in any region. Runoff coefficient is a dimensionless factor that represents the proportion of rainfall that is converted into runoff over a particular area. It is a crucial parameter in hydrological modelling, and its estimation is essential for water resource management, flood control, and irrigation planning.

In India, runoff coefficients vary significantly from region to region, depending on various factors such as topography, soil type, land use, and rainfall intensity. In this article, we will explore the different runoff coefficients in various regions and river basins in India.

Factors Affecting Runoff Coefficient

The runoff coefficient is influenced by several factors, including the following:

Topography: The slope of the land surface plays a crucial role in determining the runoff coefficient. Steeper slopes have higher runoff coefficients as water flows faster over them and does not get enough time to infiltrate into the soil.

Soil Type: The infiltration capacity of soil determines the amount of water that can be absorbed into the ground. Soils with high infiltration rates have lower runoff coefficients, while soils with low infiltration rates have higher runoff coefficients.

Land Use: Different land uses have different infiltration rates and surface roughness, which can significantly affect the runoff coefficient. For instance, forests and grasslands have lower runoff coefficients than urban areas, which have higher runoff coefficients due to the presence of impervious surfaces.

Rainfall Intensity: The intensity and duration of rainfall also play a vital role in determining the runoff coefficient. Heavy rainfall events with high intensity and short duration result in higher runoff coefficients than light rainfall events with low intensity and long duration.

Evapotranspiration: Evapotranspiration is the process by which water is returned to the atmosphere through plant transpiration and soil evaporation. Areas with high evapotranspiration rates have lower runoff coefficients as more water is absorbed into the ground.

Methods for Estimating Runoff Coefficients

Here are some methods commonly used for estimating runoff coefficients.

Soil Conservation Service Curve Number (SCS-CN) method: This method is widely used for estimating runoff coefficients in India. The SCS-CN method uses land use, soil type, and antecedent soil moisture conditions to calculate a curve number, which is then used to estimate the runoff coefficient.

Rational method: The Rational method is a widely used empirical method for estimating peak runoff rates in urban areas. This method uses the rainfall intensity, the area of the catchment, and the runoff coefficient to estimate the peak runoff rate.

Artificial Neural Network (ANN) techniques: ANN is a machine learning technique used for estimating runoff coefficients. It uses a set of input variables such as rainfall, land use, soil type, topography, and vegetation cover, to predict the output variable, which is the runoff coefficient.

Geographic Information System (GIS) techniques: GIS techniques can be used to estimate runoff coefficients by integrating data on topography, land use, soil type, and other parameters. The GIS-based methods can provide a spatially distributed estimate of runoff coefficients.

These methods are not exhaustive, and there are many other techniques available for estimating runoff coefficients. The choice of the method depends on the availability of data, the size and complexity of the catchment, and the level of accuracy required in the estimation.

Runoff Coefficient for Various Regions in India

The runoff coefficient values in India vary widely depending on the region and the factors mentioned above. Here, we will discuss the different runoff coefficients in various regions in India.

Himalayan Region

The Himalayan region is characterized by steep slopes, high rainfall, and a predominance of forests and grasslands. The runoff coefficient in this region varies from 0.2 to 0.4, depending on the slope, soil type, and land use. The higher runoff coefficients are observed in areas with steeper slopes and less vegetation cover, while lower runoff coefficients are observed in areas with flatter slopes and more vegetation cover.

Indo-Gangetic Plain

The Indo-Gangetic plain is a vast alluvial plain in northern India, covering an area of approximately 700,000 km^2. The plain is characterized by low slopes, fertile soils, and a predominance of agricultural land use. The runoff coefficient in this region varies from 0.2 to 0.3, depending on the soil type and land use. The lower runoff coefficients are observed in areas with high infiltration rates, such as sandy soils and well-drained agricultural lands.

Peninsular India

Peninsular India is a large landmass in southern India, characterized by a diverse range of topography, soil types, and land uses. The runoff coefficient in this region varies widely, ranging from 0.1 to 0.5, depending on the factors mentioned above. The higher runoff coefficients are observed in areas with steep slopes, shallow soils, and low vegetation cover, while lower runoff coefficients are observed in areas with gentle slopes, deep soils, and dense vegetation cover.

Northeast India

Northeast India is a hilly and mountainous region, characterized by high rainfall, dense forests, and steep slopes. The runoff coefficient in this region varies from 0.3 to 0.5, depending on the topography and land use. The higher runoff coefficients are observed in areas with steeper slopes and less vegetation cover, while lower runoff coefficients are observed in areas with flatter slopes and more vegetation cover.

Western Ghats

The Western Ghats are a mountain range running parallel to the western coast of India, covering an area of approximately 160,000 km^2. The region is characterized by high rainfall, dense forests, and steep slopes. The runoff coefficient in this region varies from 0.3 to 0.5, depending on the slope and land use. The higher runoff coefficients are observed in areas with steeper slopes and less vegetation cover, while lower runoff coefficients are observed in areas with flatter slopes and more vegetation cover.

Coastal Plains

The coastal plains of India are a narrow strip of land along the eastern and western coasts of India, covering an area of approximately 60,000 km^2. The region is characterized by low slopes, fertile soils, and a predominance of agricultural land use. The runoff coefficient in this region varies from 0.2 to 0.3, depending on the soil type and land use. The lower runoff coefficients are observed in areas with high infiltration rates, such as sandy soils and well-drained agricultural lands.

Central India

Central India is a large region in the middle of India, covering an area of approximately 500,000 km^2. The region is characterized by a mix of topography, soil types, and land uses. The runoff coefficient in this region varies widely, ranging from 0.1 to 0.5, depending on the factors mentioned above. The higher runoff coefficients are observed in areas with steep slopes, shallow soils, and low vegetation cover, while lower runoff coefficients are observed in areas with gentle slopes, deep soils, and dense vegetation cover.

Overall, it is important to note that the runoff coefficient for any particular region in India is not fixed and can vary over time due to changes in land use, climate, and other factors. Therefore, it is important to continually monitor and update runoff coefficients in order to accurately predict and manage water resources.

Conclusion

The runoff coefficient is a crucial parameter in hydrological modelling and water resource management. The runoff coefficient values in India vary significantly depending on the region and the factors mentioned above. The Himalayan region has a runoff coefficient of 0.2 to 0.4, the Indo-Gangetic plain has a runoff coefficient of 0.2 to 0.3, Peninsular India has a runoff coefficient of 0.1 to 0.5, Northeast India has a runoff coefficient of 0.3 to 0.5, the Western Ghats have a runoff coefficient of 0.3 to 0.5, the coastal plains have a runoff coefficient of 0.2 to 0.3, and central India has a runoff coefficient of 0.1 to 0.5. These values are critical in determining the surface water availability and water balance in these regions, and their estimation is essential for flood control, irrigation planning, and sustainable water resource management.

References

Bhattacharya, A.K., and Goyal, M.K. (2008). Runoff estimation using SCS-CN method in a Himalayan watershed. Hydrological Sciences Journal, 53(6), 1156-1168.

Ministry of Water Resources (2015). National Register of Large Dams. Central Water Commission, Government of India.

National Institute of Hydrology (2018). Hydrological Atlas of India. National Institute of Hydrology, Ministry of Water Resources, River Development and Ganga Rejuvenation, Government of India.

Singh, R., Singh, V.P., and Mishra, S.K. (2015). Rainfall-Runoff Modelling Using Artificial Neural Network Technique. Journal of Water Resource and Protection, 7, 1245-1261.

Tiwari, K.N., Singh, R., and Mishra, S.K. (2014). Estimation of Runoff and Infiltration in a Semi-arid Watershed Using SCS-CN Method. Water Resources Management, 28, 553-567.

Water Resources Information System of India (2019). Hydrological Data Book. Central Water Commission, Government of India.

Yadav, M.R., Singh, R., and Mishra, S.K. (2016). Rainfall-Runoff Modelling Using Artificial Neural Network: A Case Study of Tapi Basin, India. Journal of Water Resource and Protection, 8, 1418-1432.

Yusuf, S., and Raju, K.S. (2015). Application of SCS-CN Method for Estimation of Runoff in a Semi-arid Region: A Case Study of Sabarmati Basin, India. Water Resources Management, 29, 3029-3046.

Zaman, B., and Saraf, A.K. (2010). Estimation of Runoff Using SCS-CN Method in the Jhelum River Basin, India. Hydrological Sciences Journal, 55(4), 623-636.

6 - Managing Water Resources in Uttarakhand: Challenges and Initiatives

Abstract

Uttarakhand, known as the 'Land of Gods,' is blessed with abundant water resources, including several major rivers, glaciers, and waterfalls. However, the state faces several challenges related to water management, including overexploitation of groundwater resources, pollution of rivers, and climate change. This article discusses the challenges faced by Uttarakhand in managing its water resources and the initiatives taken by the state government to address these challenges. These initiatives include the Atal Bhujal Yojana, Namami Gange program, establishment of a water regulatory authority, and investment in water storage and distribution infrastructure. The article concludes that with the implementation of these initiatives, the state can manage its water resources more sustainably and ensure that everyone has access to safe and reliable water supply.

Introduction

Uttarakhand, also known as the "Devbhoomi" or the "Land of Gods," is a state located in the northern part of India. The state is famous for its scenic beauty, pristine rivers, and religious sites. Uttarakhand is also known for its water resources, as the state is the source of some of India's major rivers, including the Ganges, Yamuna, and Alaknanda. The state has a diverse topography, which includes mountains, hills, and plains. This article will provide an overview of Uttarakhand's water resources, including its rivers, lakes, glaciers, and waterfalls.

Rivers

Uttarakhand is the source of some of India's major rivers, including the Ganges, Yamuna, and Alaknanda. The Ganges, also known as the Ganga, is considered to be one of the holiest rivers in India. The river originates from the Gangotri glacier, which is located in the Uttarkashi district of Uttarakhand. The Ganges flows through several states, including Uttar Pradesh, Bihar, and West Bengal, before emptying into the Bay of Bengal. The river is used for various purposes, including irrigation, transportation, and religious rituals.

The Yamuna is another important river that originates from the Yamunotri glacier, which is located in the Uttarkashi district of Uttarakhand. The river flows through several states, including Uttar Pradesh, Haryana, and Delhi, before emptying into the Arabian Sea. The Yamuna is also used for various purposes, including irrigation, drinking water supply, and industrial use.

The Alaknanda is a tributary of the Ganges and originates from the Satopanth glacier, which is located in the Chamoli district of Uttarakhand. The river flows through several districts, including Chamoli, Rudraprayag, and Tehri, before merging with the Bhagirathi river to form the Ganges. The Alaknanda is used for various purposes, including hydroelectric power generation and irrigation.

Apart from these major rivers, Uttarakhand also has several other rivers, including the Kali, Ramganga, and Saryu. These rivers are also used for various purposes, including irrigation, drinking water supply, and tourism.

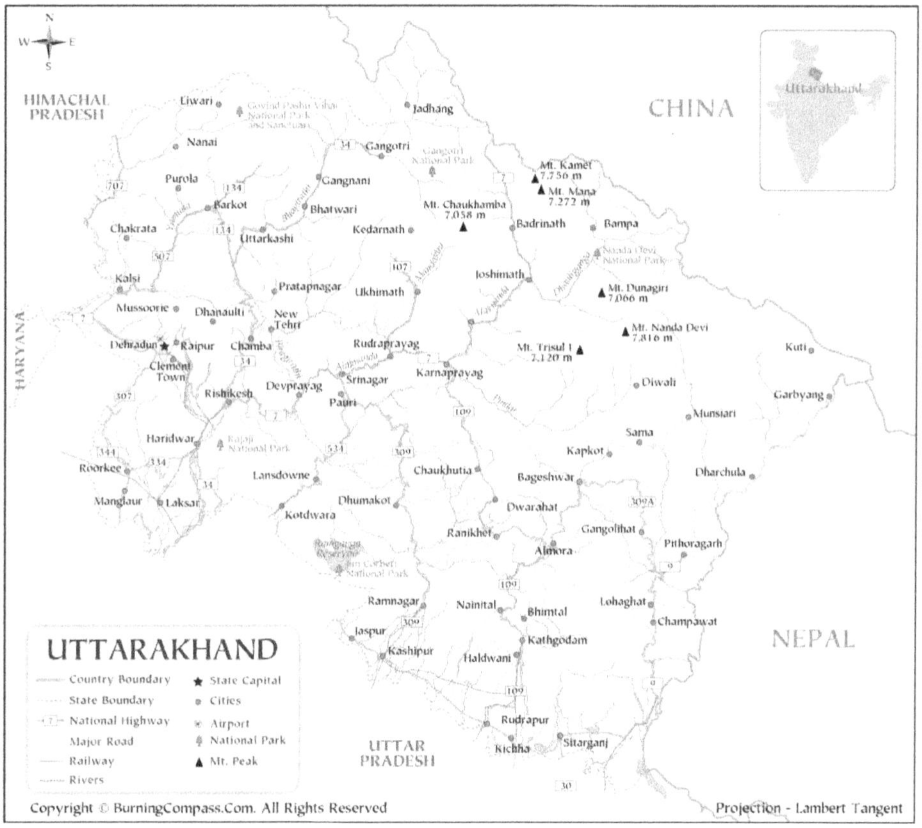

Lakes

Uttarakhand is also known for its beautiful lakes, which are located in the high-altitude regions of the state. Some of the famous lakes in Uttarakhand include the Bhimtal Lake, Nainital Lake, and Roopkund Lake.

The Bhimtal Lake is located in the Nainital district of Uttarakhand and is named after the legendary character Bhima from the Indian epic Mahabharata. The lake is surrounded by lush green hills and is a popular tourist destination.

The Nainital Lake is also located in the Nainital district of Uttarakhand and is one of the most famous lakes in India. The lake is surrounded by hills and is a popular destination for boating and other water activities.

The Roopkund Lake is located in the Chamoli district of Uttarakhand and is a high-altitude glacial lake. The lake is known for the skeletons that were found near its shores, which are believed to be of people who died in a hailstorm. The lake is a popular trekking destination and attracts adventure enthusiasts from all over the world.

Apart from these lakes, Uttarakhand also has several other lakes, including the Kedar Tal, Hemkund Sahib, and Dodital. These lakes are also popular tourist destinations and are known for their natural beauty.

Glaciers

Uttarakhand is home to several glaciers, which are an important source of water for the rivers in the state. Some of the famous glaciers in Uttarakhand include the Gangotri glacier, Yamunotri glacier, and Milam glacier.

The Gangotri glacier is located in the Uttarkashi district of Uttarakhand and is the source of the Ganges river. The glacier is around 30 kilometers long and is one of the largest glaciers in the Himalayas. The Gangotri glacier is also a popular trekking destination and attracts adventure enthusiasts from all over the world.

The Yamunotri glacier is located in the Uttarkashi district of Uttarakhand and is the source of the Yamuna river. The glacier is around 6 kilometers long and is one of the smaller glaciers in the Himalayas. The Yamunotri glacier is also a popular trekking destination and is known for its natural beauty.

The Milam glacier is located in the Pithoragarh district of Uttarakhand and is around 16 kilometers long. The glacier is located at an altitude of around 3,870 meters and is known for its natural beauty. The Milam glacier is also a popular trekking destination and is known for its panoramic views of the Himalayas.

Apart from these glaciers, Uttarakhand also has several other glaciers, including the Pindari glacier, Khatling glacier, and Gangotri group of glaciers. These glaciers are also important sources of water for the rivers in the state and are known for their natural beauty.

Waterfalls

Uttarakhand is also known for its beautiful waterfalls, which are located in the hilly regions of the state. Some of the famous waterfalls in Uttarakhand include the Kempty falls, Sahastradhara falls, and Tiger falls.

The Kempty falls are located in the Mussoorie town in the Dehradun district of Uttarakhand and are one of the most famous waterfalls in the state. The falls are around 40 feet high and are surrounded by lush green forests. The Kempty falls are a popular tourist destination and attract visitors from all over the world.

The Sahastradhara falls are located near Dehradun in Uttarakhand and are known for their medicinal properties. The falls are around 30 meters high and are believed to have therapeutic properties due to the presence of sulfur in the water. The Sahastradhara falls are a popular destination for people seeking relief from various ailments.

The Tiger falls are located in the Chakrata district of Uttarakhand and are around 50 meters high. The falls are surrounded by dense forests and are a popular destination for trekking and camping.

Apart from these waterfalls, Uttarakhand also has several other waterfalls, including the Bhatta falls, Mossy falls, and Kainchi Dham falls. These waterfalls are also popular tourist destinations and are known for their natural beauty.

Challenges

Despite having abundant water resources, Uttarakhand faces several challenges related to water management. One of the major challenges is the increasing demand for water due to the growing population and industrialization in the state. This has led to overexploitation of groundwater resources, which has resulted in the depletion of groundwater levels in several areas.

Another challenge is the pollution of rivers due to discharge of untreated sewage and industrial effluents. The Ganges, in particular, is heavily polluted due to the discharge of untreated sewage and industrial waste. This has led to several health problems for people living in the downstream areas.

Climate change is also a major challenge for water management in Uttarakhand. The state has witnessed several extreme weather events in the recent past, including floods and landslides, which have caused significant damage to infrastructure and loss of life. These extreme weather events are expected to increase in frequency and intensity in the future, which will further complicate water management in the state.

In addition to these challenges, there are also issues related to water allocation and distribution in the state. The state government is responsible for allocating water resources to different sectors, including agriculture, industry, and domestic use. However, there are often conflicts between different sectors over the allocation of water resources.

Another challenge is the lack of proper infrastructure for water storage and distribution. While the state has several dams and reservoirs, many of them are old and in need of repair. Moreover, there are several areas in the state that do not have access to piped water supply, which leads to dependence on groundwater sources.

Efforts to Address the Challenges

The state government of Uttarakhand has taken several measures to address the challenges related to water management. One of the key initiatives is the implementation of the Atal Bhujal Yojana, which is aimed at sustainable management of groundwater resources. The scheme involves the formation of water user associations and incentivizing the adoption of sustainable groundwater practices.

The state government has also launched several initiatives to improve water quality in the rivers of the state. The Namami Gange program, which was launched by the central government, has been implemented in Uttarakhand with the aim of cleaning up the Ganges river. Under this program, several sewage treatment plants have been set up in the state to treat the sewage before it is discharged into the river.

To address the issue of water allocation and distribution, the state government has set up a water regulatory authority, which is responsible for regulating the allocation and distribution of water resources. The authority also resolves disputes related to water allocation between different sectors.

The state government is also investing in the development of infrastructure for water storage and distribution. Several new dams and reservoirs are being constructed in the state to increase the storage capacity of water resources. Moreover, efforts are being made to expand the piped water supply network in the state to ensure that more people have access to safe and reliable water supply.

Conclusion

Uttarakhand is blessed with abundant water resources, including several major rivers, glaciers, and waterfalls. However, the state faces several challenges related to water management, including overexploitation of groundwater resources, pollution of rivers, and climate change. To address these challenges, the state government has launched several initiatives, including the Atal Bhujal Yojana, Namami Gange program, and the establishment of a water regulatory authority. With the implementation of these initiatives, it is hoped that the state will be able to manage its water resources more sustainably and ensure that everyone has access to safe and reliable water supply.

7 - Cracks in Buildings in Joshimath: Understanding the Impact of Land Subsidence

Abstract

This article discusses the recent development of cracks in many buildings in Joshimath, a town located in the Chamoli district of Uttarakhand, India, due to land subsidence. Land subsidence is a geological phenomenon that occurs when the ground level sinks or settles, and can be caused by various factors such as mining, oil and gas extraction, groundwater pumping, natural compaction of soil, and tectonic activity. In the case of Joshimath, the land subsidence is believed to have been caused by melting glaciers in the region, which has resulted in the weakening of the soil and the formation of underground cavities. The article highlights the impact of land subsidence on buildings, including the development of cracks that can compromise their structural integrity, weaken their resistance to natural disasters, and pose health risks to occupants. The article also discusses measures to mitigate the impact of land subsidence, such as regulating groundwater use, improving building construction quality, and undertaking structural repairs and retrofitting of affected buildings. The article concludes that taking these measures is necessary to ensure the safety of people living in affected areas.

Introduction

Joshimath is a small town located in the Chamoli district of the Indian state of Uttarakhand. The town is situated at an altitude of 1,890 meters above sea level and is known for its religious importance as it serves as a gateway to several Hindu pilgrimage sites such as Badrinath, Hemkund Sahib, and Valley of Flowers. Recently, the town has been in the news due to the cracks that have developed in several buildings. The cracks have been attributed to land subsidence, which is a geological phenomenon that can have serious consequences for the safety of the people living in the affected area.

What is Land Subsidence?

Land subsidence is a geological phenomenon that occurs when the ground level sinks or settles. This can be caused by a variety of factors such as underground mining, oil and gas extraction, excessive groundwater pumping, natural compaction of soil, and tectonic activity. In the case of Joshimath, the land subsidence is believed to have been caused by the melting of the glaciers in the region, which has resulted in the weakening of the soil and the formation of underground cavities.

Impact of Land Subsidence on Buildings

When land subsidence occurs, the ground level sinks, which can cause serious damage to buildings and other infrastructure. Buildings that are constructed on the affected land can develop cracks, which can compromise their structural integrity and pose a risk to the safety of the people living in them. In the case of Joshimath, the cracks that have developed in the buildings are a clear indication of the impact of land subsidence on the town.

The cracks in the buildings in Joshimath are not just cosmetic damage. They can have serious consequences for the safety of the people living in them. The cracks can weaken the structure of the building, making it more vulnerable to collapse during earthquakes or other natural disasters. Additionally, the cracks can also allow moisture to seep into the building, which can lead to the growth of mold and other harmful microorganisms. This can pose a serious health risk to the occupants of the building.

Measures to Mitigate the Impact of Land Subsidence

There are several measures that can be taken to mitigate the impact of land subsidence on buildings and other infrastructure. One of the most effective measures is to reduce the rate of groundwater pumping, which is a major cause of land subsidence in many parts of the world. In the case of Joshimath, it may be necessary to regulate the use of groundwater in the region to prevent further land subsidence.

Another measure that can be taken to mitigate the impact of land subsidence is to improve the quality of construction of buildings. Buildings that are constructed using high-quality materials and techniques are less likely to be affected by cracks and other forms of damage. Additionally, buildings that are designed to be earthquake-resistant are more likely to withstand the impact of natural disasters such as earthquakes.

In the case of Joshimath, it may be necessary to undertake structural repairs on the buildings that have been affected by the cracks. This will involve identifying the extent of the damage and undertaking the necessary repairs to ensure that the buildings are structurally sound. Additionally, it may be necessary to retrofit the buildings to make them more earthquake-resistant.

Conclusion

Land subsidence is a serious geological phenomenon that can have serious consequences for the safety of the people living in the affected area. In the case of Joshimath, the cracks that have developed in the buildings are a clear indication of the impact of land subsidence on the town. Measures to mitigate the impact of land subsidence on buildings and other infrastructure include reducing groundwater pumping, improving the quality of construction, and undertaking structural repairs and retrofitting. It is essential to take action to prevent further damage and ensure the safety of the people living in the affected area. As climate change and human activities continue to impact the environment, it is vital to be aware of the potential consequences and take steps to mitigate them.

Disclaimer

The views expressed in this article are based on the general perception and understanding of the subject matter. The information provided is for educational and informational purposes only and should not be considered as professional or expert advice. The author does not make any guarantees about the completeness, accuracy, reliability, suitability or availability of the information contained in the article for any purpose.

8 - Groundwater Over-Exploitation in India: Causes, Impacts, and Solutions

Abstract

Groundwater over-exploitation is a significant problem in India, affecting millions of people, especially in rural areas. This article provides an in-depth analysis of the groundwater over-exploitation crisis in India, including its causes, impacts, and solutions. Agriculture, rapid urbanization, industrialization, and climate change are among the leading causes of groundwater over-exploitation in India. The depletion of aquifers, deterioration of water quality, economic and social impacts are some of the consequences of groundwater over-exploitation. To address this issue, improving irrigation techniques, promoting crop diversification, rainwater harvesting, and implementing groundwater regulation policies are some of the solutions suggested. It is essential to address the groundwater over-exploitation crisis in India to ensure the sustainable use of this critical resource and to safeguard the livelihoods of millions of people who depend on it.

Introduction

India, with its growing population, rapid urbanization, and expanding agriculture, is experiencing severe water scarcity, particularly in its groundwater resources. Groundwater over-exploitation is a significant problem in India, affecting millions of people, especially in the rural areas. Overexploitation of groundwater leads to a range of issues such as depletion of aquifers, sinking of the land surface, deterioration of water quality, and exacerbating the impacts of climate change. This article provides an in-depth analysis of the groundwater over-exploitation crisis in India, including its causes, impacts, and solutions.

Overview of Groundwater in India

Groundwater is an essential source of water for drinking, irrigation, and industrial uses in India. It accounts for around 60% of the country's total irrigation needs, and approximately 85% of the rural domestic water requirements. The groundwater in India is mainly found in the form of shallow and deep aquifers, and it is recharged mainly by rainfall and surface water. Groundwater levels vary significantly across the country, with some regions having abundant resources, while others are severely depleted.

Groundwater Over-exploitation in India

Groundwater over-exploitation occurs when the extraction of water from aquifers exceeds the rate at which the water is naturally recharged. This leads to the depletion of the aquifer, and the water table declines, resulting in wells going dry. Overexploitation of groundwater in India is a widespread problem, and it is estimated that around 29% of the country's groundwater resources are overexploited, and a further 27% are in a critical stage.

Causes of Groundwater Over-exploitation in India

There are several reasons for the overexploitation of groundwater in India, some of which are discussed below.

Agricultural Practices: Agriculture is the largest user of groundwater in India, accounting for around 90% of the total groundwater usage. Farmers often rely on groundwater for irrigation due to the unreliability of surface water. The lack of proper irrigation techniques and management practices, combined with the high subsidies for electricity, leads to the over-extraction of groundwater.

Rapid Urbanization: Rapid urbanization has led to an increased demand for water, and groundwater is often the primary source of water in cities and towns. The rapid growth of urban areas has put immense pressure on the groundwater resources, resulting in their over-exploitation.

Industrialization: The expansion of industries in India has resulted in a higher demand for water, and groundwater is often the preferred source due to its low cost and easy accessibility. The industrial use of groundwater has increased significantly in recent years, leading to its over-extraction.

Climate Change: Climate change is also a significant factor in the overexploitation of groundwater in India. Changing precipitation patterns and increasing temperatures have led to a reduction in the recharge of groundwater, exacerbating the crisis.

Impacts of Groundwater Over-exploitation in India

The overexploitation of groundwater in India has significant impacts on the environment, economy, and social aspects of life. Some of the impacts are discussed below.

Depletion of Aquifers: Overexploitation of groundwater leads to the depletion of aquifers, and the water table declines, resulting in wells going dry. This, in turn, leads to the sinking of the land surface and damage to infrastructure such as buildings and roads.

Deterioration of Water Quality: The overexploitation of groundwater can lead to the deterioration of water quality, as the remaining water in the aquifer becomes more saline and contaminated with pollutants. This poses a risk to human health, as people may consume contaminated water, leading to water-borne diseases.

Economic Impact: The overexploitation of groundwater can have a significant economic impact, particularly in the agriculture sector, which accounts for a substantial proportion of groundwater usage. When wells go dry, farmers are forced to rely on expensive tanker water, leading to increased costs and reduced profits. This can also lead to a decline in agricultural productivity, affecting the livelihoods of farmers and exacerbating rural poverty.

Social Impact: Groundwater overexploitation can have a significant social impact, particularly on marginalized communities that rely on groundwater for their daily needs. When wells go dry, women and girls are often tasked with collecting water from further distances, leading to a loss of productive time and reduced opportunities for education and employment.

Solutions to Groundwater Over-exploitation in India

Several solutions can help to address the issue of groundwater overexploitation in India, including:

Improving Irrigation Techniques: Implementing efficient irrigation techniques such as drip irrigation and sprinkler irrigation can help to reduce groundwater usage in agriculture.

Promoting Crop Diversification: Encouraging farmers to diversify their crops can help to reduce the demand for groundwater, as some crops require less water than others. This can also lead to increased resilience to climate change and better agricultural productivity.

Regulating Groundwater Extraction: The government can regulate groundwater extraction by enforcing restrictions on the amount of water that can be extracted and the time of day when it can be extracted. This can be done through the implementation of a groundwater management policy that aims to ensure sustainable use of groundwater resources.

Promoting Rainwater Harvesting: Rainwater harvesting is a cost-effective and sustainable way to augment groundwater resources. The government can promote the use of rainwater harvesting systems in both rural and urban areas, which can help to reduce the dependence on groundwater.

Investing in Water Infrastructure: Investing in water infrastructure, such as building new reservoirs, can help to reduce the demand for groundwater and ensure that water is available for various uses. This can also help to ensure that water is distributed more equitably, particularly to marginalized communities.

Conclusion

The overexploitation of groundwater in India is a severe problem that has significant impacts on the environment, economy, and society. Addressing this issue will require a multi-pronged approach that involves improving irrigation techniques, regulating groundwater extraction, promoting crop diversification, and investing in water infrastructure. It is also essential to raise awareness among communities about the importance of conserving groundwater resources and implementing sustainable water management practices. By taking action now, India can ensure that its groundwater resources are conserved for future generations.

9 - Groundwater Quality Issues in India: Challenges and Remediation Strategies

Abstract

Groundwater quality issues in India have severe implications for human health and the environment. The deterioration of groundwater quality is mainly caused by anthropogenic activities such as agricultural practices, industrialization, urbanization, and domestic sewage disposal. Arsenic, fluoride, and nitrate contamination are the most significant groundwater quality issues in India. Remediation of groundwater contamination is a challenging task and requires a multidisciplinary approach. Several remediation techniques, such as physical, chemical, and biological methods, have been implemented to mitigate groundwater contamination in India. Community-based remediation plants have been successful in providing safe drinking water to the affected population. This article provides an overview of the groundwater quality issues in India, their causes and impacts, and various remediation strategies implemented to mitigate contamination.

Introduction

Groundwater is an essential resource for human life and the environment, and it plays a crucial role in India's economy. About 80% of India's rural water needs and 50% of urban water needs are met by groundwater. However, the quality of groundwater in India has been deteriorating over the years due to various anthropogenic activities, such as agricultural practices, industrialization, urbanization, and domestic sewage disposal.

Groundwater Quality Issues in India

Groundwater quality issues in India are widespread and have been a cause of concern for many years. The main sources of groundwater contamination in India are:

Agricultural practices: Agricultural practices such as excessive use of fertilizers and pesticides have resulted in groundwater contamination. The excessive use of fertilizers and pesticides has led to high levels of nitrate and pesticide residues in groundwater, which pose a significant risk to human health.

Industrialization: The rapid industrialization in India has led to the release of various toxic chemicals and heavy metals into the environment. The discharge of untreated industrial wastewater into rivers and other water bodies has resulted in the contamination of groundwater.

Urbanization: The rapid urbanization in India has led to increased domestic sewage disposal, which has resulted in groundwater contamination. The lack of adequate sewage treatment facilities in urban areas has led to the discharge of untreated sewage into the environment, contaminating groundwater.

Natural sources: Natural sources of groundwater contamination in India include arsenic and fluoride. Arsenic contamination is mainly found in the Ganga-Brahmaputra basin, while fluoride contamination is widespread across several states in India.

Effects of Groundwater Contamination on Human Health

Groundwater contamination can have severe effects on human health, depending on the type and extent of contamination. Some of the effects of groundwater contamination on human health are:

Acute health effects: Acute health effects of groundwater contamination include gastrointestinal illnesses, skin rashes, and respiratory problems. These health effects are typically caused by exposure to high levels of bacteria, viruses, and other microorganisms in contaminated groundwater.

Chronic health effects: Chronic health effects of groundwater contamination include various types of cancers, reproductive disorders, and neurological disorders. These health effects are typically caused by exposure to toxic chemicals and heavy metals in contaminated groundwater.

Remediation of Groundwater Contamination

Remediation of groundwater contamination is a challenging task and requires a multidisciplinary approach. The remediation of groundwater contamination involves various techniques, such as physical, chemical, and biological methods.

Physical methods: Physical methods of remediation include the use of pumps, wells, and barriers to remove or contain the contaminated groundwater. Pump and treat systems involve pumping contaminated groundwater from an aquifer to the surface, treating it, and then returning it to the aquifer. Permeable reactive barriers involve the installation of a reactive material in the groundwater flow path to treat the contaminants.

Chemical methods: Chemical methods of remediation include the use of oxidants, reducing agents, and chelating agents to treat the contaminants. Chemical oxidation involves the use of oxidants such as hydrogen peroxide and ozone to break down the contaminants. Chemical reduction involves the use of reducing agents such as zero-valent iron to convert the contaminants to less harmful forms. Chelating agents are used to immobilize heavy metals in the soil and prevent them from leaching into groundwater.

Biological methods: Biological methods of remediation involve the use of microorganisms to degrade or immobilize the contaminants. Bioremediation involves the use of microorganisms to degrade organic contaminants in the groundwater. Phytoremediation involves the use of plants to absorb and translocate contaminants from the groundwater to the atmosphere or soil.

Case Studies

Arsenic contamination in West Bengal

Arsenic contamination in West Bengal is one of the most significant groundwater quality issues in India. The contamination is mainly found in the Ganga-Brahmaputra basin, affecting millions of people. The primary source of arsenic contamination is natural, with the element leaching from the surrounding rocks and soil. The contamination is severe in the shallow aquifers, which are the primary source of drinking water for the local population.

Several remediation techniques have been implemented to mitigate the arsenic contamination in West Bengal. The most common method is the installation of community-based arsenic removal plants, which use activated alumina, iron oxide-coated sand, or other materials to remove arsenic from the water. The plants are operated and maintained by the local community and have been successful in providing safe drinking water to the affected population.

Fluoride contamination in Rajasthan

Fluoride contamination is widespread in several states in India, with Rajasthan being one of the most affected. The contamination is mainly caused by the dissolution of fluoride-rich rocks and minerals in the groundwater. The high fluoride concentration in the water leads to dental and skeletal fluorosis, a debilitating disease that affects millions of people in India.

Several remediation techniques have been implemented to mitigate the fluoride contamination in Rajasthan. The most common method is the installation of community-based defluoridation plants, which use activated alumina or bone char to remove fluoride from the water. The plants are operated and maintained by the local community and have been successful in providing safe drinking water to the affected population.

Nitrate contamination in Punjab

Nitrate contamination is a widespread issue in Punjab, mainly caused by excessive use of fertilizers and pesticides in agriculture. The contamination is severe in the shallow aquifers, which are the primary source of drinking water for the local population. High levels of nitrate in the water can cause methemoglobinemia, also known as blue baby syndrome, in infants.

Several remediation techniques have been implemented to mitigate the nitrate contamination in Punjab. The most common method is the installation of community-based nitrate removal plants, which use ion exchange resins or reverse osmosis to remove nitrate from the water. The plants are operated and maintained by the local community and have been successful in providing safe drinking water to the affected population.

Conclusion

Groundwater quality issues in India are widespread and have severe implications for human health and the environment. The deterioration of groundwater quality is mainly caused by anthropogenic activities such as agricultural practices, industrialization, urbanization, and domestic sewage disposal. The remediation of groundwater contamination is a challenging task and requires a multidisciplinary approach. Several remediation techniques, such as physical, chemical, and biological methods, have been implemented to mitigate groundwater contamination in India. Community-based remediation plants have been successful in providing safe drinking water to the affected population. However, there is a need for sustained efforts to address the groundwater quality issues in India and ensure the availability of safe drinking water for all.

10 - Challenges and Solutions for Sustainable Groundwater Management in India

Abstract

India's groundwater resources are under significant pressure due to over-extraction, contamination, and climate change. Over-extraction has resulted in declining water tables, increased energy costs, and water quality degradation. The quality of groundwater has deteriorated due to the discharge of untreated industrial and domestic wastewater, agricultural runoff, and contamination from chemicals and pesticides, posing a severe threat to public health. India's groundwater management policies have focused primarily on supply-side management, with limited emphasis on demand-side management. Potential solutions include promoting water conservation, artificial recharge of aquifers, regulating groundwater use, and improving wastewater management. Climate change is also expected to have significant impacts on groundwater resources. Sustainable groundwater management is necessary to address these challenges, and coordinated efforts by the central and state governments, local communities, and private sectors are required to address the issues surrounding groundwater resources in India.

Introduction

India is the second-most populous country in the world, with a population of over 1.3 billion. Groundwater is a critical resource for India, particularly for agriculture, which accounts for over 80% of the country's water consumption. The country's rapid population growth, along with its rapidly expanding agricultural and industrial sectors, has placed significant pressure on its groundwater resources. Groundwater is an essential resource for India's economy and livelihoods, with approximately 85% of the country's rural drinking water supply and 60% of its irrigation needs being met by groundwater. However, groundwater resources in India are under significant pressure due to overexploitation, contamination, and climate change. Over-extraction of groundwater has led to a decline in water tables, water quality degradation, and increased energy costs for pumping. In this article, we will examine the issues surrounding groundwater resources in India, the current state of groundwater management, and potential solutions to address the challenges.

Groundwater Resources Issues

Overexploitation:

One of the primary issues surrounding groundwater resources in India is over-extraction. The increasing demand for water from agriculture, industry, and domestic use has resulted in groundwater levels falling at an alarming rate. According to the Central Ground Water Board (CGWB), approximately 60% of India's groundwater blocks are in a critical condition, with the level of exploitation exceeding the rate of recharge.

Groundwater is being extracted at an unsustainable rate in many parts of India, particularly in the agricultural sector. The groundwater extraction rate in India is the highest in the world, and several regions, including Punjab, Haryana, and Rajasthan, are experiencing significant declines in groundwater levels.

The overexploitation of groundwater resources has several negative consequences, including reduced availability of water for agriculture and domestic use, increased energy costs for pumping, land subsidence, and saltwater intrusion in coastal areas.

The government of India has implemented several measures to address the overexploitation of groundwater resources, including the creation of groundwater regulatory agencies and the promotion of more efficient irrigation techniques. However, these measures have not been entirely successful in addressing the problem, and further action is needed to ensure sustainable groundwater management.

Contamination:

Another issue is the quality of groundwater. In many parts of India, the quality of groundwater has deteriorated due to the discharge of untreated industrial and domestic wastewater, agricultural runoff, and contamination from chemicals and pesticides. This poses a severe threat to public health, as the consumption of contaminated groundwater can cause various water-borne diseases.

Groundwater contamination is a significant problem in India, with a high prevalence of pollutants such as arsenic, fluoride, nitrate, and pesticides in groundwater sources. The contamination of groundwater sources poses a significant risk to human health, particularly in rural areas where groundwater is the primary source of drinking water.

The primary sources of groundwater contamination in India include industrial effluent, agricultural runoff, and inadequate sanitation infrastructure. The contamination of groundwater sources is a complex problem that requires a multifaceted approach involving regulation, enforcement, and the development of appropriate treatment technologies.

The government of India has implemented several measures to address groundwater contamination, including the establishment of the Central Ground Water Board and the National Green Tribunal. However, there is still a significant need for investment in wastewater treatment infrastructure, groundwater monitoring systems, and research into new treatment technologies.

Climate Change:

Climate change is expected to have significant impacts on groundwater resources in India, with changes in precipitation patterns, temperature, and extreme weather events likely to affect both the quantity and quality of groundwater resources.

The impact of climate change on groundwater resources is likely to be particularly severe in regions such as the Indo-Gangetic Plain, which are already experiencing high levels of groundwater stress. Climate change is likely to exacerbate the overexploitation and contamination of groundwater resources in these regions.

The government of India has recognized the need to address the impacts of climate change on groundwater resources and has implemented several measures, including the National Water Mission, which seeks to improve the efficiency of water use and promote sustainable groundwater management practices.

Current State of Groundwater Management

India's groundwater management policies are primarily focused on supply-side management, such as the construction of new infrastructure to increase water supply. However, there has been limited emphasis on demand-side management, such as promoting water conservation and regulating groundwater use. The lack of regulation has led to over-extraction of groundwater and subsequent depletion of the aquifers.

The National Water Policy of 2012 emphasized the importance of managing groundwater resources sustainably, but the implementation of policies at the state level has been slow. The implementation of the policy has also been hindered by a lack of coordination between different government departments and agencies.

Potential Solutions

To address the challenges facing India's groundwater resources, several solutions have been proposed. These include:

Promoting water conservation: Encouraging farmers to adopt water-efficient irrigation methods and promoting the use of low-water crops can help conserve groundwater resources.

Recharge of aquifers: Artificial recharge of aquifers can be used to replenish groundwater resources. Methods such as rainwater harvesting, building check dams, and groundwater recharge wells can help recharge groundwater resources.

Regulation of groundwater use: The implementation of regulations to monitor groundwater use and prevent over-extraction is necessary to sustainably manage groundwater resources.

Improved wastewater management: Effective treatment of domestic and industrial wastewater can help reduce groundwater contamination.

Conclusion

Groundwater is a vital resource for India's economy and livelihoods. The over-extraction of groundwater and subsequent depletion of aquifers, along with groundwater contamination, poses a severe threat to public health and the country's future economic development. Sustainable groundwater management is necessary to address these challenges. Addressing these challenges will require a multifaceted approach involving regulation, enforcement, investment in infrastructure, and the development of new technologies. Promoting water conservation, artificial recharge of aquifers, regulation of groundwater use, and effective wastewater management can help sustainably manage India's groundwater resources. It is necessary to have a coordinated effort by the central and state governments, local communities, and private sectors to address the issues surrounding groundwater resources in India. The government of India has implemented several measures to address these challenges, but further action is needed to ensure the long-term sustainability of groundwater resources in India.

11 - Some Thoughts on Water Resource Issues and Management in India

Background

India is endowed with a rich and vast diversity of natural resources, with water being one of them. Its development and management play a vital role in agricultural production. Integrated water management is vital for poverty reduction, environmental sustainability, and sustainable economic development. The National Water Policy of India envisages that the water resources of the country should be developed and managed in an integrated manner.

The Ministry of Jal Shakti (Government of India) was created to deal with all matters relating to water at one place in an integrated manner. It is responsible for laying down policy guidelines and programmes for the development and regulation of the country's water resources. The vision of the ministry includes optimal sustainable development, maintenance of quality, and efficient use of water resources to match the growing demands on this precious natural resource of the country.

The issue is not one of water availability but of converting that availability into supply. Tube wells, which were a boon at one point in time, have become a bane.

Many states have their own state water policy. For example, policies for ground water management, rainwater harvesting, and ground water recharge in the state of Uttar Pradesh are available at http://upgwd.gov.in/MediaGallery/Salient%20Points-English.pdf

Drinking Water

- The project authorities should reserve enough water in each reservoir for drinking purposes during hot weather, and only the remaining water available during the year should be released for irrigation and other uses. This is easily accomplished by establishing a drinking water rule curve. When the reservoir water level falls below this level, water should only be supplied to meet drinking needs.

- If the drinking water requirement is "reserved" in all reservoirs as per the National Water Policy and waste water management is made scientific and part of the accounting system, there may not be any crisis for drinking water. Many parts of the country are vulnerable to Day Zero conditions because of the unabated use of water for crops with high water requirements without leaving a small fraction for drinking water.

Crops

- We need to incentivize less water-consuming crops. Of course, water-intensive crops like rice and sugarcane may be promoted in high-rainfall areas.

- ❖ If the available water is used for less water-guzzling crops, the area under irrigation can be increased.

- ❖ Pharmaceutical wastes that enter the food chain are the most dangerous.

Subsidy

- Farmers in many states receive free or heavily subsidised electricity. So, a farmer in Punjab, who was not growing paddy before the 1980s, is now growing paddy in abundance.

- Many farmers in Punjab keep their tube well switches turned on. The government releases the supply of electricity for a few hours during the day, and one can see the water gushing out for hours.

- Everyone should pay for the use of electricity. Subsidies could be provided to farmers with small holdings by direct transfer to their bank accounts.

Rainwater Harvesting and Water Conservation

- ➢ Rainwater harvesting has two components: surface storage conservation and groundwater recharge. The choice depends on the amount of rainfall, the area available, and the availability of alternate sources. Normally, in arid areas, measures are taken to store (conserve) water. In areas with abundant rainfall, the depth of the groundwater also influences the choice. In shallow water table areas, we opt for conservation in surface water bodies. The overall goal is to reduce groundwater pressure during both monsoon and non-monsoon seasons.

- ➢ The focus should be on water conservation and storage in smaller structures. Similarly, the revival of small rivers and rivulets can be targeted.

- ➢ Rooftop rainwater harvesting systems are the best and cheapest way to overcome the water crisis in our country because we have ample rainfall.

Rivers

- o Rivers are the lifeline of our country. Unfortunately, many of our rivers are in danger of becoming extinct. The major problem being faced by our rivers is the overexploitation of surface water as well as groundwater and the increasing pollution load on them. Over-exploitation of groundwater has a direct bearing on the health of a river, as base flow or lean season flow in any river primarily depends on the flow of water from groundwater aquifers. Encroachment on flood plains is another bane for our rivers.

- o We need to rejuvenate our rivers and bring water back to them. Fortunately, a large number of smaller rivers (lower-order tributaries of well-known rivers) are being rejuvenated by the government as well as non-government organisations with the help of people's participation. If government, civil society organizations, and the community come together, they can do wonders.

Flood Management

- It is worth pondering as to why at all the need for "flood control" arises. What we term "flood" is a natural phenomenon of a river or any water body. The term "flood plain" has been born out of this only. When people started moving into the designated flood plains and settled to divert water somewhere else, preventing it from coming to its rightful place (the flood plain), we conveniently cultivated "flood control." A river should be allowed to flow longitudinally as well as laterally. Any kind of obstruction (including dams and barrages) alters its dynamics and ecological integrity. The risk of urban flooding in the event of heavy rainfall is increased by run-off quickly forming due to the lack of natural ground cover and because natural drains have been sealed off by unplanned construction.

- The real problem is the non-implementation of flood plain zoning. Cities today are more flood-prone than rivers because of habitation in floodplains combined with a lack of a proper drainage network. The Central Water Commission has circulated a model floodplain zoning bill to all states for adoption and implementation. After defining flood plain zones, at least people will know about the extent of flood risk associated with their dwellings. Accordingly, suitable structures can be planned.

- Intense population pressure on river basins and floodplains necessitates practical solutions. Given the population pressure, it may not be feasible to leave vast swaths of land uninhabited. Therefore, the option is to resort to engineering measures that will moderate floods, reduce flow velocities, ensure safe deposition of coarser sediments, and contain the floods from spreading overland and inundating low-lying areas. But jacketing rivers may not always be a good method of flood management. The jacketing of the river may cause floods on both sides of the embankment, on one side due to river flow and on the other side due to drainage congestion.

As an example, consider the Kosi River.

- ❖ If one stands on the Kosi embankment during floods, it may be difficult to differentiate between the riverside and the countryside. The problem of the Kosi is not merely the large volume of water in spate but also the large amounts of sediment, more than 50% of which is of the coarse and medium-sized detritus variety, contrary to other rivers where a larger fraction is of finer sediment. Due to its typical geography, the Kosi descends into the plains from very steep Himalayan slopes without the benefit of transitional slopes, as other rivers do.

- ❖ Consequently, when intense rainfall occurs in a high rainfall area, large volumes of water rush through the steeper slopes at great velocity, carrying with them large amounts of sediment and detritus that are deposited in the valleys. This deposit is the curse of Bihar; wherever it passes through fields, it renders them barren and unfit for cultivation. In due course of time, the river starts changing its course owing to these relatively coarse deposits. The literature shows that the migration of the Kosi takes place within a width of about 100 km of land nearer to its confluence with the Ganga.

- Sediment transport in rivers, especially Kosi, needs to be properly understood. Studies show that dams in Nepal will help only partially since rainfall in India can also cause flooding due to drainage congestion. There may be no better solution for the Kosi River than embankments in Indian territory to keep the river in a wide enough channel while preventing people from settling between the embankments. The requirement is to properly design the embankments, including provisions for lateral inflows.

Aquifer Mapping

- Aquifer mapping work needs to be completed as soon as possible. The country has approximately 25 lakh square kilometres of mappable area. Out of this, aquifer mapping over an area of 10.8 lakh sq. km. has already been completed. For the remaining area of about 14 lakh sq. km., CGWB has planned to complete aquifer mapping for areas of 5.6 lakh sq. km. and 8.4 lakh sq. km. in the next two years.

Rejuvenation

- We should not concentrate on "river cleaning," but river rejuvenation. River rejuvenation will automatically lead to river cleaning. Rather than concentrating on sewage treatment plants (STP), we need to adopt a holistic approach to river rejuvenation.

- River rejuvenation should be done by bringing more water to the river. This could be done through a series of measures such as afforestation, improving water use efficiency, ground water recharge through suitable structures under MGNREGA (the Mahatma Gandhi National Rural Employment Guarantee Act, 2005), etc.

- All over the country, we had innumerable tanks and ponds. Unfortunately, we had neglected them so much that they had almost become extinct. They are either encroached upon, silted up, or used as a garbage dump. Without the rejuvenation of tanks, ponds, and wells, there can be no water security.

Examples

- Bundelkhand is keeping the drought at bay by recharging wells.

- A strong and favourable public opinion and an equally strong and sustained political consensus and determination made the Sardar Sarovar possible in Gujarat, which has reduced water stress in the Saurashtra and Kutch regions.

- In Haryana, a state known for groundwater overexploitation, demand side management practises like drip irrigation and direct seeding are being adopted for rice.

- 50000 ha of land in Haryana's seven paddy-cultivating districts, viz., Ambala, Yamuna Nagar, Kurukshetra, Karnal, Sonipat, Jind, and Kaithal, have switched to maize and oilseeds under the "*Jal hi Jeewan Hai*" scheme of the state government. The state is supporting the crop change by providing financial assistance of Rs 5,000 per hectare and free insurance. This scheme for alternative crop diversification alone will impact crop patterns in the state, replace paddy with maize, and reduce groundwater abstraction.

- Meghalaya has inspired 2,000 villages to implement water conservation policies. It became the first state to implement such a policy that aims to ensure water security for present and future generations.

- *Lesson from Singapore*

  https://scroll.in/article/926270/singapore-taps-into-innovation-as-water-imports-dry-up

  Singapore has a national master-plan focusing on four "national taps": catching rainfall in reservoirs, recycling water, desalinating water, and imports. Singapore also invested heavily in underground drainage systems and dams. The tiny country now has 17 reservoirs that collect the rain that falls on two-thirds of its land area.

General Solutions

- We need to manage water in a participatory manner, and the focus has to be on water harvesting and water conservation.

- Drinking water is the first priority in the National Water Policy of India. Before providing water for "extra production" of sugarcane, we should ensure that drinking water reserves are maintained.

- We may need to implement some unpopular solutions like rationalising MSP (minimum support price), ending free electricity, and promoting non-water-guzzling crops instead of sugarcane and paddy.

- We must rejuvenate all our water bodies. Solar-powered pumps with micro irrigation will ensure proper utilisation of water from rejuvenated water bodies.

- We can have a scheme where we combine renovation of water bodies with a solar or electric powered pump, a piped distribution network, and micro irrigation.

- We need to fix benchmarks for water use by various industrial sectors, conduct water audits, develop a scheme like PAT (Perform, Achieve, and Trade), etc. The PAT scheme is a programme launched by the Bureau of Energy Efficiency (BEE) to reduce energy consumption and promote enhanced energy efficiency among specific energy-intensive industries in the country.

- ❖ All premises and buildings of various organisations must have functional water harvesting (rooftop rainwater harvesting).

- ❖ A target of improving water use efficiency by 20% was set when the National Water Mission was launched. The global average efficiency is said to be around 52%. Rich countries have about 75%, developing countries have about 40%, and poor countries have about 25%. In India, we estimate it to be between 30 and 40%, but only few studies have been conducted. Improving surface water efficiency will help alleviate pressure on groundwater resources. Benchmarking studies should be undertaken for a few irrigation projects.

- ❖ We have a large number of defunct tube wells all over the country. In some of the states, they seal such tube wells. These defunct tube wells should be utilised to recharge groundwater. Several farmers in Haryana are doing it from their tube wells by removing motors. Through their awareness programs, CGWB has propagated it.

- ❖ Steps for recycling and reuse by water intensive industries can also be a priority area.

- ❖ The Marathwada region should prioritise nutri-cereals (millets). These consume less water, but this will require a change in consumption patterns and also bring at least some of these under the Minimum Support Price (MSP).

Future Vision

- Wider application of nuclear and space technologies in the area of water resources management.

- Scientific re-assessment of flood-prone areas in the entire country.

- Establish a modern real-time flood forecasting and management system for Indian River basins.

Acknowledgement

The ideas and thoughts presented in this article have been inspired and compiled from the discussions in the WhatsApp group of the Ministry of Jal Shakti, Government of India.

12 - Organizations Working towards Sustainable Water Management in India

Abstract

India faces significant challenges in the management of its water resources. With a complex water governance structure and water management issues complicated by political and social factors, sustainable water management practices are crucial. This article highlights 12 organizations working towards the promotion of sustainable water management in India. These organizations conduct research, advocacy, and implementation activities related to hydrology, groundwater, water quality, and water resources management. They also provide training and capacity building programs for water resources professionals and community members. The article aims to raise awareness about the critical role these organizations play in India's water management landscape and highlight their efforts to achieve a sustainable and equitable water future.

Introduction

India is the seventh-largest country in the world by area, with a population of over 1.3 billion people. Water is a critical resource in India, and the country faces significant challenges related to hydrology, groundwater, water quality, and water resources. Due to rapid urbanization, population growth, and climate change, water scarcity and contamination have become a significant concern in India. Several organizations are working to address these issues, and this article will provide an overview of some of the organizations dealing with hydrology, groundwater, water quality, and water resources in India.

1. Central Water Commission (CWC)

The Central Water Commission (CWC) is a governmental organization under the Ministry of Jal Shakti, Government of India. CWC is responsible for the development and management of surface water resources in India. The organization conducts hydrological studies, water resources assessment, and monitoring of water quality and quantity. CWC is also responsible for the planning and implementation of major water projects such as dams, barrages, and canals.

CWC has a network of hydrological observation stations across the country to track changes in river flows, water levels, and sediment transport. The organization also provides technical support to the state governments and other stakeholders in the management of surface water resources. CWC is also responsible for flood forecasting and warning, which is critical for reducing the impact of floods in India.

2. Central Ground Water Board (CGWB)

The Central Ground Water Board (CGWB) is a governmental organization established in 1970 under the Ministry of Jal Shakti, Government of India. CGWB is responsible for developing and managing groundwater resources in India. The organization conducts hydrogeological studies, groundwater exploration, and monitoring of groundwater quality and quantity. CGWB has a network of observation wells and groundwater monitoring stations across the country to track changes in groundwater levels and quality.

CGWB also conducts research and development activities related to groundwater management. The organization provides technical support to the state governments and other stakeholders in the management of groundwater resources. CGWB is also responsible for implementing the National Aquifer Mapping and Management Program (NAQUIM), which aims to map the aquifers of India and develop a management plan for sustainable groundwater use.

3. Central Ground Water Authority

The Central Ground Water Authority (CGWA) is a statutory authority established in 1997 under the Ministry of Jal Shakti, Government of India. CGWA is responsible for regulating and controlling groundwater development in India. The organization issues permits for groundwater extraction, and it also monitors and enforces regulations related to groundwater management. CGWA provides technical support to the state governments and other stakeholders in the management of groundwater resources.

4. National Institute of Hydrology (NIH)

The National Institute of Hydrology (NIH) is a premier research organization established in 1978 under the Ministry of Jal Shakti, Government of India. NIH conducts research and development activities related to hydrology, groundwater, and water resources management. The organization also provides training and capacity building programs for water resources professionals in India.

NIH has several research divisions, including surface water hydrology, groundwater hydrology, and water quality. The organization conducts research on topics such as hydrological modelling, water resources assessment, and climate change impact on water resources.

5. Central Pollution Control Board (CPCB)

The Central Pollution Control Board (CPCB) is a governmental organization established in 1974 under the Ministry of Environment, Forest and Climate Change, Government of India. CPCB is responsible for monitoring and controlling pollution in India. The organization conducts water quality monitoring and assessment, and it also develops and enforces regulations related to water pollution.

CPCB has a network of water quality monitoring stations across the country to track changes in water quality parameters such as pH, dissolved oxygen, and total dissolved solids. The organization also provides technical support to the state governments and other stakeholders in the management of water pollution. CPCB is also responsible for implementing the National River Conservation Plan (NRCP), which aims to clean up the major rivers of India.

6. National Mission for Clean Ganga (NMCG)

The National Mission for Clean Ganga (NMCG) is a governmental organization established in 2011 under the Ministry of Jal Shakti, Government of India. NMCG is responsible for the rejuvenation and conservation of the River Ganga and its tributaries. The organization conducts river basin studies, water quality monitoring, and it also develops and implements projects related to river conservation.

NMCG has several divisions, including river basin management, water quality management, and industrial pollution control. The organization works with state governments, NGOs, and other stakeholders to implement various projects related to river conservation, such as the construction of sewage treatment plants and the promotion of eco-tourism.

7. National Institute of Oceanography (NIO)

The National Institute of Oceanography (NIO) is a premier research organization established in 1966 under the Council of Scientific and Industrial Research, Government of India. NIO conducts research and development activities related to oceanography, coastal zone management, and marine biology. The organization also provides training and capacity building programs for marine scientists and professionals.

NIO has several research divisions, including coastal processes, marine ecology, and marine geology. The organization conducts research on topics such as ocean circulation, coastal erosion, and marine biodiversity. NIO is also responsible for implementing various projects related to coastal zone management, such as the development of coastal protection structures and the promotion of sustainable coastal tourism.

8. Indian Council of Agricultural Research (ICAR)

The Indian Council of Agricultural Research (ICAR) is a governmental organization established in 1929 under the Department of Agricultural Research and Education, Government of India. ICAR is responsible for conducting research and development activities related to agriculture, including water use efficiency and crop water management.

ICAR has several research institutions and centers across the country, including the Central Soil and Water Conservation Research and Training Institute (CSWCRTI) and the Indian Institute of Water Management (IIWM). These institutions conduct research on topics such as irrigation management, soil conservation, and watershed management. ICAR also provides training and capacity building programs for farmers and water resources professionals.

9. Centre for Science and Environment (CSE)

The Centre for Science and Environment (CSE) is a non-governmental organization established in 1980. CSE is a research and advocacy organization that works on issues related to environment and sustainable development, including water resources management. The organization conducts research and policy analysis, and it also engages in public awareness and advocacy activities.

CSE has several research divisions, including water management, climate change, and sustainable cities. The organization conducts research on topics such as water governance, water pricing, and water conservation. CSE also provides training and capacity building programs for water resources professionals and NGOs.

10. WaterAid India

WaterAid India is a non-governmental organization established in 1986. WaterAid India works on issues related to water, sanitation, and hygiene (WASH) in India. The organization conducts advocacy and awareness activities, and it also implements projects related to WASH in various parts of the country.

WaterAid India has several programs, including community-based water management, school WASH, and disaster response. The organization works with communities, governments, and other stakeholders to promote access to safe water, sanitation, and hygiene. WaterAid India also conducts research and policy analysis on issues related to WASH.

11. The Energy and Resources Institute (TERI)

The Energy and Resources Institute (TERI) is a research and policy organization established in 1974. TERI conducts research and development activities related to energy, environment, and sustainable development, including water resources management. The organization also provides training and capacity building programs for professionals and policymakers.

TERI has several research divisions, including water resources management, sustainable agriculture, and climate change. The organization conducts research on topics such as water governance, water pricing, and water conservation. TERI also implements various projects related to water resources management, such as the promotion of rainwater harvesting and the development of sustainable irrigation practices.

12. Global Water Partnership - India (GWP-India)

The Global Water Partnership - India (GWP-India) is a non-governmental organization established in 2002. GWP-India works on issues related to water resources management, including water governance and water conservation. The organization conducts advocacy and awareness activities, and it also implements projects related to water resources management.

GWP-India has several programs, including water governance, water security, and climate change adaptation. The organization works with communities, governments, and other stakeholders to promote sustainable water management practices. GWP-India also provides training and capacity building programs for water resources professionals and policymakers.

Conclusion

India faces significant challenges in the management of its water resources. The country has a complex water governance structure, and water management issues are often complicated by political and social factors. However, there are several organizations in India that are working towards sustainable water management practices. These organizations conduct research, advocacy, and implementation activities related to water resources management. They also provide training and capacity building programs for water resources professionals and community members. With the help of these organizations, India can work towards a sustainable and equitable water future.

13 - Challenges and Strategies in Water Resources Management

Abstract

Water resources management is a crucial process for sustaining life on earth and ensuring the availability of water for future generations. However, water managers face numerous challenges in managing water resources effectively. The uneven distribution of water resources, increasing population, climate change, water pollution, conflicting water demands, and lack of investment are some of the major challenges faced in water resources management. To manage water resources sustainably, various strategies can be adopted, such as water conservation, integrated water resources management, water pricing, water reuse, watershed management, and climate change adaptation. This article discusses the various aspects of water resources management, the challenges faced by water managers, and the strategies that can be adopted to manage water resources effectively.

Introduction

Water is a vital natural resource that plays a crucial role in sustaining life on earth. Water resources management is a process of planning, developing, and managing water resources to meet the various needs of human beings and the environment. Water resources management includes the management of both surface water and groundwater resources. The effective management of water resources is crucial for sustainable development and for ensuring the availability of water for future generations. In this article, we will discuss the various aspects of water resources management and the challenges faced by water managers in managing water resources.

Water Resources

Water resources can be classified into two categories: surface water and groundwater. Surface water refers to the water that is available on the earth's surface, such as rivers, lakes, and reservoirs. Groundwater refers to the water that is available beneath the earth's surface, in aquifers. Both surface water and groundwater are interconnected, and the management of one resource can affect the other.

The availability of water resources varies depending on the geographical location and climatic conditions. Some regions have abundant water resources, while others face water scarcity. The uneven distribution of water resources is one of the major challenges faced by water managers.

Water Resources Management

Water resources management is a complex and multidisciplinary field that involves the integration of various disciplines, including engineering, hydrology, economics, and social sciences. Water resources management is essential for ensuring the sustainable use of water resources and for meeting the various needs of human beings and the environment. The process of water resources management involves the following steps:

Assessment of water resources: The first step in water resources management is to assess the available water resources. This includes the measurement of surface water and groundwater flows and the estimation of the quantity and quality of water resources.

Identification of water needs: The next step is to identify the various needs for water resources, including domestic, industrial, agricultural, and environmental needs.

Development of water management strategies: Based on the assessment of water resources and the identification of water needs, water management strategies are developed to optimize the use of water resources.

Implementation of water management strategies: The water management strategies are implemented, and the performance of the water management system is monitored to ensure that it is meeting the desired objectives.

Evaluation of water management strategies: The performance of the water management system is evaluated periodically to determine its effectiveness and to identify areas for improvement.

Challenges Faced in Water Resources Management

Water resources management is a challenging task due to the following reasons:

Uneven distribution of water resources: Water resources are not distributed evenly across the globe. Some regions have abundant water resources, while others face water scarcity. The uneven distribution of water resources makes it challenging to manage water resources effectively.

Increasing population: The world's population is growing rapidly, and this puts pressure on the available water resources. As the population grows, the demand for water increases, making it difficult to manage water resources sustainably.

Climate change: Climate change is affecting the availability of water resources, making it difficult to manage them effectively. Climate change has led to more frequent and severe droughts, floods, and storms, which have a significant impact on water resources.

Water pollution: Water pollution is a major problem that affects the quality of water resources. Water pollution can cause health problems and ecological damage, making it difficult to manage water resources sustainably.

Conflicting water demands: Conflicting water demands from different sectors, such as agriculture, industry, and the environment, make it challenging to manage water resources effectively.

Lack of investment: Lack of investment in water infrastructure and management systems makes it challenging to manage water resources effectively.

Strategies for Water Resources Management

To manage water resources effectively, the following strategies can be adopted:

Water conservation: Water conservation involves the implementation of various measures to reduce water use and wastage. This includes the adoption of water-efficient technologies, the implementation of water conservation policies, and the promotion of water conservation practices among individuals and businesses.

Integrated water resources management: Integrated water resources management (IWRM) is a holistic approach to water resources management that considers the interrelationships between different water users and the environment. IWRM promotes the integration of different sectors and stakeholders in the management of water resources.

Water pricing: Water pricing is an effective tool for managing water resources. By pricing water appropriately, water managers can encourage the efficient use of water resources and discourage wasteful practices.

Water reuse: Water reuse involves the treatment of wastewater and the use of recycled water for various purposes. Water reuse can reduce the demand for freshwater and alleviate the pressure on water resources.

Watershed management: Watershed management involves the management of the entire drainage basin, including both surface water and groundwater resources. This approach promotes the sustainable use of water resources and the protection of the environment.

Climate change adaptation: Climate change adaptation involves the development of strategies to adapt to the impacts of climate change on water resources. This includes the development of drought and flood management plans, the implementation of water conservation measures, and the promotion of water reuse.

Conclusion

Water resources management is a complex and challenging task that requires the integration of various disciplines and stakeholders. The effective management of water resources is essential for sustainable development and for ensuring the availability of water for future generations. Water resources management involves the assessment of water resources, the identification of water needs, the development of water management strategies, the implementation of water management strategies, and the evaluation of water management strategies. The challenges faced in water resources management include the uneven distribution of water resources, increasing population, climate change, water pollution, conflicting water demands, and lack of investment. To manage water resources effectively, various strategies can be adopted, including water conservation, integrated water resources management, water pricing, water reuse, watershed management, and climate change adaptation. The implementation of these strategies can promote the sustainable use of water resources and ensure the availability of water for future generations.

14 - Hydrological Studies: Understanding Water Resources and their Management

Abstract

This article provides an overview of hydrological studies and their importance in water resource management. Hydrological studies involve the analysis and modelling of the water cycle, including precipitation, evapotranspiration, runoff, and groundwater flow. These studies can be conducted using various methods and techniques, such as field measurements, hydrological modelling, and remote sensing. The article highlights the applications of hydrological studies in water resources planning, flood forecasting, environmental management, and disaster risk reduction. The article concludes that hydrological studies are essential in developing strategies for the sustainable management of water resources and can help stakeholders make informed decisions regarding water resource management.

Introduction

Hydrological studies are a critical aspect of water resource management, as they provide insights into the water cycle and help stakeholders make informed decisions regarding water supply, flood protection, and environmental management. In this article, we will explore the various aspects of hydrological studies and their applications in analyzing an area's water resources.

What are Hydrological Studies?

Hydrological studies refer to a systematic investigation of the water cycle and its various components, such as precipitation, evaporation, infiltration, runoff, and groundwater. These studies help in understanding the quantity and quality of water resources available in a region, and how they change over time due to natural and human-induced factors. Hydrological studies can be conducted at various scales, ranging from small catchments to entire river basins, and can cover different time frames, from daily to decadal.

The Importance of Hydrological Studies

Hydrological studies are essential for water resource management, as they provide the basis for understanding the water cycle and the various processes that contribute to the availability, quality, and distribution of water resources. These studies are used to quantify the water balance of an area, which is the difference between the amount of water that enters and leaves the system. The water balance is critical in determining the amount of water available for various uses, such as irrigation, drinking water supply, and hydropower generation.

Hydrological studies also provide insights into the dynamics of floods and droughts, which are recurring natural disasters that can cause significant damage to people and property. By analyzing historical data and modelling future scenarios, hydrologists can predict the frequency, duration, and intensity of these events and develop strategies to mitigate their impacts.

Hydrological studies are also crucial in environmental management, as they help in understanding the interactions between the water cycle and the natural ecosystems. These studies provide information on the water requirements of various plant and animal species and help in identifying areas that are vulnerable to water stress or pollution.

The Hydrological Cycle

The hydrological cycle is a continuous process that involves the movement of water between the atmosphere, land, and oceans. The cycle starts with the evaporation of water from the oceans, lakes, and rivers, which is driven by solar radiation. The water vapor rises into the atmosphere and condenses into clouds, which can precipitate as rain or snow.

The precipitation can either infiltrate into the soil or flow overland as surface runoff, which eventually reaches the rivers and oceans. Some of the water that infiltrates into the soil recharges the groundwater, while the rest is used by plants for transpiration. The groundwater also feeds the rivers and streams, which eventually flow into the oceans, where the cycle starts again.

Hydrological studies aim to quantify the various components of the hydrological cycle and how they interact with each other. These studies use various methods and techniques, such as remote sensing, hydrological modelling, and field measurements, to estimate the volume and timing of precipitation, evapotranspiration, infiltration, runoff, and groundwater recharge.

Hydrological Modelling

Hydrological modelling is a process of using mathematical equations to simulate the various components of the hydrological cycle and their interactions. These models can range from simple empirical relationships to complex physically-based models that incorporate detailed information about the terrain, land use, soil properties, and meteorological conditions.

Hydrological models can be used for various purposes, such as water resources planning, flood forecasting, and environmental impact assessment. These models require input data, such as precipitation, temperature, and land use, and can produce output data, such as streamflow, groundwater level, and water quality.

Hydrological models can also be used to evaluate the impacts of various scenarios, such as climate change, land use change, and water management strategies. These models can provide insights into the potential risks and benefits of these scenarios, and help stakeholders make informed decisions regarding water resource management.

Field Measurements

Field measurements are essential in hydrological studies, as they provide direct observations of various hydrological variables, such as precipitation, streamflow, groundwater level, and water quality. These measurements are used to calibrate and validate hydrological models, as well as to develop empirical relationships between different variables.

Field measurements can be conducted using various techniques and equipment, such as rain gauges, stream gauges, groundwater wells, and water quality sensors. These measurements can be taken at different frequencies, ranging from daily to hourly, depending on the research objectives and available resources.

Remote Sensing

Remote sensing is a technique of collecting information about the Earth's surface and atmosphere using sensors on satellites or aircraft. Remote sensing can provide valuable information about various hydrological variables, such as precipitation, evapotranspiration, and land surface temperature.

Remote sensing data can be used in conjunction with hydrological models to improve their accuracy and spatial resolution. These data can also be used to monitor and forecast natural disasters, such as floods and droughts, and to assess the impacts of land use change and climate change on water resources.

Applications of Hydrological Studies

Hydrological studies have various applications in water resource management, environmental management, and disaster risk reduction. Some of the common applications of hydrological studies are discussed below.

Water Resources Planning

Hydrological studies are essential in water resources planning, as they provide information on the availability and quality of water resources and their variations over time. Hydrological models can be used to simulate different water management scenarios, such as the construction of new dams, the implementation of water conservation measures, and the development of irrigation systems.

These models can help in assessing the potential impacts of these scenarios on the water balance, water quality, and environmental sustainability. Hydrological studies can also help in identifying areas that are vulnerable to water stress or pollution and developing strategies to address these challenges.

Flood Forecasting

Hydrological studies are crucial in flood forecasting, as they provide information on the timing, extent, and intensity of floods. Hydrological models can be used to simulate the response of rivers and streams to different rainfall scenarios, and to generate flood forecasts for different lead times.

These forecasts can be used by emergency managers and decision-makers to take appropriate measures to protect people and property from flooding. Hydrological studies can also help in identifying areas that are vulnerable to flooding and developing flood mitigation strategies, such as the construction of flood control structures and the implementation of land use regulations.

Environmental Management

Hydrological studies are essential in environmental management, as they provide information on the water requirements of various ecosystems and the impacts of water stress and pollution on these systems. Hydrological models can be used to simulate the water balance of different ecosystems, such as wetlands, forests, and agricultural lands, and to assess the impacts of different water management scenarios on these ecosystems.

Hydrological studies can also help in identifying areas that are vulnerable to water stress or pollution and developing strategies to protect these ecosystems. These strategies may include the implementation of water conservation measures, the restoration of degraded ecosystems, and the regulation of land use practices.

Conclusion

Hydrological studies are critical in water resource management, as they provide insights into the water cycle and its various components. These studies help in understanding the availability, quality, and distribution of water resources, as well as the impacts of natural and human-induced factors on these resources. Hydrological studies can be conducted using various methods and techniques, such as hydrological modelling, field measurements, and remote sensing. These studies have various applications in water resource management, environmental management, and disaster risk reduction, and can help stakeholders make informed decisions regarding the sustainable management of water resources.

15 - Assessment of Surface Water Resources: Methods, Challenges, and Implications for Sustainable Water Management

Abstract

Assessment of surface water resources is crucial for managing water resources effectively and sustainably. This article discusses the various methods and techniques used to assess water quantity and quality, including stream gauging, water level measurement, remote sensing, water sampling, biological assessment, water quality index, and real-time water quality monitoring. However, the assessment of surface water resources faces several challenges, such as limited access to water bodies, lack of data, changing climate, and land-use changes. Addressing these challenges requires collaboration among different stakeholders, such as government agencies, researchers, communities, and water users. This article highlights the implications of surface water resource assessment for sustainable water management and the importance of working together to ensure the sustainable use of surface water resources for future generations.

Introduction

Surface water resources, such as rivers, lakes, and streams, play a crucial role in providing water for various human activities, including domestic, industrial, and agricultural use. Surface water resources also support aquatic ecosystems and provide recreational opportunities. However, the availability and quality of surface water resources are threatened by various factors, such as climate change, land-use changes, and pollution. Therefore, the assessment of surface water resources is essential for managing water resources sustainably and ensuring their availability for future generations.

The assessment of surface water resources involves the measurement and analysis of various parameters, such as water quantity, water quality, and ecological health. This article discusses the different methods and techniques used to assess surface water resources, including stream gauging, water level measurement, remote sensing, water sampling, biological assessment, water quality index, and real-time water quality monitoring. The article also highlights the challenges faced in the assessment of surface water resources, such as limited access to water bodies, lack of data, changing climate, and land-use changes.

The article concludes by emphasizing the importance of surface water resource assessment for sustainable water management and the need for collaboration among different stakeholders to address the challenges faced in the assessment of surface water resources. By working together, we can ensure the sustainable use of surface water resources for future generations.

Assessment of Surface Water Resources

Surface water is defined as the water that is found on the surface of the earth in the form of rivers, lakes, ponds, and wetlands. It is a vital resource for human, animal and plant life. The availability of surface water resources is critical for many sectors, such as agriculture, industry, and domestic use. Therefore, it is important to assess the availability of surface water resources in order to manage them effectively and sustainably.

Assessment of surface water resources involves the measurement of water quantity and quality. In this article, we will discuss the methods and techniques used for the assessment of surface water resources.

Water Quantity Assessment

The assessment of water quantity is important to determine the availability of surface water resources. The following are some of the methods and techniques used for the assessment of water quantity:

Stream Gauging

Stream gauging is the most common method used for measuring the flow rate of a river or stream. It involves the installation of a gauging station at a point along the river where the flow rate is representative of the entire stream. The station consists of a streamflow measurement device, such as a current meter or a flow sensor, and a data logger that records the measurements. The data obtained from the gauging station can be used to determine the flow rate, discharge, and water level of the river or stream.

Water Level Measurement

Water level measurement is the method used to determine the water level of a lake, pond, or wetland. This method involves the installation of a water level gauge at a point on the shore of the body of water. The gauge consists of a vertical staff with a graduated scale that is marked in units of water level. The water level is measured by reading the scale at the water surface. The water level measurement is used to determine the volume and surface area of the body of water.

Remote Sensing

Remote sensing is the method used to determine the water quantity using satellite images. Remote sensing involves the use of satellite images to measure the extent and area of surface water bodies. This method is useful for large-scale assessments of water quantity in remote areas. The images obtained can be used to estimate the surface area, volume, and depth of surface water bodies.

Rainfall Measurement

Rainfall measurement is the method used to determine the amount of precipitation that falls on a particular area. This method involves the installation of rain gauges at different points in the area. The gauges measure the amount of rainfall that falls on them. The data obtained from the rain gauges can be used to estimate the amount of water that enters the surface water bodies in the area.

Water Quality Assessment

The assessment of water quality is important to determine the suitability of water for various uses. The following are some of the methods and techniques used for the assessment of water quality:

Water Sampling

Water sampling is the method used to collect water samples from surface water bodies. The samples are collected at different points in the water body and at different times of the day. The samples are then analyzed in the laboratory for various water quality parameters, such as pH, temperature, dissolved oxygen, biochemical oxygen demand (BOD), chemical oxygen demand (COD), total suspended solids (TSS), and nutrients (nitrogen and phosphorus).

Biological Assessment

Biological assessment is the method used to assess the health of the aquatic ecosystem. This method involves the collection of biological samples, such as fish, macroinvertebrates, and algae, from the water body. The samples are then analyzed in the laboratory to determine the diversity and abundance of species. The data obtained can be used to assess the ecological health of the water body.

Water Quality Index

The water quality index (WQI) is a method used to summarize the water quality data obtained from the laboratory analysis of water samples. The WQI is a numerical value that ranges from 0 to 100, with 0 representing the worst water quality and 100 representing the best water quality. The WQI is calculated by assigning weightage factors to the different water quality parameters and then combining them to obtain a single value. The WQI provides a quick and easy way to assess the overall water quality of a surface water body.

Real-time Water Quality Monitoring

Real-time water quality monitoring involves the continuous monitoring of water quality parameters in surface water bodies. This method uses sensors that are installed in the water body to measure various water quality parameters, such as temperature, dissolved oxygen, pH, and conductivity. The data obtained from the sensors are transmitted to a data logger, which records the data and sends it to a central database. Real-time water quality monitoring is useful for detecting changes in water quality over time and for providing early warning of water quality problems.

Assessment of Surface Water Availability

To estimate the volumetric quantity of surface water in rivers, lakes, and reservoirs using data/information from stream gauging, water level measurement, rainfall measurement, and remote sensing, you can follow these general steps:

Collect the necessary data: Collect data on stream gauging, water level measurement, rainfall measurement, and remote sensing for the surface water body you are interested in. This may involve accessing existing data sources, such as government or research institution databases, or collecting new data through field measurements or satellite imagery.

Analyze the data: Use statistical analysis techniques to analyze the data and identify patterns and trends. For example, you can use regression analysis to estimate the relationship between water level and flow rate, or use time series analysis to identify seasonal and long-term trends in rainfall.

Calculate water volume: Use the data and analysis to estimate the volume of water in the surface water body. For example, you can use the flow rate measurements from stream gauging to calculate the volume of water passing through a specific location over a period of time, or use water level measurements and bathymetric data to estimate the volume of water in a lake or reservoir.

Consider uncertainty and accuracy: Remember that there may be uncertainty and errors in the data and analysis, so it is important to consider the accuracy and precision of your estimates. You can use techniques such as error propagation or Monte Carlo simulation to estimate uncertainty and identify areas where more data or analysis is needed.

Evaluate water availability: Once you have estimated the volume of water in the surface water body, you can evaluate the availability of water for different uses, such as irrigation, drinking water, or hydropower generation. This may involve considering factors such as water quality, environmental regulations, and competing demands for water resources.

Overall, estimating the volumetric quantity of surface water in rivers, lakes, and reservoirs using data/information from stream gauging, water level measurement, rainfall measurement, and remote sensing requires careful data collection, analysis, and interpretation. It is important to use appropriate methods and techniques to ensure accuracy and reliability of the estimates, and to consider uncertainty and other factors that may affect water availability.

Challenges in the Assessment of Surface Water Resources

The assessment of surface water resources is a complex task that involves many challenges. The following are some of the challenges faced in the assessment of surface water resources:

Limited Access to Water Bodies

Access to surface water bodies can be limited in some areas due to physical barriers, such as rugged terrain, dense vegetation, or private property. Limited access to water bodies can make it difficult to collect water samples, install gauging stations, or conduct real-time water quality monitoring.

Lack of Data

The assessment of surface water resources requires data on water quantity and quality. However, in many areas, data on water resources are limited or not available. This can be due to the lack of monitoring infrastructure, inadequate funding, or insufficient technical capacity.

Changing Climate

Climate change can have a significant impact on surface water resources. Changes in precipitation patterns, temperature, and weather extremes can affect the quantity and quality of surface water resources. It can also affect the timing and duration of water availability, which can have implications for water management.

Land-use Changes

Land-use changes, such as deforestation, urbanization, and agricultural expansion, can have a significant impact on surface water resources. Land-use changes can alter the water balance of an area, affecting the quantity and quality of surface water resources. It can also affect the ecological health of the water bodies, which can have implications for human and animal health.

Conclusion

The assessment of surface water resources is important for managing water resources effectively and sustainably. The assessment of water quantity and quality requires the use of various methods and techniques, such as stream gauging, water level measurement, remote sensing, water sampling, biological assessment, water quality index, and real-time water quality monitoring. However, the assessment of surface water resources also faces many challenges, such as limited access to water bodies, lack of data, changing climate, and land-use changes. Addressing these challenges requires the collaboration of different stakeholders, such as government agencies, researchers, communities, and water users. By working together, we can ensure the sustainable use of surface water resources for future generations.

16 - Surface Water Balance: Understanding the Water Cycle and Managing Water Resources

Abstract

This article provides an overview of surface water balance, which is a key tool for understanding the water cycle and managing water resources. Surface water balance involves measuring and analyzing the different components of the water balance equation, such as precipitation, evapotranspiration, runoff, total storage, and change in storage, to gain insights into the availability and distribution of water resources over time. The article discusses the various applications of surface water balance, including water resources management, flood forecasting, drought monitoring, climate change studies, and environmental studies. The challenges and limitations of surface water balance are also discussed, such as data availability and quality, spatial and temporal variability, and modelling complexity and uncertainty. Despite these challenges, surface water balance remains an important tool for understanding and managing water resources, and advancements in technology and modelling techniques are helping to improve its accuracy and effectiveness.

Introduction

Surface water balance is a critical concept in hydrology that refers to the quantitative assessment of the amount of water that enters and leaves a given catchment or surface water system. It involves the measurement of various components of the water cycle, such as precipitation, evapotranspiration, infiltration, and runoff, to determine the overall water budget of a particular watershed or water system. Understanding the surface water balance is essential for various water management and planning activities, including irrigation, urban water supply, flood control, and environmental conservation. This article aims to provide an in-depth discussion of the surface water balance, including its components, measurements, and applications.

Components of the Surface Water Balance

The surface water balance can be conceptualized as a simple equation that expresses the total amount of water that enters a given catchment or surface water system and the total amount that leaves it. Mathematically, the water balance equation is expressed as:

$$P = E + T + R + \Delta S$$

where P is the precipitation, E is the evapotranspiration, T is the total amount of water that is stored in the catchment or surface water system, R is the runoff, and ΔS is the change in the amount of water stored in the catchment or surface water system over time. The different components of the water balance equation are discussed in detail below.

Precipitation (P)

Precipitation is the primary input into the surface water system. It includes all forms of moisture that fall from the atmosphere onto the earth's surface, such as rain, snow, sleet, and hail. Precipitation can be measured using various instruments, including rain gauges, snow pillows, and weather radars. Rain gauges are the most common instruments used to measure precipitation, and they work by collecting the amount of rainwater that falls within a specific area. Snow pillows are used to measure the snow-water equivalent, which is the amount of water that would be produced if the snow were to melt. Weather radars use electromagnetic waves to estimate the amount of precipitation in a given area.

Evapotranspiration (E)

Evapotranspiration is the process by which water is returned to the atmosphere from the earth's surface. It involves the combined effects of evaporation and transpiration. Evaporation is the process by which water is converted from its liquid state to its gaseous state, while transpiration is the process by which plants release water vapor through their leaves. Evapotranspiration is influenced by various factors, including temperature, humidity, wind speed, and vegetation cover. It can be measured using various techniques, including the lysimeter method, the Bowen ratio method, and the eddy covariance method.

Total Storage (T)

The total storage component of the water balance equation refers to the amount of water that is stored in the catchment or surface water system over time. It includes water that is stored in various forms, such as groundwater, soil moisture, and surface water. The amount of water stored in the catchment or surface water system is influenced by various factors, including climate, geology, topography, and land use. It can be measured using various techniques, including the water table observation method, the neutron probe method, and the soil moisture probe method.

Runoff (R)

Runoff is the portion of the precipitation that flows over the land surface and eventually enters a stream, river, lake, or other surface water body. It is influenced by various factors, including rainfall intensity, soil type, vegetation cover, and topography. Runoff can be measured using various techniques, including the stream gauging method, the velocity-area method, and the tracer method.

Change in Storage (ΔS)

The change in storage component of the water balance equation refers to the change in the amount of water stored in the catchment or surface water system over time. It can be either positive or negative, indicating whether the system is gaining or losing water over time. The change in storage component can be measured using various techniques, including the water balance method, the remote sensing method, and the hydrological modelling method.

Measurement of Surface Water Balance Components

The different components of the surface water balance can be measured using various techniques and methods, depending on the availability of data, the spatial and temporal scales of interest, and the level of accuracy required. Some of the common methods used for measuring the different components of the surface water balance are discussed below.

Measurement of Precipitation

Precipitation can be measured using various techniques, including rain gauges, weather radars, and satellite-based sensors. Rain gauges are the most common instruments used to measure precipitation, and they work by collecting the amount of rainwater that falls within a specific area. Weather radars use electromagnetic waves to estimate the amount of precipitation in a given area. Satellite-based sensors, such as the Tropical Rainfall Measuring Mission (TRMM) and the Global Precipitation Measurement (GPM), provide global coverage of precipitation with high spatial and temporal resolution.

Measurement of Evapotranspiration

Evapotranspiration can be measured using various techniques, including the lysimeter method, the Bowen ratio method, and the eddy covariance method. The lysimeter method involves measuring the amount of water that is lost through evapotranspiration from a specific area of land enclosed in a container called a lysimeter. The Bowen ratio method involves measuring the temperature and humidity gradients between the soil surface and the atmosphere to estimate the rate of evapotranspiration. The eddy covariance method involves measuring the turbulent fluxes of heat, water vapor, and carbon dioxide between the surface and the atmosphere using specialized instruments.

Measurement of Total Storage

The total storage component of the water balance equation can be measured using various techniques, including the water table observation method, the neutron probe method, and the soil moisture probe method. The water table observation method involves measuring the depth of the water table at specific locations in the catchment or surface water system. The neutron probe method involves measuring the amount of moisture in the soil using a neutron probe that emits and detects neutrons. The soil moisture probe method involves measuring the amount of moisture in the soil using a probe that measures the electrical conductivity of the soil.

Measurement of Runoff

Runoff can be measured using various techniques, including the stream gauging method, the velocity-area method, and the tracer method. The stream gauging method involves measuring the water level and flow rate of a stream or river using specialized instruments, such as a stream gauge. The velocity-area method involves measuring the velocity of the water and the cross-sectional area of the stream or river to estimate the flow rate. The tracer method involves adding a known amount of a tracer substance, such as a dye or a radioactive isotope, to the water and measuring its concentration downstream to estimate the flow rate.

Measurement of Change in Storage

The change in storage component of the water balance equation can be measured using various techniques, including the water balance method, the remote sensing method, and the hydrological modelling method. The water balance method involves accounting for the different components of the water balance equation and comparing the inputs and outputs over a specific period to estimate the change in storage. The remote sensing method involves using satellite-based sensors, such as the Gravity Recovery and Climate Experiment (GRACE) mission, to estimate the change in the amount of water stored in a catchment or surface water system over time. The hydrological modelling method involves using mathematical models to simulate the different components of the water cycle and estimate the change in storage over time.

Applications of Surface Water Balance

The surface water balance has various applications in hydrology, water resources management, and environmental studies. Some of the key applications of surface water balance are discussed below.

Water resources management: Surface water balance is an important tool for managing water resources in a catchment or surface water system. It helps to identify the sources and sinks of water within the system and provides information on the availability and distribution of water resources over time. This information can be used to plan and implement various water management strategies, such as water conservation, irrigation, and flood control.

Flood forecasting: Surface water balance can also be used for flood forecasting by providing information on the volume and timing of water flows in a catchment or surface water system. This information can be used to issue early warnings and alerts to communities and authorities to take appropriate measures to mitigate the impacts of floods, such as evacuations and flood-proofing measures.

Drought monitoring: Surface water balance can also be used for monitoring drought conditions by providing information on the amount and distribution of water resources in a catchment or surface water system over time. This information can be used to assess the severity and extent of drought conditions and to develop strategies to mitigate the impacts of drought, such as water conservation and crop management practices.

Climate change studies: Surface water balance can also be used for studying the impacts of climate change on the water cycle and water resources. By analyzing changes in the different components of the water balance equation, such as precipitation, evapotranspiration, and runoff, over time, scientists can gain insights into how climate change is affecting the water cycle and water resources in a specific region or globally.

Environmental studies: Surface water balance can also be used for studying the impacts of human activities and land use changes on the water cycle and water resources. By analyzing changes in the different components of the water balance equation, such as land cover, precipitation, evapotranspiration, and runoff, over time, scientists can gain insights into how human activities and land use changes are affecting the water cycle and water resources in a specific region or ecosystem.

Challenges and Limitations of Surface Water Balance

Despite its many applications and benefits, surface water balance also has several challenges and limitations that need to be considered when using it as a tool for water resources management and environmental studies. Some of the key challenges and limitations of surface water balance are discussed below.

Data availability and quality: One of the main challenges of surface water balance is the availability and quality of data required to measure and analyze the different components of the water balance equation. In many regions, especially in developing countries, there is a lack of data on precipitation, evapotranspiration, and other water balance components, making it difficult to accurately estimate water resources availability and distribution.

Spatial and temporal variability: Another challenge of surface water balance is the spatial and temporal variability of the different water balance components within a catchment or surface water system. Precipitation, evapotranspiration, and runoff can vary significantly over space and time, making it difficult to accurately estimate water resources availability and distribution.

Modelling complexity and uncertainty: Surface water balance also involves the use of mathematical models to simulate the different components of the water cycle and estimate the change in storage over time. These models can be complex and require significant expertise and computing resources to develop and run. Additionally, there is often uncertainty associated with the models' outputs, making it difficult to accurately predict water resources availability and distribution.

Conclusion

Surface water balance is an important tool for managing water resources and studying the water cycle and its impacts on the environment. By measuring and analyzing the different components of the water balance equation, such as precipitation, evapotranspiration, runoff, total storage, and change in storage, scientists and water managers can gain insights into the availability and distribution of water resources over time. However, surface water balance also has several challenges and limitations that need to be considered, including data availability and quality, spatial and temporal variability, and modelling complexity and uncertainty.

Despite these challenges, surface water balance remains an important tool for understanding and managing water resources. Advances in remote sensing technology and data processing techniques are improving the accuracy and availability of data on water balance components, making it easier to estimate water resources availability and distribution over time. Additionally, the development of more advanced and sophisticated modelling techniques is helping to reduce the uncertainty associated with surface water balance modelling.

Overall, surface water balance is a valuable tool for understanding and managing water resources in a changing climate and environment. By using this tool effectively, scientists and water managers can develop more effective strategies for conserving and managing water resources, mitigating the impacts of floods and droughts, and protecting the environment.

17 - Determination of Optimal Cropping Pattern: A Guide for Agricultural Planning

Abstract

Choosing the right cropping pattern is critical for agricultural productivity and sustainability. It involves identifying the crops that can be grown in a particular area, determining their suitability based on climate, soil type, water availability, and market demand, and devising a plan that maximizes yield and profitability. This article provides a comprehensive guide to determining the optimal cropping pattern, including key factors to consider, methods of analysis, and strategies for planning and implementation. We discuss various approaches to crop selection, including agronomic, economic, and ecological considerations, and highlight the importance of diversification and crop rotation. We also explore the use of tools such as crop simulation models, remote sensing, and GIS mapping in optimizing crop selection and management. By providing a holistic approach to agricultural planning, this article aims to help farmers and policymakers make informed decisions that lead to sustainable and profitable agricultural practices.

Introduction

Agriculture is the backbone of the world's economy, providing food, fiber, and fuel for the ever-growing population. However, with the increasing demands of food security, the effects of climate change, and the need for sustainable practices, the importance of choosing the right cropping pattern cannot be overstated. A cropping pattern refers to the specific combination of crops that a farmer grows in a particular area, and it plays a crucial role in determining the productivity and sustainability of agriculture. A well-designed cropping pattern can help maximize yield, optimize resource use, minimize risks, and improve the quality of soil and water resources.

Determining the optimal cropping pattern requires careful consideration of several factors, including climate, soil type, water availability, market demand, and crop suitability. With the help of various analytical tools and approaches, farmers and policymakers can make informed decisions that lead to sustainable and profitable agricultural practices. In this article, we provide a comprehensive guide to determining the optimal cropping pattern, including key factors to consider, methods of analysis, and strategies for planning and implementation.

Factors to Consider in Determining Optimal Cropping Pattern

Several factors need to be considered when determining the optimal cropping pattern. These factors are briefly discussed below.

Climate: Climate is one of the most critical factors that influence the choice of crops. Different crops have different temperature and rainfall requirements, and the suitability of a particular crop for a particular area depends on the prevailing climatic conditions. The length of the growing season, frost-free period, and the amount of sunlight available are also critical factors to consider.

Soil Type: Soil type plays a crucial role in crop selection, as it determines the availability of nutrients, water, and other essential elements required for plant growth. Different crops have different soil requirements, and the soil's physical and chemical properties can significantly affect crop yield and quality.

Water Availability: Water availability is another essential factor to consider when choosing a cropping pattern. Different crops have different water requirements, and the availability of water can significantly affect crop yield and quality. The type of irrigation system and the timing and amount of water application are also critical factors to consider.

Market Demand: Market demand for a particular crop is an essential factor that determines its profitability. The demand for crops can vary depending on the region, season, and consumer preferences, and choosing crops with high demand can lead to better returns.

Crop Suitability: Crop suitability refers to the ability of a crop to grow and produce under specific conditions. The crop's adaptability to the local environment, pest and disease resistance, and the crop's ability to produce high yields are essential factors to consider.

Methods of Analysis

Several analytical tools and approaches can be used to determine the optimal cropping pattern. The most commonly used methods are briefly discussed below.

Agronomic Considerations: Agronomic considerations involve evaluating the soil and climatic conditions of a particular area to determine which crops are most suitable for cultivation. Agronomic considerations take into account factors such as crop nutrient requirements, soil type, water availability, and climate. This approach involves selecting crops based on their adaptability to local conditions and the availability of necessary resources. Agronomic considerations also involve crop rotation and intercropping, which can help optimize resource use and minimize soil degradation.

Economic Considerations: Economic considerations involve evaluating the profitability of different crops based on market demand, production costs, and expected yields. This approach involves selecting crops that have a high demand and are profitable to grow. Economic considerations also involve crop diversification and crop rotation, which can help minimize risks and optimize returns.

Ecological Considerations: Ecological considerations involve selecting crops that promote ecological sustainability, including biodiversity, soil health, and water conservation. This approach involves selecting crops that are adapted to local environmental conditions and can contribute to ecosystem services such as soil fertility and water infiltration. Ecological considerations also involve crop rotation and intercropping, which can help promote soil health and minimize environmental degradation.

Remote Sensing and GIS Mapping: Remote sensing and GIS mapping are powerful tools that can be used to determine the optimal cropping pattern. Remote sensing can be used to monitor crop growth, identify areas of soil degradation, and map the distribution of different crops. GIS mapping can be used to analyze the spatial relationships between different crops, soil type, water availability, and other factors to determine the most optimal cropping pattern.

Strategies for Planning and Implementation

After selecting the optimal cropping pattern, the next step is to develop a plan for implementation. The following strategies can be used to ensure successful implementation of the cropping pattern.

Crop Rotation: Crop rotation involves growing different crops in a particular area in a planned sequence to optimize soil fertility, minimize pest and disease problems, and promote long-term sustainability. Crop rotation involves selecting crops that have different nutrient requirements, growth patterns, and pest and disease resistance.

Intercropping: Intercropping involves growing two or more crops in the same area to optimize resource use and minimize risk. Intercropping can help improve soil health, increase yields, and reduce pest and disease problems.

Sustainable Soil Management: Sustainable soil management involves adopting practices that promote soil health and minimize soil degradation. This includes practices such as conservation tillage, cover cropping, and soil testing.

Irrigation Management: Irrigation management involves optimizing water use and minimizing waste. This includes selecting the appropriate irrigation system, timing and amount of water application, and soil moisture monitoring.

Conclusion

The optimal cropping pattern plays a crucial role in determining the productivity and sustainability of agriculture. Determining the optimal cropping pattern requires careful consideration of several factors, including climate, soil type, water availability, market demand, and crop suitability. Various analytical tools and approaches can be used to determine the optimal cropping pattern, including agronomic, economic, and ecological considerations, as well as remote sensing and GIS mapping. Strategies for planning and implementation of the cropping pattern include crop rotation, intercropping, sustainable soil management, and irrigation management. By following a holistic approach to agricultural planning, farmers and policymakers can make informed decisions that lead to sustainable and profitable agricultural practices.

18 - Managing Soil Moisture Balance for Sustainable Agriculture: Strategies and Implications

Abstract

The soil moisture balance is a critical aspect of agriculture and environmental sustainability, as it determines the water availability for crops and affects the health of ecosystems. The soil moisture balance is the difference between the amount of water that enters the soil and the amount of water that leaves the soil through evaporation, transpiration, and drainage. A positive soil moisture balance is desirable for crop growth, while a negative soil moisture balance can lead to water stress and reduced yields. Managing the soil moisture balance involves various strategies such as irrigation, soil management, crop management, and climate-smart agriculture. These strategies can help improve the water-holding capacity of soil, reduce water loss, and increase agricultural productivity. This article provides an overview of the soil moisture balance, its implications for agriculture, and the strategies for managing it.

Introduction

Soil moisture balance is an important aspect of agricultural and environmental sciences. The balance of soil moisture is critical for crop growth, as plants require a certain amount of water to thrive. Soil moisture balance is the difference between the amount of water that enters the soil and the amount of water that leaves the soil through evaporation, transpiration, and drainage. A positive soil moisture balance indicates that the soil is wetter than it was previously, while a negative soil moisture balance indicates that the soil is drier than it was previously.

Soil Moisture

Soil moisture is the water that is held in the pores and spaces between soil particles. The amount of water that soil can hold is determined by its texture, structure, and composition. Soils with a high clay content can hold more water than soils with a high sand content because clay particles are smaller and have more surface area to hold onto water molecules.

Soil moisture is critical for plant growth because it is necessary for many biological processes, such as photosynthesis, respiration, and nutrient uptake. The amount of water that plants require varies depending on the species, growth stage, and environmental conditions.

Soil Moisture Balance Components

The soil moisture balance is the difference between the amount of water that enters the soil and the amount of water that leaves the soil. The components of the soil moisture balance include precipitation, evapotranspiration, and drainage.

Precipitation

Precipitation is the primary source of water for soil. It is the process by which water falls from the atmosphere to the surface of the Earth. Precipitation can be in the form of rain, snow, or hail, and its amount and intensity vary depending on location, season, and weather conditions.

The amount of water that enters the soil through precipitation is called infiltration. Infiltration depends on the intensity and duration of precipitation, soil texture, and soil structure. If precipitation intensity is high, soil structure is poor, or soil is compacted, infiltration can be reduced, leading to surface runoff.

Evapotranspiration

Evapotranspiration is the process by which water is lost from the soil through evaporation and transpiration. Evaporation is the process by which water changes from a liquid to a gas and enters the atmosphere. Transpiration is the process by which water is absorbed by plant roots, travels through the plant, and is released into the atmosphere through small openings in the leaves called stomata.

The amount of water lost through evapotranspiration depends on environmental factors such as temperature, humidity, wind, and solar radiation, as well as plant characteristics such as leaf area, stomatal conductance, and root depth.

Drainage

Drainage is the process by which excess water leaves the soil and enters groundwater or surface water. Drainage occurs when the amount of water entering the soil exceeds the amount of water that can be stored or used by plants. The amount of drainage depends on soil properties such as texture, structure, and permeability.

If the soil is too permeable, water may drain too quickly, leading to a negative soil moisture balance. On the other hand, if the soil is too impermeable, water may accumulate, leading to waterlogging and other problems.

Soil Moisture Balance Equation

The soil moisture balance equation is a simple accounting tool that is used to estimate the amount of water in the soil at any given time. The equation is based on the principle of conservation of mass, which states that the amount of water in the soil cannot change unless there is a change in the inputs or outputs of the system.

The soil moisture balance equation is as follows:

$\Delta S = P - ET - D$

Where:

ΔS = change in soil moisture storage

P = precipitation
ET = evapotranspiration
D = drainage

The change in soil moisture storage (ΔS) is the difference between the amount of water that enters the soil through precipitation (P), and the amount of water that leaves the soil through evapotranspiration (ET) and drainage (D).

If the soil moisture balance is positive ($\Delta S > 0$), it means that the soil is gaining water, and the soil moisture content is increasing. If the soil moisture balance is negative ($\Delta S < 0$), it means that the soil is losing water, and the soil moisture content is decreasing.

Interpreting the Soil Moisture Balance

The soil moisture balance is an important tool for farmers, agronomists, and environmental scientists. By monitoring the soil moisture balance, it is possible to determine the water requirements of crops and predict the risk of drought or waterlogging.

Positive Soil Moisture Balance

A positive soil moisture balance is desirable for crop growth, as it indicates that the soil is wetter than it was previously. A positive soil moisture balance can occur when:

- There is sufficient rainfall to meet the water requirements of crops
- The soil has good water-holding capacity and can store excess water
- The rate of evapotranspiration is low due to low temperatures or high humidity

Negative Soil Moisture Balance

A negative soil moisture balance is undesirable for crop growth, as it indicates that the soil is drier than it was previously. A negative soil moisture balance can occur when:

- There is insufficient rainfall to meet the water requirements of crops
- The soil has poor water-holding capacity and cannot store excess water
- The rate of evapotranspiration is high due to high temperatures or low humidity

Managing Soil Moisture Balance

Managing the soil moisture balance is critical for crop growth and environmental sustainability. Farmers and agronomists can use several strategies to manage the soil moisture balance, including:

Irrigation

Irrigation is the process of adding water to crops artificially. Irrigation can be done using surface water, groundwater, or recycled water. Irrigation can help maintain a positive soil moisture balance and provide water to crops during dry periods.

Soil Management

Soil management practices can help improve the water-holding capacity of soil and reduce the risk of waterlogging. Soil management practices include:

- Adding organic matter to the soil
- Reducing soil compaction
- Improving soil structure

Crop Management

Crop management practices can help reduce water loss through evapotranspiration and improve the efficiency of water use. Crop management practices include:

- Planting drought-tolerant crops
- Using crop rotation to manage soil moisture
- Reducing tillage to conserve soil moisture

Climate Smart Agriculture

Climate-smart agriculture is an approach to farming that focuses on reducing greenhouse gas emissions, adapting to climate change, and increasing agricultural productivity. Climate-smart agriculture practices can help manage the soil moisture balance by:

- Improving water-use efficiency
- Reducing soil erosion and nutrient loss
- Increasing soil carbon sequestration

Conclusion

The soil moisture balance is an important aspect of agricultural and environmental sciences. The balance of soil moisture is critical for crop growth, as plants require a certain amount of water to thrive. The soil moisture balance is the difference between the amount of water that enters the soil and the amount of water that leaves the soil through evaporation, transpiration, and drainage. Positive soil moisture balance is desirable for crop growth, while negative soil moisture balance is undesirable. Managing the soil moisture balance is critical for crop growth and environmental sustainability. Farmers and agronomists can use several strategies to manage the soil moisture balance, including irrigation, soil management, crop management, and climate-smart agriculture.

19 - Estimation of Soil Erosion: Methods, Challenges, and Applications

Abstract

Soil erosion is a natural process that occurs due to various factors such as rainfall, wind, and human activities. It poses a significant threat to soil fertility, biodiversity, and the environment in general. Therefore, estimation of soil erosion is essential for the sustainable management of land resources. Several methods have been developed to estimate soil erosion, ranging from field-based to modelling approaches. Each method has its advantages and limitations, and the choice of method depends on the available resources and the specific requirements of the study. This article provides an overview of the various methods used for estimating soil erosion, their applications, and the challenges associated with each method. Additionally, the article highlights the importance of soil erosion estimation for sustainable land management and the need for continued research in this field.

Introduction

Soil erosion is the process of detachment, transport, and deposition of soil particles from one location to another due to the influence of external factors such as rainfall, wind, and human activities. Soil erosion is a natural phenomenon that can occur in both agricultural and non-agricultural areas. However, it poses a significant threat to soil fertility, biodiversity, and the environment in general, especially when it occurs at a higher rate than the soil's ability to regenerate. Soil erosion can lead to decreased crop productivity, soil degradation, and water pollution, among other adverse effects.

Estimation of soil erosion is essential for the sustainable management of land resources. It provides information on the extent and rate of soil loss and enables the development of appropriate soil conservation measures. Several methods have been developed for estimating soil erosion, ranging from field-based to modelling approaches. Each method has its advantages and limitations, and the choice of method depends on the available resources and the specific requirements of the study.

This article provides an overview of the various methods used for estimating soil erosion, their applications, and the challenges associated with each method. Additionally, the article highlights the importance of soil erosion estimation for sustainable land management and the need for continued research in this field.

Field-based Methods

Field-based methods involve direct measurement of soil loss from a specific area using erosion plots, sediment traps, or erosion pins. These methods are relatively simple, inexpensive, and provide accurate estimates of soil loss at the field scale. However, they require a considerable amount of labor and time, and their applicability is limited to small areas.

Erosion Plots

Erosion plots are small areas of land, typically 2-10 square meters, where soil erosion is measured by direct observation of the soil surface. The soil surface is typically covered with vegetation or mulch to simulate natural conditions. The amount of soil loss is measured by weighing the soil removed from the plot or by measuring the depth of the soil removed. Erosion plots are useful for evaluating the effectiveness of soil conservation practices such as contour farming, terracing, and cover cropping.

Sediment Traps

Sediment traps are structures that capture and measure sediment that is transported downslope by runoff. Sediment traps consist of a collection basin that is placed at the bottom of a sloping field or drainage area. The basin is designed to capture sediment-laden runoff and measure the sediment's weight or volume. Sediment traps are useful for estimating soil loss in areas where erosion is driven by surface runoff.

Erosion Pins

Erosion pins are metal rods that are inserted into the soil at a specific depth and orientation. The pins are installed before the onset of erosion and are retrieved after a specified period. The amount of soil loss is estimated by measuring the length of the exposed pin above the soil surface. Erosion pins are useful for estimating soil loss in areas where erosion is driven by rainfall or wind.

Modelling Methods

Modelling methods are used to estimate soil erosion over large areas and extended periods. These methods use mathematical models to simulate the various processes involved in soil erosion, such as rainfall, runoff, infiltration, and sediment transport. Modelling methods are useful for assessing soil erosion at a regional or watershed scale and for predicting soil erosion under different land use and management scenarios. However, they require a considerable amount of data and computational resources and may be less accurate than field-based methods.

Empirical models: Empirical models are based on field measurements of soil erosion and use statistical relationships between soil erosion and environmental factors such as rainfall, slope, soil texture, and land use. The most commonly used empirical model for estimating soil erosion is the Universal Soil Loss Equation (USLE) and its revised version, the Revised Universal Soil Loss Equation (RUSLE). The USLE and RUSLE estimate soil erosion as a function of rainfall erosivity, soil erodibility, slope length and steepness, cover management factor, and support practice factor. These factors can be estimated using field measurements or secondary data sources such as soil maps and climate data.

Process-based models: Process-based models simulate the physical processes that cause soil erosion and predict soil loss rates based on the physical properties of soil and vegetation, as well as environmental factors such as rainfall, wind, and slope. The most widely used process-based model for soil erosion estimation is the Water Erosion Prediction Project (WEPP) model. The WEPP model simulates soil erosion by integrating soil hydrology, erosion mechanics, and plant growth. The WEPP model requires input data on soil properties, vegetation cover, climate, and topography.

Geospatial models: Geospatial models use spatially explicit data sources, such as satellite imagery, remote sensing, and geographic information systems (GIS), to estimate soil erosion at regional or global scales. Geospatial models combine environmental factors such as rainfall, slope, soil type, and land use with spatial data on soil erosion rates to estimate soil erosion. Examples of geospatial models for soil erosion estimation include the Revised Universal Soil Loss Equation GIS (RUSLE GIS) model and the Soil and Water Assessment Tool (SWAT) model. The RUSLE GIS model combines the USLE with GIS data on soil properties, land use, and topography to estimate soil erosion rates. The SWAT model simulates the hydrology and water quality of large watersheds and includes soil erosion as one of its output variables.

In summary, empirical models, process-based models, and geospatial models are all useful tools for estimating soil erosion. Each type of model has its strengths and limitations, and the choice of model should depend on the scale of interest, data availability, and accuracy requirements.

Challenges and Applications

Estimating soil erosion is a challenging task due to the complexity and variability of the factors that influence erosion. Several challenges need to be addressed to improve the accuracy and applicability of soil erosion estimation methods. These challenges include:

Data Availability

Accurate estimation of soil erosion requires detailed input data on soil properties, topography, climate, and land use. However, data availability may be limited, especially in developing countries, where data collection and management systems are weak. Therefore, efforts should be made to improve data collection and sharing mechanisms to facilitate soil erosion estimation.

Model Validation

The accuracy of soil erosion estimation models depends on their ability to reproduce observed erosion rates. However, model validation is often challenging due to the lack of long-term erosion data and the complexity of the factors that influence erosion. Therefore, efforts should be made to collect long-term erosion data and to validate models using independent data sets.

Scale

Soil erosion occurs at different scales, ranging from individual fields to watersheds and regions. Therefore, the choice of method for estimating soil erosion depends on the scale of interest.

However, methods that are applicable at a small scale may not be applicable at a larger scale, and vice versa. Therefore, efforts should be made to develop methods that are applicable at different scales.

Estimation of soil erosion has several applications in sustainable land management. These applications include:

Soil Conservation

Estimation of soil erosion is essential for developing appropriate soil conservation measures, such as contour farming, terracing, and cover cropping. Soil conservation measures can reduce soil loss, improve soil fertility, and increase crop productivity.

Watershed Management

Estimation of soil erosion is useful for watershed management, where it can be used to identify areas with high erosion risk and to prioritize soil conservation measures. Soil conservation measures can reduce sediment delivery to rivers and streams, thus improving water quality and aquatic habitat.

Climate Change Mitigation

Estimation of soil erosion can contribute to climate change mitigation efforts by reducing the amount of carbon released from eroded soils. Soil erosion can result in the loss of soil organic matter, which is a significant source of carbon. Therefore, soil conservation measures can help to sequester carbon in soils and reduce greenhouse gas emissions.

Conclusion

Estimation of soil erosion is an essential task for sustainable land management. Soil erosion can result in the loss of soil fertility, reduced crop productivity, and degraded ecosystem services. Therefore, efforts should be made to develop and apply accurate and reliable methods for estimating soil erosion. Several methods, including field-based methods, modelling methods, and geospatial methods, are available for estimating soil erosion. Each method has its strengths and limitations and should be chosen based on the scale of interest, data availability, and accuracy requirements. Improving data collection and sharing mechanisms and validating models using independent data sets can help to address the challenges associated with estimating soil erosion. Estimation of soil erosion has several applications in soil conservation, watershed management, and climate change mitigation, and can contribute to achieving sustainable land management practices.

20 - Drought Risk Assessment and Management: Strategies for Mitigating the Impact of Drought

Abstract

This article explores the importance of drought risk assessment and management in mitigating the impact of drought on communities and ecosystems. Drought risk assessment involves identifying areas that are vulnerable to drought and evaluating the potential impact of a drought on those areas. Drought management involves implementing measures to reduce the impact of drought and increase the resilience of communities to its effects. This article discusses various methods for drought risk assessment, including drought indices, remote sensing, hydrological modelling, and vulnerability assessment. Drought management strategies discussed in this article include water conservation, drought monitoring and early warning systems, drought contingency planning, agricultural management, ecosystem management, and public education and outreach. Additionally, the article addresses the challenges and limitations of drought risk assessment and management, such as limited data availability, resource constraints, lack of coordination and cooperation, and the impact of climate change. Overall, this article emphasizes the need for a coordinated and collaborative effort among different stakeholders to reduce the impact of drought and increase the resilience of communities and ecosystems to its effects.

Introduction

Drought is a natural phenomenon that can result in severe social, economic, and environmental consequences. It occurs when the amount of rainfall or precipitation in a given area is less than what is needed to sustain normal vegetation, water supply, and other uses. Droughts are a common occurrence in many parts of the world and can have devastating effects on agriculture, water resources, and human health.

Drought risk assessment and management are essential components of drought preparedness and mitigation strategies. Drought risk assessment involves identifying areas that are vulnerable to drought and evaluating the potential impact of a drought on those areas. Drought management involves implementing measures to reduce the impact of drought and increase the resilience of communities to its effects.

This article provides an overview of drought risk assessment and management, including its importance, methods, and strategies.

Importance of Drought Risk Assessment and Management

Drought risk assessment and management are critical because they enable decision-makers to understand the potential impact of drought on communities and ecosystems. By identifying areas that are vulnerable to drought, decision-makers can prioritize resources and implement measures to reduce the impact of drought.

Drought risk assessment is also essential because it enables decision-makers to plan for droughts before they occur. This proactive approach is much more effective than a reactive approach that relies on emergency measures after a drought has already occurred.

Drought management is also crucial because it helps to increase the resilience of communities to droughts. By implementing measures that reduce the impact of drought, decision-makers can help communities recover more quickly from the effects of drought.

Drought Risk Assessment

Drought risk assessment involves evaluating the potential impact of a drought on a given area. This assessment is typically based on several factors, including the climate and weather patterns, the availability of water resources, and the vulnerability of communities and ecosystems. There are several methods for drought risk assessment, including:

Drought indices

Drought indices are tools that use climate and weather data to evaluate the severity and duration of droughts. These indices are used to monitor drought conditions and to predict the potential impact of drought on communities and ecosystems. Some of the most common drought indices include:

- Palmer Drought Severity Index (PDSI)
- Standardized Precipitation Index (SPI)
- Crop Moisture Index (CMI)

Remote sensing

Remote sensing involves the use of satellite imagery to monitor vegetation health and water availability. This method can be used to detect drought conditions and to evaluate the potential impact of drought on ecosystems and agricultural lands.

Hydrological modelling

Hydrological modelling involves using computer models to simulate the movement of water through a watershed or river basin. This method can be used to evaluate the potential impact of drought on water resources and to identify areas that are vulnerable to drought.

Vulnerability assessment

Vulnerability assessment involves evaluating the vulnerability of communities and ecosystems to the effects of drought. This assessment typically includes an evaluation of the social, economic, and environmental factors that can increase vulnerability to drought.

Drought Management

Drought management involves implementing measures to reduce the impact of drought and increase the resilience of communities and ecosystems to its effects. Some of the most common drought management strategies include:

Water conservation

Water conservation involves reducing water use to ensure that available water resources are used efficiently. This can include measures such as water reuse, leak detection, and the implementation of water-efficient technologies.

Drought monitoring and early warning systems

Drought monitoring and early warning systems involve using drought indices and other tools to monitor drought conditions and predict the potential impact of drought on communities and ecosystems. This enables decision-makers to take proactive measures to reduce the impact of drought.

Drought contingency planning

Drought contingency planning involves developing plans to manage the effects of drought. This can include measures such as water rationing, the implementation of emergency water supplies, and the prioritization of water uses.

Agricultural management

Agricultural management involves implementing measures to reduce the impact of drought on agricultural lands. This can include measures such as crop diversification, the use of drought-resistant crops, and improved irrigation practices.

Ecosystem management

Ecosystem management involves implementing measures to reduce the impact of drought on ecosystems. This can include measures such as the restoration of wetlands, the implementation of water conservation measures in protected areas, and the establishment of drought-resistant plant species.

Public education and outreach

Public education and outreach are essential components of drought management. Educating the public about the impacts of drought and the measures that can be taken to reduce its effects can help to increase awareness and promote behavior change.

Challenges and Limitations of Drought Risk Assessment and Management

Despite the importance of drought risk assessment and management, there are several challenges and limitations that must be addressed. Some of the most common challenges and limitations include:

Limited data availability

Drought risk assessment and management rely heavily on climate and weather data. However, in many regions, there is a lack of reliable data on precipitation, temperature, and other climatic variables. This can limit the accuracy of drought risk assessments and the effectiveness of drought management strategies.

Resource constraints

Implementing effective drought management strategies requires significant resources, including financial resources, technical expertise, and institutional capacity. In many regions, these resources are limited, making it difficult to implement effective drought management strategies.

Lack of coordination and cooperation

Drought risk assessment and management require coordination and cooperation among different stakeholders, including government agencies, non-governmental organizations, and local communities. However, in many regions, there is a lack of coordination and cooperation among these stakeholders, which can limit the effectiveness of drought management strategies.

Climate change

Climate change is expected to increase the frequency and severity of droughts in many regions. This can make it more difficult to implement effective drought management strategies and increase the vulnerability of communities and ecosystems to the effects of drought.

Conclusion

Drought risk assessment and management are essential components of drought preparedness and mitigation strategies. Drought risk assessment involves identifying areas that are vulnerable to drought and evaluating the potential impact of a drought on those areas. Drought management involves implementing measures to reduce the impact of drought and increase the resilience of communities to its effects.

There are several methods for drought risk assessment, including drought indices, remote sensing, hydrological modelling, and vulnerability assessment. Drought management strategies include water conservation, drought monitoring and early warning systems, drought contingency planning, agricultural management, ecosystem management, and public education and outreach.

Despite the importance of drought risk assessment and management, there are several challenges and limitations that must be addressed, including limited data availability, resource constraints, lack of coordination and cooperation, and the impact of climate change.

Addressing these challenges and limitations will require a coordinated and collaborative effort among different stakeholders, including government agencies, non-governmental organizations, and local communities. By working together, we can reduce the impact of drought and increase the resilience of communities and ecosystems to its effects.

21 - Flood Risk Assessment and Management: Strategies for Mitigating the Impacts of Floods

Abstract

Floods are one of the most common and destructive natural disasters, causing significant economic and social damages and loss of life worldwide. Effective flood risk assessment and management are essential to mitigate the impacts of floods and improve the resilience of communities to future flood events. This article provides an overview of flood risk assessment and management, including the process of flood risk assessment, the development and implementation of flood risk management policies and strategies, and the role of different stakeholders in flood risk management. It also highlights the challenges of flood risk management and the multidisciplinary approach required to address them. By working together, governments, communities, NGOs, and private sector organizations can develop and implement effective flood risk management policies and strategies that can reduce the impact of floods and improve the resilience of communities to future flood events.

Introduction

Flood risk assessment and management are critical components of disaster risk reduction. Floods are one of the most common and destructive natural disasters, causing significant economic and social damages, and loss of life worldwide. Floods can be caused by various factors, including heavy rainfall, snowmelt, coastal storm surges, and riverine flooding, among others. Effective flood risk assessment and management are essential to mitigate the impacts of floods and improve the resilience of communities to future flood events.

This article provides an overview of flood risk assessment and management, including its definition, importance, and methods. It also highlights the challenges of flood risk management and the role of different stakeholders in addressing flood risk.

What is Flood Risk Assessment?

Flood risk assessment is the process of evaluating the probability of flooding and its potential consequences. It involves the identification of flood hazards, vulnerabilities, and impacts, as well as the assessment of the likelihood and severity of future flooding. The objective of flood risk assessment is to provide decision-makers with information that can be used to inform flood risk management strategies and policies.

Flood risk assessment involves several steps, including the collection and analysis of data, the identification of flood hazards, the evaluation of vulnerability, and the estimation of potential impacts. It also involves the development of flood risk maps, which provide a visual representation of the areas that are most at risk of flooding.

The Importance of Flood Risk Assessment

Flood risk assessment is critical for several reasons. First, it helps decision-makers to identify areas that are most at risk of flooding and to prioritize flood risk management efforts. This ensures that limited resources are allocated to areas where they are most needed, improving the effectiveness of flood risk management.

Second, flood risk assessment can inform land-use planning and development decisions. It can help to identify areas that are unsuitable for development or that require specific flood-resistant building codes. This can prevent the construction of buildings and infrastructure in areas that are vulnerable to flooding and reduce the potential impact of future floods.

Third, flood risk assessment can inform emergency preparedness and response plans. By identifying areas that are most at risk of flooding, emergency responders can develop plans to evacuate residents, protect critical infrastructure, and respond to flood emergencies effectively.

Methods for Flood Risk Assessment

Flood risk assessment can be conducted using various methods, including qualitative, semi-quantitative, and quantitative methods.

Qualitative methods involve the identification of flood hazards and the evaluation of vulnerability using expert judgment and qualitative data. This method is typically used when data is limited or when a quick assessment is required.

Semi-quantitative methods involve the use of qualitative data and limited quantitative data to evaluate flood risk. This method is typically used when some data is available but not enough for a full quantitative analysis.

Quantitative methods involve the use of statistical and mathematical models to estimate the likelihood and potential impact of future floods. This method is typically used when sufficient data is available and a more detailed analysis is required.

Flood risk assessment can also be conducted using different types of models, including hydrological models, hydraulic models, and flood inundation models.

Hydrological models are used to simulate the flow of water in rivers and other waterways. They can be used to estimate the amount and timing of water flowing into a particular area and to predict the likelihood of flooding.

Hydraulic models are used to simulate the flow of water in urban areas, including the effects of buildings and other infrastructure on floodwater flow. They can be used to predict the depth and velocity of floodwater in specific areas and to identify areas that are most at risk of flooding.

Flood inundation models are used to simulate the extent and severity of flooding in specific areas. They can be used to predict the depth and velocity of floodwater in different areas and to identify areas that are most at risk of flooding.

Challenges of Flood Risk Management

Flood risk management is a complex and challenging task that requires the involvement of multiple stakeholders, including governments, communities, NGOs, and private sector organizations. One of the main challenges of flood risk management is the limited availability of data and information. Many developing countries lack the necessary resources and infrastructure to collect and analyze data, making it challenging to conduct accurate flood risk assessments.

Another challenge of flood risk management is the complexity of flood risk. Flood risk is influenced by various factors, including climate change, land-use changes, and population growth. As a result, effective flood risk management requires a multidisciplinary approach that considers the social, economic, and environmental factors that contribute to flood risk.

Finally, flood risk management is often complicated by political and institutional factors. Decision-making processes can be slow and bureaucratic, and different stakeholders may have conflicting interests and priorities. This can make it challenging to implement effective flood risk management policies and strategies.

The Role of Different Stakeholders in Flood Risk Management

Effective flood risk management requires the involvement of multiple stakeholders, including governments, communities, NGOs, and private sector organizations.

Governments have a critical role to play in flood risk management. They are responsible for developing and implementing flood risk management policies and strategies, including the provision of flood protection infrastructure, land-use planning and regulation, and emergency preparedness and response planning.

Communities also have an important role to play in flood risk management. They can contribute to flood risk assessment by providing local knowledge and information about flood hazards and vulnerabilities. They can also participate in the development of flood risk management policies and strategies, including the implementation of flood-resistant building codes and the provision of early warning systems.

NGOs can also play a significant role in flood risk management. They can provide technical assistance and expertise to governments and communities, including the development of flood risk maps, the implementation of community-based flood risk management programs, and the provision of emergency relief and recovery assistance.

Finally, the private sector can also contribute to flood risk management. Private sector organizations can provide funding and resources for flood risk management initiatives, including the construction of flood protection infrastructure and the development of flood-resistant buildings and infrastructure.

Conclusion

Flood risk assessment and management are critical components of disaster risk reduction. Floods are one of the most common and destructive natural disasters, causing significant economic and social damages, and loss of life worldwide. Effective flood risk assessment and management are essential to mitigate the impacts of floods and improve the resilience of communities to future flood events.

Flood risk assessment involves the evaluation of the probability of flooding and its potential consequences. It involves the identification of flood hazards, vulnerabilities, and impacts, as well as the assessment of the likelihood and severity of future flooding. Flood risk management involves the development and implementation of policies and strategies to reduce flood risk.

Flood risk assessment and management are challenging tasks that require the involvement of multiple stakeholders, including governments, communities, NGOs, and private sector organizations. Effective flood risk management requires a multidisciplinary approach that considers the social, economic, and environmental factors that contribute to flood risk. By working together, stakeholders can develop and implement effective flood risk management policies and strategies that can reduce the impact of floods and improve the resilience of communities to future flood events.

22 - Managing Floods: An Overview of Flood Routing Techniques and Their Challenges

Abstract

This article provides an overview of flood routing techniques, which are a critical tool for managing floods and protecting human life and property. Flood routing techniques involve the use of mathematical models and data to predict the timing and magnitude of floods, and to manage the impact of floods on the surrounding area. The article discusses the different types of flood routing techniques, including hydrological routing, hydraulic routing, and integrated routing, and their applications in flood management. Additionally, the article highlights the challenges that flood routing techniques face, such as limited data availability, uncertainty, changing climate, and limited resources. Overall, this article emphasizes the importance of effective flood management and the need to continue to develop and refine flood routing techniques based on new data and technologies.

Introduction

Floods are among the most destructive natural disasters that can occur in any location worldwide. Flood routing techniques are used to predict the flow of water in a channel during a flood. These techniques can be used to estimate the height and the speed of the water flow in a river or stream. This information can be used to help plan for the possible impact of a flood, to manage the floodwaters and to limit the damage to property and people.

Flood routing techniques involve the use of mathematical models that are designed to simulate the flow of water in a river or stream. There are several different types of flood routing techniques that can be used, including hydrological models, hydraulic models, and empirical models. In this article, we will discuss the different types of flood routing techniques and how they are used in flood prediction and management.

Types of Flood Routing Techniques

Hydrological models

Hydrological models are mathematical models that are designed to simulate the water cycle in a river basin. These models take into account various factors such as rainfall, evapotranspiration, soil moisture, and land use to predict the amount of water that will flow into a river or stream during a flood. Hydrological models can be used to estimate the magnitude and timing of floods in a particular area. There are two types of hydrological models that can be used in flood routing:

a. *Conceptual models*:

Conceptual models are simple models that use a set of equations to simulate the hydrological cycle in a river basin. These models are easy to use, but they are less accurate than other types of hydrological models.

b. *Physically-based models*:

Physically-based models are more complex than conceptual models, and they use detailed equations to simulate the hydrological cycle in a river basin. These models are more accurate than conceptual models, but they require more data and resources to use.

Hydraulic models

Hydraulic models are mathematical models that are designed to simulate the flow of water in a river or stream. These models take into account various factors such as the shape of the riverbed, the roughness of the riverbed, and the gradient of the riverbed to predict the height and speed of the water flow during a flood.

There are two types of hydraulic models that can be used in flood routing:

a. *1D models*: One-dimensional (1D) models are used to simulate the flow of water in a river or stream in one direction only. These models are easy to use and require less data and resources than other types of hydraulic models.

b. *2D models*: Two-dimensional (2D) models are used to simulate the flow of water in a river or stream in two directions. These models are more complex than 1D models, and they require more data and resources to use. However, they are more accurate than 1D models because they take into account the shape of the riverbed in two directions.

Empirical models

Empirical models are mathematical models that are based on observed data rather than on theoretical equations. These models are used to estimate the flow of water in a river or stream during a flood based on past flood events.

Empirical models are simple to use, but they are less accurate than hydrological or hydraulic models because they do not take into account the physical properties of the river or stream. However, they can be useful in areas where there is limited data available, or where other types of models are not suitable.

Modelling Software

There are various software options for different types of models. Here are some examples:

Hydrological models:

HEC-HMS: This software is developed by the US Army Corps of Engineers, and it can be used to perform hydrological modelling, including watershed analysis, floodplain mapping, and flood routing.

MIKE SHE: This is a hydrological model developed by DHI, which can be used to simulate the water balance of a catchment, including groundwater and surface water interactions.

SWMM: The Storm Water Management Model is an open-source hydrological model developed by the US EPA, which can be used to simulate the quality and quantity of stormwater runoff.

Hydraulic models:

HEC-RAS: This software is developed by the US Army Corps of Engineers, and it can be used to perform hydraulic modelling, including river and floodplain hydraulics, bridge and culvert hydraulics, and sediment transport.

MIKE 21: This is a hydraulic model developed by DHI, which can be used to simulate waves, currents, and sediment transport in coastal and marine environments.

LISFLOOD-FP: This is a 2D flood inundation model developed by the University of Bristol, which can be used to simulate flood events and produce flood inundation maps.

Empirical models:

USGS StreamStats: This is a web-based application developed by the US Geological Survey, which can be used to estimate the flow characteristics of a stream or river based on digital elevation data and other spatial data.

Flood Frequency Analysis (FFA): This is a statistical method used to estimate flood frequency and magnitude based on historical flood data. The FFA can be performed using statistical software such as R, SAS, or SPSS.

Flood Routing Techniques

Flood routing techniques are used to predict the flow of water in a river or stream during a flood. There are several different flood routing techniques that can be used, including:

Muskingum-Cunge method

The Muskingum-Cunge method is a hydrological model that is used to predict the flow of water in a river or stream during a flood. This method is based on the principle of conservation of mass and energy, and it takes into account the storage and routing of water in the river channel.

The Muskingum-Cunge method uses a set of equations to predict the water flow in a river or stream. These equations take into account the inflow of water into the river or stream, the outflow of water from the river or stream, and the storage of water in the river or stream. The method is used to estimate the water flow at a particular point in the river or stream, as well as the timing and magnitude of the peak flow.

Kinematic wave method

The kinematic wave method is a hydraulic model that is used to predict the flow of water in a river or stream during a flood. This method is based on the principle of conservation of mass and momentum, and it takes into account the slope and roughness of the riverbed.

The kinematic wave method uses a set of equations to predict the water flow in a river or stream. These equations take into account the slope and roughness of the riverbed, as well as the inflow and outflow of water. The method is used to estimate the height and speed of the water flow at a particular point in the river or stream.

Muskingum method

The Muskingum method is a hydrological model that is used to predict the flow of water in a river or stream during a flood. This method is based on the principle of conservation of mass and energy, and it takes into account the storage and routing of water in the river channel.

The Muskingum method uses a set of equations to predict the water flow in a river or stream. These equations take into account the inflow of water into the river or stream, the outflow of water from the river or stream, and the storage of water in the river or stream. The method is used to estimate the water flow at a particular point in the river or stream, as well as the timing and magnitude of the peak flow.

Hydrodynamic models

Hydrodynamic models are hydraulic models that are used to predict the flow of water in a river or stream during a flood. These models take into account the complex physical properties of the river or stream, such as turbulence, eddies, and sediment transport.

Hydrodynamic models use a set of equations to predict the water flow in a river or stream. These equations take into account the slope and roughness of the riverbed, as well as the inflow and outflow of water. The models are used to estimate the height and speed of the water flow at a particular point in the river or stream, as well as the timing and magnitude of the peak flow.

Applications of Flood Routing Techniques

Flood routing techniques are used in a variety of applications, including flood prediction, flood management, and flood mitigation. These techniques can be used to estimate the height and speed of the water flow in a river or stream during a flood, as well as the timing and magnitude of the peak flow.

Flood prediction

Flood routing techniques are used to predict the occurrence of floods in a particular area. These techniques can be used to estimate the timing and magnitude of a flood, as well as the potential impact of the flood on the surrounding area. Flood prediction can help emergency managers to prepare for a flood, to evacuate people from areas that are at risk, and to mobilize resources to respond to the flood.

Flood management

Flood routing techniques are used to manage the impact of floods on the surrounding area. These techniques can be used to control the flow of water in a river or stream, to protect infrastructure and property from flood damage, and to manage the water supply during a flood. Flood management can include the construction of levees and flood walls, the installation of flood control structures such as dams and weirs, and the development of flood warning systems.

Flood mitigation

Flood routing techniques can also be used to mitigate the impact of floods on the surrounding area. This can include the development of floodplain management strategies, the creation of natural flood protection measures such as wetlands and floodplain forests, and the restoration of rivers and streams to their natural state. Flood mitigation can help to reduce the risk of flood damage, to protect human life and property, and to preserve natural ecosystems.

Challenges in Flood Routing Techniques

Despite the many benefits of flood routing techniques, there are several challenges that must be addressed in order to effectively manage floods. These challenges include:

Limited data availability: Flood routing techniques rely heavily on data, including hydrological data, hydraulic data, and meteorological data. However, in many areas, data is limited or not available, making it difficult to accurately predict the timing and magnitude of floods.

Uncertainty: Flood routing techniques involve a high degree of uncertainty, due to the complexity of the natural systems involved. This uncertainty can make it difficult to accurately predict the impact of a flood, and can result in errors in flood predictions and management.

Changing climate: Climate change is expected to have a significant impact on the frequency and intensity of floods in many areas. This can make it difficult to develop effective flood management strategies, as historical data may no longer accurately predict the timing and magnitude of floods.

Limited resources: Effective flood management requires significant resources, including funding, personnel, and equipment. In many areas, these resources are limited, making it difficult to implement effective flood management strategies.

Conclusion

Flood routing techniques are a critical tool for managing floods and protecting human life and property. These techniques can be used to predict the timing and magnitude of floods, to manage the impact of floods on the surrounding area, and to mitigate the risk of flood damage. However, these techniques also face several challenges, including limited data availability, uncertainty, changing climate, and limited resources. To effectively manage floods, it is important to address these challenges and to continue to develop and refine flood routing techniques based on new data and technologies.

23 - Regulating Development in Flood-Prone Areas: Benefits and Challenges of Flood Plain Zoning

Abstract

Flood plain zoning is a land use planning technique that aims to regulate development within flood-prone areas to reduce the risk of damage caused by floods. This article provides an overview of the history of flood plain zoning, including the creation of the National Flood Insurance Program in the United States. It also describes the various methods for flood plain zoning, such as flood hazard maps, elevation requirements, flood proofing, and acquisition and relocation. The benefits of flood plain zoning, including reduced risk of flood damage, improved public safety, cost savings, and environmental benefits, are discussed. However, the article also highlights some challenges, such as economic impact, enforcement, political challenges, and equity concerns. Finally, the article discusses the future prospects of flood plain zoning, including its importance in ensuring public safety and reducing the risk of flood damage, and the potential for technology and public education to improve its effectiveness.

Introduction

Flood plain zoning is a land use planning technique that involves regulating development within flood-prone areas. Flood plain zoning aims to reduce the risk of damage caused by floods by designating areas of land that are prone to flooding and regulating how these areas are used. This article provides an in-depth analysis of flood plain zoning, including its history, benefits, challenges, and future prospects.

History of Flood Plain Zoning

The history of flood plain zoning can be traced back to the early 20th century when the United States was experiencing widespread flooding that caused significant damage to property and loss of life. The federal government, through the National Flood Control Act of 1936, created the National Flood Insurance Program (NFIP), which aimed to reduce the impact of floods by encouraging flood plain zoning.

The NFIP provided insurance for homes and businesses in flood-prone areas and established minimum standards for local communities to participate in the program. These standards included the creation of flood plain management plans, the mapping of flood-prone areas, and the implementation of zoning regulations.

Methods for Flood Plain Zoning

There are several methods for flood plain zoning, which can vary depending on local conditions, land use patterns, and regulatory frameworks. The following are some common methods for flood plain zoning:

Flood Hazard Maps: One of the most common methods of flood plain zoning is through the use of flood hazard maps. These maps identify areas at risk of flooding and are used to establish flood plain boundaries, which can then be used to regulate land use and development within those areas.

Flood Risk Assessment: Flood risk assessment is another method for flood plain zoning. It involves evaluating the probability and potential consequences of flooding in a given area, and using this information to determine appropriate land use regulations.

Elevation Requirements: Elevation requirements are another common method for flood plain zoning. This involves requiring new or existing buildings to be constructed at a certain height above the base flood elevation, to minimize flood damage.

Flood Proofing: Flood proofing is a technique that can be used in conjunction with other flood plain zoning methods. It involves retrofitting buildings with features such as waterproof barriers, to reduce the risk of flood damage.

Acquisition and Relocation: In some cases, it may be necessary to acquire and relocate buildings located in flood-prone areas. This method of flood plain zoning involves purchasing properties at risk of flooding and relocating them to safer areas.

Natural Flood Management: Natural flood management is an approach that involves using natural features, such as wetlands or trees, to mitigate flooding. This method of flood plain zoning can be effective in reducing flood risk and providing other environmental benefits.

These methods can be used in various combinations to create a customized flood plain zoning plan that is appropriate for local conditions and regulatory frameworks. It is important to consider the specific challenges and opportunities of each approach when designing a flood plain zoning plan.

Benefits of Flood Plain Zoning

Flood plain zoning has several benefits, including:

Reduced Risk of Flood Damage

One of the primary benefits of flood plain zoning is the reduction of the risk of flood damage. By regulating development in flood-prone areas, flood plain zoning ensures that structures are built to withstand flooding, reducing the likelihood of damage to property and loss of life.

Improved Public Safety

Flood plain zoning can also improve public safety by ensuring that structures are built to code and designed to withstand flooding. This can prevent buildings from collapsing during floods, reducing the risk of injury or death.

Cost Savings

Flood plain zoning can also lead to cost savings for property owners and taxpayers. By reducing the risk of flood damage, property owners may be able to secure lower insurance premiums, while taxpayers may save money on disaster relief and recovery efforts.

Environmental Benefits

Flood plain zoning can also have environmental benefits. By preventing development in flood-prone areas, flood plain zoning can help preserve wetlands and other natural areas, which can provide critical habitat for wildlife and help prevent erosion and flooding.

Challenges of Flood Plain Zoning

Despite its many benefits, flood plain zoning can also pose several challenges, including:

Economic Impact

Flood plain zoning can have a significant economic impact on property owners and local communities. Property owners may face restrictions on the use of their land, which can affect property values and limit development opportunities. Local communities may also face economic challenges, as flood plain zoning can limit tax revenue from development.

Enforcement

Flood plain zoning can be difficult to enforce, particularly in areas where there is limited local capacity or resources. Effective enforcement requires a commitment to regular monitoring and inspection, which can be costly and time-consuming.

Political Challenges

Flood plain zoning can also face political challenges, particularly in areas where property owners or developers may resist regulation. Local officials may also be hesitant to implement flood plain zoning regulations, fearing backlash from constituents or pressure from developers.

Equity Concerns

Flood plain zoning can also raise equity concerns, particularly in areas where low-income communities are disproportionately impacted by flooding. These communities may not have the resources to relocate or rebuild after a flood, leading to a cycle of poverty and vulnerability.

Future Prospects of Flood Plain Zoning

Despite its many challenges, flood plain zoning is likely to continue to play an important role in reducing the impact of flooding in the future. As climate change leads to more frequent and severe flooding, flood plain zoning will become increasingly important in ensuring public safety and reducing the risk of flood damage.

Advances in technology, including improved flood mapping and modelling tools, can also help improve the effectiveness of flood plain zoning. These tools can help local officials identify areas at high risk of flooding, and make informed decisions about land use and development.

In addition, public awareness and education about the risks of flooding and the benefits of flood plain zoning can help build support for this approach. Educating property owners and local communities about the benefits of flood plain zoning can help reduce resistance to regulation.

Finally, incorporating equity considerations into flood plain zoning can help ensure that vulnerable communities are not disproportionately impacted by flooding. This may involve providing resources and support to low-income communities to help them relocate or rebuild after a flood, or prioritizing these communities in flood mitigation and prevention efforts.

Conclusion

Flood plain zoning is an essential land use planning technique that can help reduce the risk of flood damage and improve public safety. The history of flood plain zoning in the United States demonstrates its effectiveness in regulating development within flood-prone areas. However, flood plain zoning also faces various challenges, including economic impact, enforcement, political challenges, and equity concerns. Despite these challenges, the future of flood plain zoning is promising, as it will continue to play an important role in reducing the impact of flooding, especially as climate change leads to more frequent and severe flooding. Advances in technology and public education can also help improve the effectiveness of flood plain zoning. It is essential to balance the benefits of flood plain zoning with its potential challenges to ensure that it remains a useful tool in reducing the risk of flood damage and improving public safety.

24 - Reservoir Operation: Principles, Objectives, Methods, and Challenges

Abstract

This article provides an overview of reservoir operation, which involves managing the storage and release of water from reservoirs to meet various demands while ensuring the safety of the dam and downstream areas. The article discusses the basic principles of reservoir operation, the different objectives of reservoir operation, the various methods of reservoir operation, and the challenges associated with reservoir operation. The objectives of reservoir operation are categorized into economic, environmental, and social objectives, and the methods of reservoir operation include rule-based, optimization-based, and real-time control methods. The challenges associated with reservoir operation include uncertainty in inflow and demand, conflicting objectives, infrastructure constraints, and stakeholder participation. By understanding these principles, objectives, methods, and challenges, stakeholders can make informed decisions about reservoir operation that balance the various demands and minimize negative impacts on the environment and society.

Introduction

Reservoirs are an important component of water resource systems, used for a variety of purposes such as water supply, irrigation, hydropower generation, flood control, and recreation. Reservoirs can store water during wet periods and release it during dry periods, providing a reliable source of water for various purposes. Reservoir operation refers to the process of managing the storage and release of water from reservoirs to meet various demands, while ensuring the safety of the dam and downstream areas.

In this article, we will discuss the basic principles of reservoir operation, the different objectives of reservoir operation, the various methods of reservoir operation, and the challenges associated with reservoir operation.

Basic Principles of Reservoir Operation

The operation of a reservoir involves managing the inflow of water into the reservoir, the outflow from the reservoir, and the storage of water in the reservoir. The basic principles of reservoir operation are:

Water balance: The inflow of water into the reservoir must be balanced by the outflow from the reservoir, and the storage of water in the reservoir. The inflow of water into the reservoir can be from precipitation, surface runoff, groundwater recharge, or diversion from other water sources.

Demand management: The outflow from the reservoir must meet the various demands for water, such as water supply, irrigation, hydropower generation, flood control, and recreation.

Safety: The operation of the reservoir must ensure the safety of the dam and downstream areas. The dam must be operated within its design limits to avoid overtopping or failure, and the downstream areas must be protected from flooding.

Objectives of Reservoir Operation

The objectives of reservoir operation can be broadly classified into three categories:

Economic objectives: The economic objectives of reservoir operation are to maximize the benefits derived from the use of the reservoir. These benefits can be in the form of hydropower generation, irrigation, water supply, flood control, and recreation.

Environmental objectives: The environmental objectives of reservoir operation are to minimize the negative impacts of reservoir operation on the environment. These impacts can be in the form of water quality degradation, loss of biodiversity, and alteration of natural flow regimes.

Social objectives: The social objectives of reservoir operation are to ensure that the benefits of reservoir operation are distributed equitably among different stakeholders. This can involve ensuring access to water for marginalized communities, ensuring that the downstream areas are not adversely affected by the reservoir operation, and ensuring that the recreational benefits of the reservoir are accessible to all.

Methods of Reservoir Operation

The methods of reservoir operation depend on the objectives of the reservoir operation and the characteristics of the water resource system. The methods can be broadly classified into three categories:

Rule-based methods: Rule-based methods involve the use of predetermined rules to determine the release of water from the reservoir. These rules can be based on historical data, hydrological models, or expert judgment. Rule-based methods are simple and easy to implement, but they may not be optimal for all situations.

Optimization-based methods: Optimization-based methods involve the use of mathematical models to determine the optimal release of water from the reservoir. These models can be used to balance the various objectives of reservoir operation, such as maximizing the benefits derived from the reservoir while minimizing the negative impacts on the environment. Optimization-based methods require detailed data on the water resource system and can be computationally intensive.

Real-time control methods: Real-time control methods involve the use of feedback control systems to adjust the release of water from the reservoir in real-time based on current conditions. These systems can be based on sensors that measure the water level and flow rate in the reservoir and downstream areas. Real-time control methods can be effective in responding to changing conditions, such as sudden changes in inflow or demand, but they require sophisticated control algorithms and communication infrastructure.

Challenges Associated with Reservoir Operation

Reservoir operation is a complex process that involves balancing multiple objectives and managing uncertainty. Some of the challenges associated with reservoir operation are:

Uncertainty in inflow and demand: The inflow of water into the reservoir and the demand for water can be highly uncertain, particularly in regions with variable climate and hydrology. This uncertainty can make it difficult to optimize reservoir operation and may require the use of robust decision-making approaches.

Conflicting objectives: The different objectives of reservoir operation, such as maximizing economic benefits, minimizing environmental impacts, and ensuring social equity, can sometimes be in conflict with each other. Resolving these conflicts requires careful consideration of the trade-offs between the different objectives.

Infrastructure constraints: The operation of a reservoir is constrained by the design and capacity of the dam and other infrastructure. The operation of the reservoir must be within these constraints to ensure the safety of the dam and downstream areas.

Stakeholder participation: Reservoir operation can have significant impacts on different stakeholders, such as farmers, communities, and recreational users. The participation of these stakeholders in the decision-making process can help ensure that their needs and perspectives are taken into account.

Conclusion

Reservoir operation is an important component of water resource management, and its effective implementation can help ensure a reliable supply of water for various purposes while minimizing the negative impacts on the environment and ensuring the safety of the dam and downstream areas. Reservoir operation involves balancing multiple objectives and managing uncertainty, and requires the use of appropriate methods and approaches that are tailored to the characteristics of the water resource system and the objectives of reservoir operation. The challenges associated with reservoir operation can be addressed through careful planning, stakeholder participation, and the use of robust decision-making approaches.

25 - Understanding Water Quality Index: Assessing and Monitoring the Health of Water Bodies

Abstract

Water is an essential natural resource, and its quality is critical for maintaining a healthy environment and supporting various human activities. However, water bodies are under increasing pressure from pollution, climate change, and population growth. To assess and monitor the health of water bodies, scientists and environmental agencies use a tool called the Water Quality Index (WQI). The WQI is a numerical rating system that combines multiple water quality parameters, including physical, chemical, and biological factors, to provide an overall assessment of water quality. This article provides an in-depth understanding of the WQI, including its components, calculation methods, and interpretation, and its applications in monitoring and managing water quality. It also highlights the importance of maintaining good water quality for public health, ecosystem services, and sustainable development.

Introduction

Water is a vital resource that sustains life and supports various human activities such as drinking, agriculture, industry, and recreation. However, the quality of water is under threat from various sources, including human activities, natural processes, and climate change. Water pollution is a global issue that affects the health of ecosystems and public health. Polluted water bodies can cause waterborne diseases, harm aquatic life, and degrade ecosystem services, including water supply and purification, flood control, and habitat for wildlife. To ensure the health of water bodies and protect public health, it is crucial to assess and monitor water quality continuously.

Water quality monitoring involves measuring and analyzing various water quality parameters, including physical, chemical, and biological factors. These parameters provide insights into the health of water bodies and the potential risks they pose to human and environmental health. However, analyzing each parameter independently can be time-consuming and complicated, and interpreting the results can be challenging. Therefore, scientists and environmental agencies use a tool called the Water Quality Index (WQI) to provide a comprehensive assessment of water quality.

What is Water Quality Index (WQI)?

The Water Quality Index (WQI) is a tool used to assess and monitor the health of water bodies. It is a numerical rating system that combines multiple water quality parameters to provide an overall assessment of water quality. The WQI assigns a score to each parameter based on its relative importance to water quality and its potential risks to human and environmental health. The scores are then weighted and combined to produce an overall score that represents the water quality index.

The WQI is designed to be a simple and easy-to-use tool for environmental managers, policymakers, and the public to understand water quality. It provides a standardized approach to water quality assessment, allowing for comparison across different water bodies and regions. The WQI also enables the identification of specific water quality parameters that require improvement and helps prioritize management actions to improve water quality.

Components of Water Quality Index

The Water Quality Index (WQI) includes several components that provide a comprehensive assessment of water quality. The components include physical, chemical, and biological factors that are critical for maintaining healthy water bodies. The following are the primary components of the WQI:

Physical Parameters

Physical parameters are the properties of water that can be measured without altering its chemical composition. These parameters include temperature, color, turbidity, total dissolved solids (TDS), and total suspended solids (TSS). Temperature is an essential physical parameter that affects aquatic life, as different species require different temperature ranges for survival. High temperatures can reduce dissolved oxygen levels in water, making it difficult for aquatic organisms to breathe. Color and turbidity affect the aesthetic quality of water and can also impact aquatic life by reducing light penetration and photosynthesis. TDS and TSS are indicators of the presence of dissolved and suspended solids in water, respectively, which can impact water clarity and aquatic life.

Chemical Parameters

Chemical parameters are the properties of water that result from the presence of various chemical substances. These parameters include pH, dissolved oxygen (DO), biochemical oxygen demand (BOD), chemical oxygen demand (COD), nutrients, heavy metals, and pesticides. pH is a measure of the acidity or alkalinity of water, which affects the ability of aquatic organisms to survive and reproduce. DO is a critical parameter for aquatic life, as it is required for respiration. BOD and COD are measures of the organic matter present in water, which can consume oxygen and deplete DO levels, leading to a decrease in aquatic life. Nutrients such as nitrogen and phosphorus can cause eutrophication, which leads to excessive algae growth and a decrease in DO levels. Heavy metals and pesticides can be toxic to aquatic life and can accumulate in the food chain, posing a risk to human health.

Biological Parameters

Biological parameters are the properties of water that result from the presence of living organisms. These parameters include the presence of bacteria, algae, and other microorganisms, as well as the abundance and diversity of aquatic life. The presence of fecal coliform bacteria is an indicator of water contamination with human or animal waste, which can cause waterborne diseases. Algae growth can indicate eutrophication, while a lack of aquatic life can indicate degraded water quality.

Calculation Methods of Water Quality Index

The calculation of the Water Quality Index (WQI) involves several steps. First, the data for each water quality parameter is collected and recorded. Next, the data is standardized by assigning a score based on a scale of 0-100. The scoring system is based on the relative importance of each parameter to water quality and the potential risks it poses to human and environmental health. The higher the score, the better the water quality for that parameter.

After assigning scores to each parameter, the scores are weighted based on their relative importance to water quality. The weights are determined based on the specific characteristics of the water body being assessed and the goals of the assessment. For example, if the water body is used for drinking water supply, parameters such as pH, DO, and fecal coliform bacteria may be given more weight than other parameters.

Once the scores are weighted, they are combined to produce an overall score for the Water Quality Index. The overall score is usually on a scale of 0-100, with higher scores indicating better water quality. The final score represents the overall health of the water body and provides a useful tool for comparing water quality across different regions and over time.

Interpretation of Water Quality Index

The interpretation of the Water Quality Index (WQI) depends on the specific goals of the assessment and the characteristics of the water body being assessed. Generally, a WQI score of 80-100 indicates excellent water quality, 60-79 indicates good water quality, 40-59 indicates fair water quality, 20-39 indicates poor water quality, and 0-19 indicates very poor water quality.

A WQI score below 60 indicates a need for management actions to improve water quality. The specific management actions will depend on the water quality parameters that require improvement and the goals of the assessment. For example, if the WQI score is low due to high nutrient levels, management actions may include reducing nutrient inputs from agricultural or urban runoff, upgrading wastewater treatment facilities, or implementing best management practices to reduce fertilizer use.

Applications of Water Quality Index

The Water Quality Index (WQI) has several applications in monitoring and managing water quality. Some of the main applications include:

Monitoring Water Quality

The WQI is a valuable tool for monitoring water quality over time. By assessing water quality periodically, environmental managers and policymakers can identify trends and changes in water quality and take appropriate management actions to maintain or improve water quality.

Comparing Water Quality

The WQI provides a standardized method for comparing water quality across different regions and over time. This allows policymakers and environmental managers to identify areas where water quality is particularly good or poor and target management actions accordingly.

Communicating Water Quality

The WQI is a useful tool for communicating water quality to the public. By providing a simple and understandable score, the WQI can help to raise awareness about water quality issues and motivate stakeholders to take action to protect water resources.

Prioritizing Management Actions

The WQI can help to prioritize management actions to improve water quality. By identifying the water quality parameters that require improvement and the management actions that are most effective at addressing these parameters, the WQI can guide decision-making and resource allocation.

Limitations of Water Quality Index

While the Water Quality Index (WQI) is a valuable tool for monitoring and managing water quality, there are some limitations to its use. Some of the main limitations include:

Data Availability and Quality

The accuracy of the WQI depends on the availability and quality of data for each water quality parameter. In many cases, data may be limited or unreliable, particularly in developing countries or in areas with limited monitoring capacity.

Parameter Selection and Weighting

The selection and weighting of parameters can vary depending on the specific goals of the assessment and the characteristics of the water body being assessed. This can lead to variations in the WQI scores across different regions and over time.

Lack of Biological Parameters

The WQI focuses primarily on physical and chemical parameters and does not include many biological parameters, such as the abundance and diversity of aquatic life. This can limit its ability to accurately reflect the health of aquatic ecosystems.

Lack of Context

The WQI provides a standardized score for water quality, but it does not provide context for the specific conditions of the water body being assessed. This can limit its usefulness for making management decisions that are tailored to the specific characteristics of the water body.

Conclusion

The Water Quality Index (WQI) is a valuable tool for monitoring and managing water quality. By providing a simple and understandable score for water quality, the WQI can help to raise awareness about water quality issues and motivate stakeholders to take action to protect water resources. However, the WQI has some limitations, including data availability and quality, parameter selection and weighting, lack of biological parameters, and lack of context. Despite these limitations, the WQI remains a useful tool for comparing water quality across different regions and over time, prioritizing management actions, and communicating water quality to the public.

26 - Dam Break Analysis: Importance, Methods, and Applications

Abstract

Dam break analysis is a critical tool used to assess the potential consequences of a dam failure and to develop emergency plans to mitigate the risk. In this article, we discuss the importance of dam break analysis in dam safety management, emergency planning, and environmental impact assessment. We also describe the various methods used for dam break analysis, including empirical, analytical, and numerical modelling. Each method has its advantages and limitations, and the choice of method depends on the available data and the complexity of the dam and its associated structures. Overall, dam break analysis is essential to evaluate the performance of a dam and its associated structures, assess the risk of dam failure, and develop strategies to mitigate the risk and minimize the potential consequences of a failure.

Introduction

Dam break analysis is an important tool used to predict the potential consequences of a dam failure. Dams are structures built to store water and generate hydroelectric power. However, if a dam fails, it can cause catastrophic flooding downstream, resulting in loss of life and property damage. Therefore, dam break analysis is a critical step in dam safety management. In this article, we will discuss the various aspects of dam break analysis, including its importance, methods, and applications.

Importance of Dam Break Analysis

Dam break analysis is essential to assess the risk of dam failure and the potential consequences of a failure. The consequences of a dam failure can be severe, including loss of life, property damage, and environmental damage. Therefore, it is crucial to identify potential failure scenarios and their consequences to develop emergency plans and mitigate the risk.

Dam break analysis is also essential to evaluate the performance of a dam and its associated structures. The analysis can identify potential failure modes and assess the effectiveness of the dam's design and maintenance strategies. This information can help dam owners and operators make informed decisions about the dam's management and prioritize maintenance and repair efforts.

Methods of Dam Break Analysis

There are several methods used for dam break analysis, each with its advantages and limitations. The most common methods are empirical, analytical, and numerical modelling.

Empirical Methods

Empirical methods are based on historical data and observations of dam failures. These methods use statistical analysis to estimate the probability of dam failure and the resulting flood wave characteristics. Empirical methods are relatively simple and require less data than other methods. However, they are limited by the availability and quality of historical data and cannot account for the unique characteristics of individual dams.

Analytical Methods

Analytical methods are based on mathematical equations that describe the physical processes involved in dam failure and flood wave propagation. These methods can provide detailed information about the dam failure and flood wave characteristics. Analytical methods are relatively simple and require less computational resources than numerical methods. However, they are limited by the assumptions and simplifications made in the mathematical models and cannot account for complex geometries and boundary conditions.

Numerical Methods

Numerical methods use computer simulations to model the dam failure and flood wave propagation. These methods can account for complex geometries, boundary conditions, and physical processes, providing detailed information about the dam failure and flood wave characteristics. Numerical methods are the most accurate and flexible methods but require significant computational resources and expertise.

Applications of Dam Break Analysis

Dam break analysis is used in various applications, including dam safety management, emergency planning, and environmental impact assessment.

Dam Safety Management

Dam break analysis is a critical component of dam safety management. The analysis can identify potential failure scenarios and assess the effectiveness of the dam's design and maintenance strategies. This information can help dam owners and operators make informed decisions about the dam's management and prioritize maintenance and repair efforts.

Emergency Planning

Dam break analysis is also essential for emergency planning. The analysis can predict the flood wave characteristics and the areas likely to be affected by the dam failure. This information can help emergency responders develop evacuation plans and identify critical infrastructure that may be at risk.

Environmental Impact Assessment

Dam break analysis can also be used to assess the environmental impact of a dam failure. The analysis can predict the flood wave characteristics and the areas likely to be affected by the dam failure, allowing for the identification of critical habitats and species that may be at risk.

Conclusion

Dam break analysis is a critical tool used to assess the risk of dam failure and the potential consequences of a failure. The analysis can identify potential failure scenarios, evaluate the performance of a dam and its associated structures, and develop emergency plans to mitigate the risk and minimize the potential consequences of a failure. The choice of method depends on the available data and the complexity of the dam and its associated structures. Empirical, analytical, and numerical modelling methods all have their advantages and limitations. However, the ultimate goal of dam break analysis is to ensure the safety of people, property, and the environment. By using the appropriate method and analyzing the potential consequences of a dam failure, dam owners and operators can make informed decisions about the dam's management, prioritize maintenance and repair efforts, and develop effective emergency plans.

27 - Assessment of Environmental Flows: Challenges and Opportunities for Effective Management

Abstract

Environmental flows are essential for the health of aquatic ecosystems and the sustainability of water resources for human and non-human use. However, the assessment of environmental flows is a complex process that involves various methods and faces several challenges. These challenges include data availability, lack of consensus, conflicting objectives, climate change, and political and institutional constraints. This article explores the challenges and opportunities for the assessment and management of environmental flows. It highlights the importance of improved data collection, collaborative approaches, adaptive management, and the incorporation of climate change in developing effective management strategies. By addressing these challenges and leveraging these opportunities, we can ensure the sustainability of water resources and the health of aquatic ecosystems for future generations.

Introduction

Environmental flows refer to the quantity, quality, and timing of water flows necessary to maintain the ecological functions of rivers, wetlands, and other aquatic ecosystems. Environmental flows are essential to the survival of aquatic ecosystems, and they help to maintain the quality of water resources for human and non-human use. The assessment of environmental flows is, therefore, a crucial aspect of water resource management.

In this article, we will discuss the assessment of environmental flows, including the methods used, the challenges faced, and the future of environmental flow assessments.

Methods of Environmental Flow Assessment

Environmental flow assessment involves the use of various methods to determine the quantity, quality, and timing of water flows necessary to maintain the ecological functions of rivers, wetlands, and other aquatic ecosystems. Some of the commonly used methods include:

Hydrological Models

Hydrological models are used to simulate the hydrological processes in a river basin, including the flow of water, sediment, and nutrients. The models are based on various inputs, such as precipitation, temperature, and land use, and they help to predict the flow of water in rivers and other water bodies. Hydrological models are useful in assessing the impacts of human activities, such as dam construction, on the natural flow of rivers and the ecological functions of aquatic ecosystems.

Habitat Suitability Models

Habitat suitability models are used to identify the characteristics of aquatic ecosystems that are essential for the survival of aquatic species. The models consider factors such as water quality, flow velocity, temperature, and depth, among others, to determine the suitability of a habitat for different species. Habitat suitability models are useful in assessing the impacts of environmental flows on the survival of aquatic species and identifying the minimum flows necessary to maintain suitable habitats.

Hydrodynamic Models

Hydrodynamic models are used to simulate the flow of water in rivers, wetlands, and other aquatic ecosystems. The models consider factors such as channel geometry, roughness, and bed slope to predict the flow of water and sediment in rivers. Hydrodynamic models are useful in assessing the impacts of human activities, such as dam construction, on the flow of water in rivers and the ecological functions of aquatic ecosystems.

Ecological Response Models

Ecological response models are used to predict the response of aquatic ecosystems to changes in environmental flows. The models consider factors such as water quality, habitat suitability, and the life history of aquatic species to predict the impacts of changes in environmental flows on aquatic ecosystems. Ecological response models are useful in assessing the impacts of human activities, such as dam construction, on the ecological functions of aquatic ecosystems and identifying the minimum flows necessary to maintain the ecological health of these systems.

Challenges of Environmental Flow Assessment

The assessment of environmental flows is a complex process that involves many challenges. Some of the challenges include:

Data Availability

Environmental flow assessment requires data on the hydrological, ecological, and socio-economic conditions of a river basin. However, data availability is often limited, especially in developing countries, where data collection is often insufficient or nonexistent. This makes it difficult to accurately assess environmental flows and develop effective management strategies.

Lack of Consensus

There is often a lack of consensus among stakeholders on the definition of environmental flows and the methods used to assess them. This can lead to conflicting interests and hinder the development of effective environmental flow management strategies.

Conflicting Objectives

The management of environmental flows often involves balancing the needs of aquatic ecosystems with the needs of human activities, such as irrigation, hydropower generation, and domestic use. These conflicting objectives can make it difficult to develop management strategies that effectively protect aquatic ecosystems while meeting the needs of human activities.

Climate Change

Climate change is expected to have significant impacts on the hydrological cycle and the ecological functions of aquatic ecosystems. The uncertainty surrounding the impacts of climate change makes it challenging to accurately assess environmental flows and develop effective management strategies.

Political and Institutional Constraints

Environmental flow assessment and management often involve political and institutional constraints that can hinder the implementation of effective strategies. This includes issues such as conflicting policies, institutional fragmentation, and insufficient funding.

Future of Environmental Flow Assessment

Despite the challenges facing environmental flow assessment, there are several opportunities to improve the process and develop more effective management strategies. Some of the key opportunities include:

Improved Data Collection

Improving data collection on the hydrological, ecological, and socio-economic conditions of river basins is essential for accurately assessing environmental flows and developing effective management strategies. This includes investing in monitoring networks, remote sensing technologies, and citizen science initiatives.

Collaborative Approaches

Collaborative approaches involving stakeholders from different sectors and levels of governance can help to overcome conflicts and develop management strategies that balance the needs of aquatic ecosystems with human activities. This includes the use of multi-stakeholder platforms, co-management arrangements, and participatory approaches.

Adaptive Management

Adaptive management involves using a flexible and iterative approach to environmental flow management, where management strategies are continuously revised based on monitoring and evaluation. This approach recognizes the uncertainty and complexity of environmental flows and allows for the development of more effective and responsive management strategies.

Incorporation of Climate Change

Incorporating climate change into environmental flow assessments and management strategies is essential for developing strategies that are resilient to future changes. This includes the use of climate models, scenario planning, and adaptation measures.

Conclusion

Environmental flows are essential for the survival of aquatic ecosystems and the quality of water resources for human and non-human use. The assessment of environmental flows is a complex process that involves the use of various methods and faces several challenges, including data availability, lack of consensus, conflicting objectives, climate change, and political and institutional constraints.

However, there are several opportunities to improve the assessment and management of environmental flows, including improved data collection, collaborative approaches, adaptive management, and incorporation of climate change. These opportunities can help to develop more effective management strategies that balance the needs of aquatic ecosystems with the needs of human activities and ensure the sustainability of water resources for future generations.

28 - Managing Surface Water Resources: Challenges, Strategies, and Opportunities

Abstract

Surface water is a vital resource for sustaining life and promoting economic development. However, effective management of surface water resources is a complex and challenging task that involves many different stakeholders and factors. This article explores the various aspects of surface water management, including its importance, challenges, and strategies for effective management. Key strategies for effective surface water management include water conservation, water pollution control, integrated water resources management, watershed management, climate adaptation measures, and international cooperation. Advanced technologies and public awareness initiatives are also critical in promoting sustainable water use practices. By adopting a holistic and integrated approach to surface water management, we can ensure the sustainable use of this vital resource for generations to come.

Introduction

Surface water is one of the most precious resources available to us. It is essential for sustaining life, and its importance cannot be overstated. However, the management of surface water resources is a complex and challenging task that requires careful planning, monitoring, and regulation. In this article, we will explore the various aspects of surface water management, including its importance, challenges, and strategies for effective management.

Importance of Surface Water Resources

Surface water is a critical resource for human survival, and its importance cannot be overstated. It is used for drinking, irrigation, industrial purposes, and many other activities. Surface water resources also play a crucial role in maintaining ecological balance, supporting aquatic life, and providing recreational opportunities.

However, surface water resources are under increasing pressure due to population growth, urbanization, and climate change. As a result, effective management of surface water resources has become essential to ensure their sustainable use.

Challenges in Surface Water Management

Surface water management is a complex and challenging task that involves many different stakeholders and factors. Some of the key challenges in surface water management include:

Water scarcity: Many regions of the world are facing water scarcity due to population growth, urbanization, and climate change. This can lead to conflicts over water resources and make it challenging to manage surface water effectively.

Water pollution: Surface water resources are often polluted by industrial effluents, agricultural runoff, and untreated sewage. This can lead to health hazards and damage to the ecosystem.

Climate change: Climate change is altering the hydrological cycle, leading to more frequent and severe droughts, floods, and storms. This makes it challenging to manage surface water resources effectively.

Infrastructure development: Infrastructure development such as dams, reservoirs, and irrigation systems can alter the natural flow of surface water, leading to ecological imbalances and conflicts over water resources.

Strategies for Effective Surface Water Management

Effective management of surface water resources requires a multi-pronged approach that involves different stakeholders and factors. Some of the key strategies for effective surface water management include:

Water conservation: Water conservation is essential to ensure the sustainable use of surface water resources. This can be achieved through measures such as rainwater harvesting, water-efficient irrigation systems, and the promotion of water-saving practices in households and industries.

Water pollution control: Water pollution control is essential to maintain the quality of surface water resources. This can be achieved through measures such as the regulation of industrial effluents and the treatment of sewage before discharge into surface water bodies.

Integrated water resources management: Integrated water resources management (IWRM) is a holistic approach to water management that considers the social, economic, and environmental aspects of water use. It involves the participation of all stakeholders in decision-making processes and the integration of different water uses such as irrigation, domestic, and industrial.

Watershed management: Watershed management involves the management of the land and water resources within a particular watershed. It aims to maintain the ecological balance of the watershed by reducing soil erosion, preventing deforestation, and promoting reforestation.

Climate adaptation measures: Climate adaptation measures are essential to manage the impacts of climate change on surface water resources. This can include measures such as the development of drought-resistant crops, the promotion of water-saving technologies, and the construction of flood-resistant infrastructure.

International cooperation: International cooperation is essential to manage surface water resources that cross international borders. This can involve the negotiation of treaties, agreements, and protocols that promote equitable and sustainable use of transboundary water resources.

Conclusion

Effective management of surface water resources is essential to ensure their sustainable use. It requires a multi-pronged approach that involves different stakeholders and factors. Some of the key strategies for effective surface water management include water conservation, water pollution control, integrated water resources management, watershed management, climate adaptation measures, and international cooperation.

In addition, the role of technology cannot be ignored in surface water management. Advanced technologies such as remote sensing, geographic information systems (GIS), and data analytics can provide valuable insights into surface water resources, and help in making informed decisions.

Furthermore, public awareness and education are also critical in promoting sustainable water use practices. This can involve awareness campaigns, training programs, and community-based initiatives.

In conclusion, effective management of surface water resources is essential for sustaining life and promoting economic development. While there are many challenges in surface water management, there are also many opportunities for innovation and collaboration. By adopting a holistic and integrated approach to surface water management, we can ensure the sustainable use of this vital resource for generations to come.

29 - Managing Surface Water Resources in the Face of Climate Change: Impacts and Adaptation Measures

Abstract

This article discusses the impact of climate change on surface water resources, including changes in water availability, quality, and ecosystem function, as well as an increased risk of extreme events. The article provides adaptation measures that can be taken to ensure sustainable management of these resources, including water conservation, watershed management, restoration of wetlands and riparian zones, infrastructure adaptation, and ecosystem-based adaptation. A holistic approach to water management is emphasized, recognizing the interconnected nature of water resources. The adoption of these adaptation measures can help to enhance the resilience of surface water resources to climate change impacts.

Introduction

Climate change refers to long-term changes in the Earth's climate, including changes in temperature, precipitation patterns, and other weather-related phenomena, that are primarily caused by human activities, particularly the burning of fossil fuels and deforestation. These human activities release large amounts of greenhouse gases, such as carbon dioxide and methane, into the atmosphere, which trap heat and cause the Earth's atmosphere to warm. This warming trend has numerous impacts on the environment, including melting glaciers and sea ice, rising sea levels, changes in weather patterns and precipitation, and more frequent and severe weather events.

Surface water resources, such as rivers, lakes, and wetlands, are critical for human well-being and ecosystem health. However, climate change is expected to have significant impacts on surface water resources, including changes in water availability, quality, and ecosystem function. This article will discuss the impact of climate change on surface water resources and the adaptation measures that can be taken to ensure the sustainable management of these resources.

Impact of Climate Change on Surface Water Resources

Changes in Water Availability

Climate change is expected to result in changes in the timing, amount, and distribution of water availability in surface water resources. Changes in precipitation patterns can result in increased flooding in some areas and increased drought in others. These changes can have significant impacts on water availability for human uses and ecosystem function.

Changes in Water Quality

Climate change can also impact water quality in surface water resources. Increased temperatures can lead to increased algal blooms, which can result in decreased water quality and harm to aquatic ecosystems. Changes in precipitation patterns can also result in changes in nutrient and pollutant runoff, which can impact water quality.

Impacts on Ecosystem Function

Surface water resources support a wide range of ecosystem functions, including habitat for fish and other aquatic species, water purification, and nutrient cycling. Climate change can impact these functions, leading to the loss of biodiversity and changes in ecosystem services.

Increased Risk of Extreme Events

Climate change can also result in increased risk of extreme events, such as floods and droughts, which can have significant impacts on surface water resources. These events can lead to increased erosion, sedimentation, and changes in water quality.

Adaptation Measures

Water Conservation

Water conservation measures can help to reduce the demand for surface water resources, ensuring that water resources are used efficiently. These measures can include the use of water-efficient technologies, the promotion of water conservation practices, and the use of drought-tolerant crops.

Watershed Management

Watershed management can help to ensure the sustainable management of surface water resources by addressing the interconnected nature of water resources. This approach can include the development of watershed management plans, the establishment of monitoring programs, and the implementation of best management practices to reduce nutrient and pollutant runoff.

Restoration of Wetlands and Riparian Zones

Wetlands and riparian zones play critical roles in supporting surface water resources, including water purification, flood control, and habitat for aquatic species. Restoration of these ecosystems can help to improve the resilience of surface water resources to climate change impacts.

Infrastructure Adaptation

Climate-resilient water supply infrastructure can help to ensure the sustainable management of surface water resources. This includes the development of infrastructure that is resilient to climate change impacts, such as increased flooding and drought. Infrastructure adaptation measures can include the development of water storage facilities, the use of green infrastructure, and the development of climate-resilient water supply systems.

Ecosystem-based Adaptation

Ecosystem-based adaptation measures can help to enhance the resilience of surface water resources to climate change impacts. This approach includes the use of natural ecosystems, such as wetlands and riparian zones, to enhance water management and ecosystem function.

Conclusion

Climate change is expected to have significant impacts on surface water resources, including changes in water availability, quality, and ecosystem function. The adoption of adaptation measures, such as water conservation, watershed management, restoration of wetlands and riparian zones, infrastructure adaptation, and ecosystem-based adaptation, can help to ensure the sustainable management of surface water resources in the face of climate change. It is important to consider the interconnected nature of water resources and to adopt a holistic approach to water management to ensure the sustainable management of surface water resources.

30 - Groundwater: Challenges, Assessment, and Management

Abstract

This article covers various aspects related to groundwater, including groundwater problems, groundwater assessment, groundwater management, and the management of aquifer recharge. Groundwater is a vital resource used for drinking water, irrigation, and industrial processes. However, overuse, contamination, and other factors can threaten its sustainability and quality, leading to serious consequences for human health, the environment, and the economy. The article discusses common groundwater problems, such as overexploitation, contamination, saltwater intrusion, land subsidence, and the effects of climate change. The article also covers the steps involved in groundwater assessment, including identifying the location and extent of the groundwater resources, evaluating the quantity and quality of the groundwater, and developing management strategies. Effective groundwater management requires a multi-disciplinary approach that considers the needs of various stakeholders, including water users, communities, and ecosystems. The article outlines various measures, including groundwater monitoring, water allocation and planning, groundwater recharge, water conservation, and groundwater quality management, that can be used to manage groundwater effectively. Finally, the article explains the management of aquifer recharge, including the construction of artificial recharge structures and sustainable land use and agricultural practices.

Introduction

Groundwater is a critical natural resource that provides drinking water, irrigation, and industrial processes. However, it faces several challenges, including overexploitation, contamination, saltwater intrusion, land subsidence, and the impacts of climate change. To address these challenges, effective groundwater assessment and management strategies are necessary. Groundwater assessment involves evaluating the quantity and quality of groundwater through geological and hydrogeological mapping, field measurements, laboratory analysis, and computer modeling. Groundwater management involves the planning, development, and implementation of strategies and techniques to ensure the effective and sustainable use of groundwater, including groundwater monitoring and assessment, water allocation and planning, groundwater recharge, water conservation, and groundwater quality management. By managing groundwater resources in a sustainable and equitable manner, we can help to ensure that this vital resource is available for future generations.

Groundwater Problems

Groundwater problems are a major concern in many regions of the world. Groundwater is a vital resource that provides drinking water, irrigation, and industrial processes. However, overuse, contamination, and other factors can threaten the sustainability and quality of this resource, which can have serious consequences for human health, the environment, and the economy.

Some of the common problems associated with groundwater include:

Overexploitation: Groundwater is often overused, leading to depletion of aquifers and reduced water availability. This can lead to reduced crop yields, increased competition for water resources, and other economic and environmental problems.

Contamination: Groundwater can become contaminated by a variety of sources, including agricultural and industrial activities, municipal waste, and natural processes. Contaminants can include nitrates, pesticides, heavy metals, and other pollutants that can pose a risk to human health and the environment.

Saltwater intrusion: Saltwater intrusion is the process by which seawater enters freshwater aquifers, reducing the availability and quality of freshwater resources.

Land subsidence: Over-pumping of groundwater can cause the land to sink, leading to subsidence. This can cause damage to buildings, infrastructure, and other structures.

Climate change: Climate change can affect groundwater resources by altering precipitation patterns, reducing recharge rates, and increasing the likelihood of extreme weather events.

To address groundwater problems, a combination of management strategies is needed. These may include reducing water demand through conservation measures, regulating groundwater withdrawals, promoting the use of alternative water sources, and implementing measures to reduce contamination.

Overall, groundwater problems are a significant challenge that requires careful management and conservation to ensure the sustainable use of this vital resource.

Groundwater Assessment

Groundwater assessment is the process of evaluating the quantity and quality of groundwater in a particular area. Groundwater is an important natural resource that is used for a variety of purposes such as drinking water, irrigation, and industrial processes. However, groundwater resources can be overused and contaminated, leading to long-term consequences for both the environment and human health.

Groundwater assessment typically involves several steps. The first step is to identify the location and extent of the groundwater resources in the area. This is done through geological and hydrogeological mapping, which can help to identify the type and distribution of rock formations, soils, and other features that affect groundwater flow.

Once the location and extent of the groundwater resources have been determined, the next step is to evaluate the quantity and quality of the groundwater. This is typically done through a combination of field measurements, laboratory analysis of water samples, and computer modeling. Field measurements may include measurements of water levels, flow rates, and water quality parameters such as temperature, pH, dissolved oxygen, and dissolved solids.

Laboratory analysis of water samples can provide more detailed information about the chemical composition of the groundwater and the presence of contaminants such as heavy metals, pesticides, and other pollutants. Computer modeling can be used to simulate groundwater flow and assess the long-term sustainability of the groundwater resources.

Based on the results of the groundwater assessment, management strategies can be developed to protect and sustainably manage the groundwater resources. This may include regulations on groundwater use, programs to reduce pollution and contamination, and measures to promote water conservation and recharge. Overall, groundwater assessment is an important tool for ensuring the long-term availability and quality of this vital resource.

Groundwater Management

Groundwater management involves the planning, development, and implementation of strategies and techniques to ensure the effective and sustainable use of groundwater for human populations, agriculture, and the natural environment. Groundwater is a critical source of freshwater for many regions of the world, but it is a finite resource that must be carefully managed to avoid depletion and degradation.

Effective groundwater management requires a multi-disciplinary approach that considers the needs of various stakeholders, including water users, communities, and ecosystems. This approach typically involves a range of measures, including:

Groundwater monitoring and assessment: This involves monitoring the quantity and quality of groundwater resources to assess their health and identify any changes or impacts. This information is used to inform management decisions and adjust management strategies as needed.

Water allocation and planning: This involves developing and implementing water allocation plans that balance the needs of various users and stakeholders, including the environment. These plans typically include guidelines for groundwater use, recharge, and management practices.

Groundwater recharge: This includes developing and implementing techniques to recharge groundwater resources, such as artificial recharge systems or land-use practices that promote infiltration and recharge.

Water conservation: This includes promoting water conservation practices such as fixing leaks, using water-efficient appliances and fixtures, and limiting outdoor water use.

Groundwater quality management: This includes developing and implementing strategies to protect and improve the quality of groundwater resources, such as implementing best management practices for agriculture and other land uses, and implementing programs to reduce the impacts of urbanization and development on water resources.

Overall, effective groundwater management requires ongoing collaboration between government agencies, communities, and other stakeholders. It also requires a long-term perspective, considering the potential impacts of climate change and the need for ongoing adaptation and resilience-building efforts. By managing groundwater resources in a sustainable and equitable manner, we can help to ensure that this vital resource is available for future generations.

Management of Aquifer Recharge

Aquifer recharge refers to the process of replenishing underground water sources, or aquifers, with water from various sources such as rainfall, surface water, or treated wastewater. The management of aquifer recharge involves the use of various techniques and strategies to optimize the quantity and quality of water that enters the aquifers.

One of the primary methods of managing aquifer recharge is the construction of artificial recharge structures such as infiltration basins, percolation ponds, and recharge wells. These structures can be designed to capture and divert excess surface water to the aquifer through natural filtration processes or direct injection into the aquifer.

Another approach to aquifer recharge management is the implementation of sustainable land use and agricultural practices that reduce water usage and improve the infiltration of water into the soil. This can involve using crops with lower water requirements, implementing water-efficient irrigation systems, and minimizing the use of fertilizers and pesticides that can contaminate groundwater.

The management of aquifer recharge also involves monitoring and regulation of groundwater use and groundwater quality. This includes measuring and tracking changes in groundwater levels, monitoring water quality for contamination, and implementing appropriate management strategies to prevent over-extraction or pollution of the aquifer.

In summary, the management of aquifer recharge is a critical aspect of sustainable water resource management. It involves the implementation of a range of techniques and strategies to ensure the efficient and sustainable replenishment of underground water sources, which are essential for supporting human populations and ecosystems.

Groundwater Quality Management

Groundwater quality management involves the planning, development, and implementation of strategies and techniques to ensure the effective and sustainable use of groundwater for human populations, agriculture, and the natural environment. Groundwater quality is a critical component of overall water quality and is impacted by a range of human activities and natural factors.

Effective management of groundwater quality requires a multi-disciplinary approach that considers the needs of various stakeholders, including water users, communities, and ecosystems. This approach typically involves a range of measures, including:

Groundwater quality monitoring and assessment: This involves monitoring the quality of groundwater resources to assess their health and identify any changes or impacts. This information is used to inform management decisions and adjust management strategies as needed.

Source water protection: This includes developing and implementing strategies to protect the quality of groundwater resources, such as protecting the land use around recharge areas, implementing best management practices for agriculture and other land uses, and reducing the impacts of urbanization and development on water resources.

Water treatment: This includes developing and implementing water treatment systems to remove contaminants and improve water quality for human consumption and other uses.

Pollution prevention: This includes developing and implementing strategies to prevent pollution from entering groundwater resources, such as implementing regulations and incentives for industries to reduce their pollution discharge and promoting public education campaigns to increase awareness of water pollution prevention.

Water reuse and recycling: This includes developing and implementing water reuse and recycling systems that treat and reuse wastewater for non-potable uses such as irrigation or industrial processes, thereby reducing the overall demand on groundwater resources.

Overall, effective management of groundwater quality requires ongoing collaboration between government agencies, communities, and other stakeholders. It also requires a long-term perspective, considering the potential impacts of climate change and the need for ongoing adaptation and resilience-building efforts. By managing groundwater quality in a sustainable and equitable manner, we can help to ensure that this vital resource is available for future generations.

Seawater Intrusion

Seawater intrusion is the process by which saltwater enters freshwater aquifers, contaminating them and reducing their quality and quantity. This can occur in coastal areas where the freshwater aquifers are in hydraulic connection with the sea and the pressure of the saltwater exceeds that of the freshwater.

Seawater intrusion can have a significant impact on the availability of freshwater resources, which can lead to a range of environmental and socio-economic problems. When seawater enters freshwater aquifers, it reduces the volume of fresh water available for use, which can lead to increased competition for water resources and the need to find alternative sources of water.

Seawater intrusion can also have a negative impact on the quality of freshwater resources, making them unsuitable for drinking or other uses. This can result in increased health risks, reduced agricultural productivity, and damage to aquatic ecosystems.

The factors that can influence seawater intrusion include the geological and hydrological conditions of the area, as well as the rate and magnitude of groundwater withdrawals. Coastal aquifers that are over-pumped or subjected to other forms of stress are more vulnerable to seawater intrusion.

There are several strategies that can be used to prevent or mitigate the effects of seawater intrusion. These include the regulation of groundwater withdrawals, the promotion of water conservation and management practices, the development of alternative water sources such as desalination and the use of artificial recharge methods to increase the amount of freshwater in the aquifer.

Overall, seawater intrusion is a serious threat to freshwater resources in coastal areas. Proper management and conservation of these resources are necessary to ensure their sustainable use and prevent negative impacts on the environment and society.

Conclusion

In conclusion, groundwater is a vital resource that provides drinking water, irrigation, and industrial processes, but it is facing several problems such as overuse, contamination, and climate change impacts. These issues can have severe consequences for human health, the environment, and the economy, making it crucial to manage and conserve groundwater resources sustainably.

Groundwater assessment is an essential tool for evaluating the quantity and quality of groundwater in a particular area. It involves several steps such as geological and hydrogeological mapping, field measurements, laboratory analysis of water samples, and computer modeling. Based on the results of the assessment, management strategies can be developed to protect and sustainably manage groundwater resources.

Effective groundwater management requires a multi-disciplinary approach that considers the needs of various stakeholders and involves measures such as groundwater monitoring and assessment, water allocation and planning, groundwater recharge, water conservation, and groundwater quality management.

Furthermore, the management of aquifer recharge involves the use of various techniques and strategies to optimize the quantity and quality of water that enters the aquifers, such as artificial recharge structures and sustainable land use practices.

By managing groundwater resources sustainably and equitably, we can ensure that this vital resource is available for future generations. It requires ongoing collaboration and a long-term perspective that considers the potential impacts of climate change and the need for ongoing adaptation and resilience-building efforts.

31 - Groundwater Recharge: Methods, Factors, and Challenges for Sustainable Resource Management

Abstract

Groundwater recharge is a vital process for maintaining sustainable groundwater resources, supporting ecosystems, and meeting human needs. This article explores the methods used for natural and artificial groundwater recharge, including infiltration basins, injection wells, recharge trenches, and spreading basins. The factors influencing groundwater recharge, such as climate, soil type, vegetation cover, and land use, are also discussed. However, the implementation of groundwater recharge projects faces various challenges, including water availability, land use conflicts, water quality concerns, and aquifer overexploitation. It is essential to address these challenges and develop effective recharge strategies to ensure the sustainability of groundwater resources.

Introduction

Groundwater recharge is the process by which water is added to the groundwater reservoir through infiltration of precipitation, snowmelt, or surface water into the soil. Groundwater is an essential resource for both humans and the environment. It is a major source of drinking water, irrigation water, and industrial water, and it supports many wetlands, streams, and rivers. However, groundwater resources are under threat due to over-exploitation, climate change, land-use changes, and pollution. To sustain groundwater resources, it is necessary to recharge them.

This article discusses the concept of groundwater recharge, its importance, methods of recharge, factors affecting recharge, and challenges associated with recharge.

Importance of Groundwater Recharge

Groundwater recharge is essential to maintain the water table and sustain groundwater resources. Groundwater recharge helps to:

Replenish groundwater resources: Groundwater is a finite resource, and its availability is declining due to over-pumping and climate change. Groundwater recharge helps to replenish the groundwater resources and maintain the water table.

Improve water quality: Groundwater recharge helps to improve the quality of groundwater by diluting contaminants and reducing salinity.

Sustain ecosystems: Groundwater recharge helps to sustain wetlands, streams, and rivers, which are important habitats for many species of plants and animals.

Support agriculture: Groundwater recharge provides irrigation water for agriculture, which is critical for food production.

Provide drinking water: Groundwater is a major source of drinking water for many communities, and groundwater recharge helps to ensure a sustainable supply of safe drinking water.

Methods of Groundwater Recharge

There are several methods of groundwater recharge, which can be broadly classified into two categories:

Natural recharge: Natural recharge occurs through infiltration of precipitation, snowmelt, or surface water into the soil. Natural recharge is influenced by factors such as soil type, vegetation cover, land use, and topography.

Artificial recharge: Artificial recharge involves intentional augmentation of the natural recharge process. Artificial recharge is achieved by diverting surface water or treated wastewater into the ground through infiltration basins, injection wells, or recharge trenches.

The choice of recharge method depends on various factors, such as the availability of water, land use, soil type, and aquifer characteristics.

Natural recharge:

Natural recharge occurs through several processes, such as:

Infiltration: Infiltration is the process by which water enters the soil surface. Infiltration rate depends on soil type, vegetation cover, and land use. Soil with high permeability allows for more infiltration and recharge than soils with low permeability.

Percolation: Percolation is the downward movement of water through the soil. Percolation rate depends on soil type, texture, and structure. Fine-textured soils with high clay content have low percolation rates, while coarse-textured soils with high sand content have high percolation rates.

Seepage: Seepage is the lateral movement of water through the soil. Seepage rate depends on the slope of the land, soil type, and vegetation cover.

Runoff: Runoff is the surface water that flows over the land and enters streams, rivers, and lakes. Runoff can contribute to groundwater recharge by recharging streams and rivers, which in turn recharge groundwater.

Artificial recharge:

Artificial recharge is a deliberate process of augmenting the natural recharge process. Artificial recharge is done by various methods, such as:

Infiltration basins: Infiltration basins are shallow depressions in the ground designed to collect surface water and allow it to infiltrate into the soil. Infiltration basins are used in areas with permeable soils and low runoff.

Injection wells: Injection wells are drilled wells that inject water directly into the aquifer. Injection wells are used in areas with impermeable soils or where the water table is deep.

Recharge trenches: Recharge trenches are excavated trenches filled with coarse gravel or crushed stone to enhance infiltration. Recharge trenches are used in areas with shallow groundwater and permeable soils.

Spreading basins: Spreading basins are large earthen structures designed to store and percolate surface water into the ground. Spreading basins are used in areas with low permeability soils and high runoff.

Factors affecting Groundwater Recharge

Groundwater recharge is influenced by various factors, such as:

Climate: Climate influences the amount and timing of precipitation, which affects groundwater recharge. Areas with high rainfall and low evapotranspiration rates have higher groundwater recharge rates.

Soil type: Soil type affects the infiltration rate and percolation rate, which affects groundwater recharge. Sandy soils with high permeability allow for faster infiltration and recharge, while clayey soils with low permeability have lower infiltration rates and recharge.

Vegetation cover: Vegetation cover affects the infiltration rate, evapotranspiration rate, and soil moisture content, which affects groundwater recharge. Areas with dense vegetation cover have higher infiltration rates and recharge than areas with sparse vegetation cover.

Land use: Land use affects the infiltration rate, runoff rate, and soil moisture content, which affects groundwater recharge. Urban areas with impervious surfaces have low infiltration rates and recharge, while agricultural areas with permeable soils have higher infiltration rates and recharge.

Challenges associated with Groundwater Recharge

Groundwater recharge is not without challenges. Some of the challenges associated with groundwater recharge include:

Water availability: Groundwater recharge requires a sufficient supply of water. In areas with limited water availability, it may be difficult to implement recharge projects.

Land use conflicts: Groundwater recharge projects may conflict with other land uses, such as agriculture or urban development. Such conflicts can delay or prevent the implementation of recharge projects.

Water quality concerns: Recharge projects may introduce contaminants into the aquifer if the water source is not properly treated. This can lead to degradation of groundwater quality.

Aquifer overexploitation: Recharge projects can increase the demand for groundwater, leading to overexploitation of the aquifer. This can result in a decline in the water table and depletion of the groundwater resource.

Conclusion

Groundwater recharge is a critical process for sustaining groundwater resources and supporting ecosystems, agriculture, and human needs. Natural and artificial recharge methods can be used to enhance groundwater recharge, but the choice of method depends on various factors such as water availability, soil type, and aquifer characteristics. Factors such as climate, soil type, vegetation cover, and land use influence groundwater recharge. However, groundwater recharge is not without challenges, including water availability, land use conflicts, water quality concerns, and aquifer overexploitation. To ensure the sustainability of groundwater resources, it is necessary to implement effective recharge strategies that balance the needs of human and environmental systems.

32 - Groundwater Balance: Understanding the Components and Factors Affecting Sustainable Use of Groundwater Resources

Abstract

Groundwater balance is a critical component of the hydrological cycle, determining the availability and sustainability of groundwater resources. This article discusses the various components of groundwater balance, including precipitation, groundwater recharge, evapotranspiration, and change in groundwater storage, and the factors affecting them, such as climate, land use, soil type, vegetation cover, and groundwater withdrawals. Understanding groundwater balance is essential in developing effective management strategies to ensure sustainable use of groundwater resources. Groundwater balance can also help in identifying areas with high recharge potential and developing strategies to enhance recharge rates. This article emphasizes the importance of groundwater balance in water resource management and environmental sustainability.

Introduction

Groundwater is a crucial component of the earth's hydrological cycle. It is an essential source of water for many communities, industries, and agriculture worldwide. Groundwater balance refers to the state of equilibrium between the input and output of groundwater in a given area. The input includes precipitation, infiltration, and surface water runoff, while the output includes groundwater discharge, evapotranspiration, and surface water recharge. Understanding groundwater balance is essential in managing groundwater resources sustainably. This article will explore the concept of groundwater balance, its components, factors affecting it, and its importance in water resource management.

Components of Groundwater Balance

The groundwater balance equation represents the balance between the inputs and outputs of groundwater in a given area. The equation is expressed as:

$$P = Q + ET + \Delta S$$

where P represents precipitation, Q represents groundwater recharge, ET represents evapotranspiration, and ΔS represents the change in groundwater storage.

Precipitation

Precipitation is the primary source of water for groundwater recharge. Precipitation can occur as rain or snow and is dependent on factors such as location, climate, and season. In areas with high precipitation rates, such as tropical regions, there is a high potential for groundwater recharge. On the other hand, in arid regions with low precipitation rates, groundwater recharge is limited.

Groundwater Recharge

Groundwater recharge refers to the process by which precipitation infiltrates the ground and percolates to the groundwater table. The rate of recharge is dependent on several factors, such as the soil type, permeability, land use, and vegetation cover. In areas with sandy soils and high permeability, such as coastal plains, recharge rates are higher than in areas with impermeable soils such as clay.

Evapotranspiration

Evapotranspiration (ET) is the combined process of evaporation and transpiration. Evaporation refers to the process by which water changes from liquid to vapor and enters the atmosphere. Transpiration refers to the process by which water is absorbed by plants, transported through the plant, and then released into the atmosphere through pores on the leaf surface. ET is an essential process in the water cycle, as it contributes to the atmospheric water vapor that eventually falls as precipitation.

Change in Groundwater Storage

The change in groundwater storage represents the difference between the amount of water entering and leaving the aquifer. It is influenced by factors such as the rate of recharge and discharge, pumping, and natural discharge to surface water bodies. If the change in groundwater storage is positive, it indicates that groundwater recharge exceeds the amount of groundwater being withdrawn from the aquifer. A negative change in groundwater storage, on the other hand, indicates that groundwater withdrawal exceeds the recharge rate.

Factors Affecting Groundwater Balance

Several factors can affect groundwater balance, including climate, land use, soil type, vegetation cover, and groundwater withdrawals.

Climate

Climate is one of the most significant factors affecting groundwater balance. Precipitation is the primary source of groundwater recharge, and its distribution and variability significantly impact groundwater availability. Climate change can affect precipitation patterns, leading to changes in groundwater recharge rates. For example, a decrease in precipitation rates can reduce groundwater recharge rates, leading to a decline in groundwater availability.

Land Use

Land use is another critical factor affecting groundwater balance. Human activities, such as urbanization and agriculture, can alter the natural landscape and affect the recharge and discharge of groundwater. Land use changes can increase the amount of impervious surfaces such as roads and buildings, reducing the amount of water that can infiltrate the ground and recharge the aquifer. Land use changes can also lead to increased water demand, leading to over-pumping of groundwater and depletion of aquifer storage.

Soil Type

Soil type is a critical factor in determining the rate of groundwater recharge. Soil properties such as texture, structure, and permeability influence the rate of infiltration and percolation of water through the soil. In general, sandy soils have a higher infiltration rate than clay soils, which have a lower infiltration rate due to their small pore sizes. Therefore, soil type can affect groundwater recharge rates and the availability of groundwater.

Vegetation Cover

Vegetation cover can also affect groundwater balance. Plants can reduce runoff by absorbing and storing water in their root systems. Vegetation cover can increase infiltration rates, leading to increased groundwater recharge. However, excessive vegetation cover can also reduce recharge rates by intercepting rainfall and reducing the amount of water that reaches the soil surface.

Groundwater Withdrawals

Groundwater withdrawals for domestic, agricultural, and industrial uses can affect groundwater balance. Over-pumping can lead to groundwater depletion and reduced availability. In some cases, groundwater withdrawals can exceed recharge rates, leading to a negative change in groundwater storage. Groundwater withdrawals can also affect surface water bodies, as they can reduce streamflow and affect aquatic ecosystems.

Importance of Groundwater Balance

Groundwater balance is essential for sustainable water resource management. Understanding the components of groundwater balance and the factors affecting it can help in the development of effective management strategies. Some of the reasons why groundwater balance is important include:

Water Availability

Groundwater is a critical source of water for many communities, industries, and agriculture worldwide. Understanding groundwater balance is essential in managing groundwater resources sustainably, ensuring that there is enough water for current and future needs.

Environmental Sustainability

Groundwater is also an essential component of the environment, supporting ecosystems and maintaining water quality. Groundwater-dependent ecosystems, such as wetlands and riparian zones, are important habitats for many plant and animal species. Maintaining groundwater balance is crucial for ensuring the sustainability of these ecosystems.

Groundwater Management

Groundwater management involves balancing the competing demands for groundwater, such as agricultural, industrial, and domestic uses, while ensuring that the aquifer is not overexploited. Understanding groundwater balance is essential in developing effective management strategies to ensure sustainable use of groundwater resources.

Groundwater Recharge

Groundwater recharge is essential for maintaining groundwater balance. Understanding the factors affecting groundwater recharge, such as land use, vegetation cover, and soil type, can help in identifying areas with high recharge potential and developing strategies to enhance recharge rates.

Conclusion

Groundwater balance is a critical component of the hydrological cycle, determining the availability and sustainability of groundwater resources. The components of groundwater balance, including precipitation, groundwater recharge, evapotranspiration, and change in groundwater storage, are influenced by various factors such as climate, land use, soil type, vegetation cover, and groundwater withdrawals. Understanding groundwater balance is essential in developing effective management strategies to ensure sustainable use of groundwater resources. Groundwater balance can also help in identifying areas with high recharge potential and developing strategies to enhance recharge rates.

33 - Methodology and Challenges in Estimating Groundwater Resources

Abstract

Groundwater is a crucial source of freshwater, especially for agriculture and drinking water. Estimating groundwater resources is crucial for its management and sustainable use. The methodology for estimating groundwater resources involves the collection of hydrogeological data, calculation of recharge, discharge, and storage, and estimation of groundwater availability. However, there are several challenges in estimating groundwater resources, including the variability in groundwater availability, limited data availability, complex hydrogeological systems, and human-induced changes. This article discusses the methodology and challenges in estimating groundwater resources to ensure the sustainable management of groundwater resources.

Introduction

Groundwater is a critical resource that provides drinking water to millions of people worldwide, and its estimation is essential for effective management and planning. The estimation of groundwater resources involves determining the quantity and quality of water that can be sustainably extracted from an aquifer. This article discusses the methodology for estimating groundwater resources, including data collection and analysis, calculation of recharge, discharge, and storage, and estimation of groundwater availability. It also highlights the challenges involved in estimating groundwater resources and the importance of accurate data collection and analysis in the process.

Methodology for Estimating Groundwater Resources

The estimation of groundwater resources involves determining the quantity and quality of water that can be sustainably extracted from a particular aquifer. The methodology for groundwater resource estimation is based on the principles of hydrogeology, which involves the study of the movement, distribution, and quality of groundwater in aquifers.

The first step in groundwater resource estimation is to conduct a hydrogeological study of the area under consideration. This involves mapping the geological formations and identifying the aquifers in the area. The hydrogeological study also involves the measurement of groundwater levels and the collection of data on rainfall, surface water, and other hydrological parameters.

The next step is to conduct pumping tests to determine the aquifer's hydraulic characteristics, such as permeability and transmissivity. Pumping tests involve pumping water from the aquifer at a constant rate and measuring the rate of water level decline over time. This data is used to calculate the hydraulic conductivity of the aquifer, which is a measure of its ability to transmit water.

The final step is to use the data collected in the hydrogeological study and pumping tests to estimate the groundwater resources in the area. This involves developing a numerical groundwater model that simulates the movement of water in the aquifer and predicts the quantity and quality of water that can be sustainably extracted.

The methodology for estimating groundwater resources involves the following steps:

Step 1: Data Collection and Analysis

Accurate data collection and analysis are essential for groundwater resource estimation. The data collected should include information on the hydrogeology of the area, groundwater levels, pumping test data, and hydrological data such as rainfall and surface water availability.

Data collection is typically done through a combination of field surveys, drilling of boreholes, and installation of observation wells. Field surveys involve mapping the geological formations and identifying the aquifers in the area. Boreholes are drilled to collect data on the hydraulic characteristics of the aquifer, and observation wells are installed to measure groundwater levels over time. The data collected includes the thickness and depth of the aquifer, the permeability of the rock, and the water quality parameters.

Data analysis involves the use of statistical and numerical methods to interpret the data collected. This includes the use of computer models to simulate the movement of water in the aquifer and predict the quantity and quality of water that can be sustainably extracted.

Hydrogeological Survey: A detailed hydrogeological survey of the area should be conducted to identify the aquifers and their characteristics.

Geophysical Survey: Geophysical methods such as resistivity and seismic surveys should be used to supplement the hydrogeological survey.

Borehole Drilling: Boreholes should be drilled to collect data on the aquifer's hydraulic characteristics.

Pumping Tests: Pumping tests should be conducted to determine the aquifer's hydraulic conductivity.

Groundwater Level Monitoring: Groundwater levels should be monitored continuously in the area under consideration.

Rainfall Monitoring: Rainfall should be monitored regularly to determine the recharge potential of the aquifer.

Surface Water Availability: Surface water availability should be assessed to determine its impact on groundwater recharge.

Water Quality Analysis: Water quality analysis should be conducted to determine the suitability of groundwater for various uses.

Step 2: Calculation of Recharge

The next step in estimating groundwater resources is to calculate the recharge to the aquifer. Recharge is the amount of water that enters the aquifer through precipitation, infiltration, or other sources. The recharge to the aquifer can be estimated using various methods, including the water balance method, the chloride mass balance method, and the groundwater modelling method. The water balance method involves calculating the difference between the precipitation and the evapotranspiration in the area. The chloride mass balance method involves measuring the concentration of chloride in the groundwater and comparing it with the concentration in the surface water. The groundwater modelling method involves simulating the movement of water in the aquifer using mathematical models.

Step 3: Calculation of Discharge

The next step in estimating groundwater resources is to calculate the discharge from the aquifer. Discharge is the amount of water that flows out of the aquifer through springs, rivers, or other sources. The discharge from the aquifer can be estimated using various methods, including the water balance method, the baseflow separation method, and the groundwater modelling method. The water balance method involves calculating the difference between the inflow and the outflow of water in the area. The baseflow separation method involves separating the baseflow from the total flow in a river. The groundwater modelling method involves simulating the movement of water in the aquifer and its interaction with surface water using mathematical models.

Step 4: Calculation of Storage

The next step in estimating groundwater resources is to calculate the storage in the aquifer. Storage is the amount of water that is stored in the aquifer at a particular time. The storage in the aquifer can be estimated using various methods, including the water balance method and the groundwater modelling method. The water balance method involves calculating the difference between the recharge and the discharge in the area. The groundwater modelling method involves simulating the movement of water in the aquifer and its interaction with surface water using mathematical models.

Step 5: Estimation of Groundwater Availability

The final step in estimating groundwater resources is to estimate the groundwater availability. Groundwater availability is the amount of water that can be withdrawn from the aquifer without causing depletion or degradation of the resource. The groundwater availability can be estimated using various methods, including the safe yield method and the groundwater modelling method.

The safe yield method involves calculating the maximum amount of water that can be withdrawn from the aquifer without causing depletion or degradation of the resource. The groundwater modelling method involves simulating the movement of water in the aquifer and its interaction with surface water using mathematical models to estimate the sustainable yield.

Importance of Proper Data Collection and Analysis

Proper data collection and analysis help in:

- Developing sustainable groundwater management strategies that ensure the long-term availability of groundwater resources.
- Identifying potential groundwater-related problems such as over-pumping, seawater intrusion, and land subsidence.
- Evaluating the impact of climate change on groundwater resources and developing appropriate adaptation strategies.
- Ensuring equitable distribution of groundwater resources among various users.

Challenges in Estimating Groundwater Resources

Groundwater is a vital resource for human consumption, agriculture, industry, and ecosystems. However, estimating groundwater resources is a challenging task due to various natural and anthropogenic factors. Some of the challenges involved in estimating groundwater resources are:

Variability in groundwater availability: Groundwater availability varies from place to place and season to season, depending on various factors like rainfall, geological characteristics, and human activities.

Limited data availability: Groundwater monitoring networks are not extensive enough to cover all the aquifers, and data may not be available for a long enough period to make accurate assessments.

Complex hydrogeological systems: Hydrogeological systems can be highly complex, with interactions between surface water and groundwater, and the presence of multiple aquifers and confining layers.

Human-induced changes: Human activities like land-use changes, groundwater pumping, and contamination can alter groundwater availability, quality, and recharge processes.

Handling the Challenges

Handling the challenges in groundwater resource estimation and its sustainable management requires a multi-pronged approach that involves improving data collection, developing hydrological models, implementing sustainable groundwater management practices, promoting water conservation, addressing pollution sources, and raising public awareness.

Improve groundwater monitoring networks: Expand groundwater monitoring networks to cover all major aquifers, and increase the frequency of monitoring to capture seasonal variability.

Use remote sensing and geophysical techniques: Utilize remote sensing and geophysical techniques like satellite imagery and ground-penetrating radar to gather information about the subsurface hydrogeological systems.

Develop hydrological models: Develop hydrological models that integrate different data sources and simulate groundwater recharge, discharge, and flow processes.

Implement sustainable groundwater management practices: Implement sustainable groundwater management practices like conjunctive use of surface water and groundwater, rainwater harvesting, and artificial recharge to increase groundwater recharge and reduce over-pumping.

Promote water conservation: Promote water conservation measures like drip irrigation, water reuse, and use of drought-tolerant crops to reduce water demand and preserve groundwater resources.

Address pollution sources: Address pollution sources like industrial effluent discharge, agricultural runoff, and septic tank leakage to prevent contamination of groundwater resources.

Raise public awareness: Educate the public about the importance of groundwater resources, and encourage them to adopt water-saving practices and avoid activities that can harm groundwater quality.

Conclusion

Groundwater is an important source of freshwater. The estimation of groundwater resources is crucial for its management. The availability of reliable data is crucial for developing effective groundwater management strategies that ensure the long-term sustainability of this vital resource. The methodology for estimating groundwater resources involves the collection of hydrogeological data, the calculation of recharge, discharge, and storage, and the estimation of groundwater availability. However, there are several challenges in estimating groundwater resources, including the variability in groundwater availability, limited data availability, complex hydrogeological systems, and human-induced changes. Therefore, it is important to improve the collection of hydrogeological data and to develop more accurate methods for estimating groundwater resources to ensure its sustainable management.

34 - Mapping Groundwater Potential Zones: Techniques, Applications, and Challenges

Abstract

Groundwater is a critical resource for human activities and ecosystems, particularly in regions where surface water resources are limited or unreliable. Mapping groundwater potential zones is essential for identifying areas with high potential for sustainable groundwater development and management. This article provides an overview of the techniques used to map groundwater potential zones, including remote sensing, geospatial analysis, and hydrogeological modelling. It also discusses the applications of groundwater potential mapping in various sectors, such as agriculture, industry, and urban water supply. Additionally, the article highlights the challenges associated with mapping groundwater potential zones, including data availability, scale mismatch, and uncertainty. By understanding the techniques, applications, and challenges of mapping groundwater potential zones, stakeholders can make informed decisions regarding groundwater management and sustainability.

Introduction

Groundwater is an essential resource for many human activities, including agriculture, industry, and domestic water supply. In arid and semi-arid regions, groundwater is often the primary source of water for irrigation and other purposes. However, groundwater resources are under increasing pressure due to population growth, urbanization, and climate change. To ensure the sustainability of groundwater resources, it is essential to identify areas with high potential for groundwater development and management. Mapping groundwater potential zones can provide valuable information for stakeholders, including water managers, policymakers, and farmers.

Mapping groundwater potential zones involves identifying areas where groundwater recharge is high and where aquifers are capable of storing and transmitting water. The process of mapping groundwater potential zones involves integrating various data sources, including geological, hydrogeological, climatic, and socioeconomic data. Remote sensing, geospatial analysis, and hydrogeological modelling are commonly used techniques for mapping groundwater potential zones.

This article provides an overview of the techniques used to map groundwater potential zones, including remote sensing, geospatial analysis, and hydrogeological modelling. It also discusses the applications of groundwater potential mapping in various sectors, such as agriculture, industry, and urban water supply. Additionally, the article highlights the challenges associated with mapping groundwater potential zones, including data availability, scale mismatch, and uncertainty.

Techniques for Mapping Groundwater Potential Zones

Remote Sensing

Remote sensing involves the use of satellite or airborne sensors to capture images of the Earth's surface. Remote sensing data can be used to map various hydrological variables, such as land use, land cover, soil moisture, and vegetation cover, which are essential for mapping groundwater potential zones. For example, the Normalized Difference Vegetation Index (NDVI) can be used to estimate vegetation cover, which is an indicator of groundwater recharge. Areas with high NDVI values indicate high vegetation cover, which suggests high potential for groundwater recharge. Similarly, soil moisture data can be used to estimate the amount of water that infiltrates into the soil and recharges the aquifers.

Geospatial Analysis

Geospatial analysis involves the use of Geographic Information Systems (GIS) to integrate and analyze various spatial datasets. GIS allows users to overlay and analyze different layers of spatial data, such as geological maps, hydrological data, and land use maps. Geospatial analysis can be used to identify areas with high potential for groundwater recharge based on the geological and hydrological characteristics of the area. For example, areas with high permeability and porosity of the aquifer, high elevation, and low slope are likely to have high potential for groundwater recharge.

Hydrogeological Modelling

Hydrogeological modelling involves the use of numerical models to simulate the movement of water through the subsurface. Hydrogeological models can be used to estimate groundwater recharge, groundwater flow, and groundwater storage capacity. These models can be used to identify areas with high potential for sustainable groundwater development and management. For example, groundwater models can be used to estimate the impact of different scenarios, such as land use changes or climate variability, on groundwater resources and identify areas that may be particularly vulnerable to depletion or contamination.

Hydrogeological Modelling Software

Hydrogeological modelling software plays an important role in mapping groundwater potential zones. Here are some popular hydrogeological modelling software:

MODFLOW: MODFLOW (Modular Groundwater Flow Model) is one of the most widely used hydrogeological modelling software for simulating groundwater flow and transport. It is a three-dimensional numerical model that uses finite-difference methods to solve the groundwater flow equations.

GMS (Groundwater Modelling System): GMS is a comprehensive groundwater modelling software that includes tools for building, calibrating, and visualizing groundwater models. It can be used to simulate both steady-state and transient groundwater flow and transport.

FEFLOW: FEFLOW (Finite Element subsurface FLOW system) is a versatile hydrogeological modelling software that uses finite element methods to simulate groundwater flow and transport. It can be used to simulate complex hydrogeological systems, including coupled surface water-groundwater systems.

MODPATH: MODPATH is a particle-tracking tool that can be used in conjunction with MODFLOW to simulate groundwater flow and transport. It can be used to track the movement of groundwater particles through a hydrogeological system, which can be useful for identifying potential groundwater flow paths and contamination sources.

SEAWAT: SEAWAT is a software package that combines the capabilities of MODFLOW and MT3DMS to simulate variable-density groundwater flow and transport. It can be used to simulate saltwater intrusion in coastal aquifers, as well as other hydrogeological systems where density-dependent flow and transport are important.

Visual MODFLOW: Visual MODFLOW is a hydrogeological modelling software that includes tools for building, calibrating, and visualizing groundwater models. It uses finite-difference methods to solve the groundwater flow equations and can be used to simulate both steady-state and transient groundwater flow and transport.

There are many other hydrogeological modelling software available, and the choice of software will depend on the specific needs of the user and the complexity of the hydrogeological system being studied.

Applications of Mapping Groundwater Potential Zones

Agriculture

Agriculture is the largest user of groundwater globally, accounting for around 70% of global groundwater withdrawals. Mapping groundwater potential zones can help identify areas with high potential for sustainable groundwater irrigation, which can support food production and improve rural livelihoods. By mapping groundwater potential zones, farmers can identify areas with high potential for groundwater recharge and locate areas where irrigation is likely to be most efficient. In addition, mapping groundwater potential zones can help identify areas where groundwater recharge is vulnerable to land use changes, such as deforestation, which can lead to reduced recharge and lower groundwater availability for irrigation.

Industry

Industry is another important user of groundwater, particularly in regions where surface water resources are limited. Mapping groundwater potential zones can help identify areas where groundwater resources can be sustainably developed for industrial purposes, such as manufacturing or mining. Mapping groundwater potential zones can also help identify areas where industrial activities may pose a risk to groundwater resources, such as through contamination or excessive withdrawals.

Urban Water Supply

Urban areas are growing rapidly worldwide, and demand for water is increasing. Mapping groundwater potential zones can help identify areas where groundwater resources can be sustainably developed to meet urban water demand. Mapping groundwater potential zones can also help identify areas where urban activities, such as land use changes or groundwater withdrawals, may impact groundwater recharge and availability.

Challenges Associated with Mapping Groundwater Potential Zones

Data Availability

Mapping groundwater potential zones requires access to a variety of data sources, including geological, hydrological, climatic, and socioeconomic data. However, data availability can be a significant challenge, particularly in developing countries or regions with limited resources. In addition, data quality and accuracy can vary widely between different sources and can impact the reliability of groundwater potential maps.

Scale Mismatch

Mapping groundwater potential zones requires data at multiple spatial scales, from global to local. However, data at different scales may not always be compatible, and there can be a mismatch between the scale of data available and the scale of analysis required. For example, global-scale climate data may not be suitable for mapping groundwater potential zones at the local level.

Uncertainty

Groundwater potential mapping involves a significant degree of uncertainty due to the complexity of groundwater systems and the limitations of available data. Uncertainty can arise from a variety of sources, such as errors in data collection or modelling assumptions. As a result, groundwater potential maps should be interpreted with caution, and stakeholders should be aware of the uncertainty associated with the maps.

Conclusion

Mapping groundwater potential zones is essential for identifying areas with high potential for sustainable groundwater development and management. Remote sensing, geospatial analysis, and hydrogeological modelling are commonly used techniques for mapping groundwater potential zones. Mapping groundwater potential zones has important applications in various sectors, such as agriculture, industry, and urban water supply. However, mapping groundwater potential zones is not without challenges, such as data availability, scale mismatch, and uncertainty. By understanding the techniques, applications, and challenges of mapping groundwater potential zones, stakeholders can make informed decisions regarding groundwater management and sustainability.

35 - Exploring Aquifer Storage and Recovery: A Sustainable Solution for Water Management

Abstract

This article provides a comprehensive overview of the Aquifer Storage and Recovery (ASR) process, which involves injecting water into an underground aquifer for storage and future use. It covers the history of ASR, including its development and the technology behind it. The article describes the process of injection and extraction of water, including water treatment, and the benefits of using ASR, such as increased water supply, water quality, cost-effectiveness, flexibility, and sustainability. The article also outlines the drawbacks of ASR, such as limited aquifer capacity, water quality concerns, infrastructure requirements, and technical challenges. Finally, the article highlights several case studies of ASR implementation across the world.

Introduction

Aquifer storage and recovery (ASR) is a process that involves the storage of water in an underground aquifer for future use. This technique is gaining popularity across the globe as a reliable water supply method, especially in arid and semi-arid regions where water scarcity is a significant issue. ASR involves the injection of water into an aquifer during times of plenty, and then extraction of the same water when needed. The process of ASR is a combination of storage and recharge, as the water injected into the aquifer helps to replenish the groundwater resources.

This article will provide a detailed overview of the ASR process, including its history, the technology behind it, and the benefits and drawbacks of using ASR for water storage.

History of Aquifer Storage and Recovery

The concept of ASR dates back to the early 1900s when the need for water storage was first realized. The first recorded ASR project was implemented in Berlin, Germany, in 1903, where a well was used to recharge an aquifer for future use. The idea of ASR was further developed in the United States in the 1930s, where it was used to store water in the Floridan aquifer in Florida. ASR technology continued to develop throughout the 20th century, and today it is widely used across the world.

ASR Technology

The ASR process involves the injection of water into an aquifer, followed by the extraction of the same water at a later time. The technology behind ASR is relatively straightforward, but the injection and extraction process must be carefully managed to ensure the water is of good quality and safe for use.

Injection Wells

The injection of water into an aquifer is done through an injection well. The injection well is drilled into the ground and is designed to allow water to be injected into the aquifer without any contamination. The well is lined with a casing to prevent the water from coming into contact with any soil or rock formations that may contaminate the water.

The water is typically injected into the aquifer under pressure, which helps to distribute the water evenly throughout the aquifer. The pressure must be carefully managed to avoid damaging the aquifer and to ensure that the water is distributed evenly.

Extraction Wells

The extraction of water from the aquifer is done through an extraction well. The extraction well is drilled into the ground and is designed to allow water to be extracted from the aquifer without any contamination. The well is lined with a casing to prevent the water from coming into contact with any soil or rock formations that may contaminate the water.

The water is typically extracted from the aquifer using a pump. The pump must be carefully managed to avoid damaging the aquifer and to ensure that the water is of good quality and safe for use.

Water Treatment

The water injected into the aquifer must be of good quality and safe for use. The water must be treated to ensure that it is free of contaminants and safe for injection into the aquifer. The treatment process may involve filtration, chemical treatment, or other processes, depending on the quality of the water and the requirements of the aquifer.

The water extracted from the aquifer must also be treated to ensure that it is safe for use. The treatment process may involve filtration, chemical treatment, or other processes, depending on the quality of the water and the requirements for its intended use.

Benefits of Aquifer Storage and Recovery

There are many benefits to using ASR for water storage. The most significant benefits include:

Increased Water Supply: ASR provides an additional source of water that can be used during times of drought or water scarcity. The stored water can be used for drinking, irrigation, and other purposes, which helps to reduce the demand for freshwater resources.

Increased Water Quality: The injection of water into an aquifer can help to improve the quality of the groundwater by diluting any contaminants present in the aquifer. This process also helps to protect the aquifer from further contamination by reducing the concentration of pollutants.

Cost-Effective: ASR is often more cost-effective than building new reservoirs or dams. The technology used in ASR is relatively simple, and the infrastructure required is minimal compared to other water storage methods.

Flexible: ASR is a flexible water storage method that can be easily scaled up or down to meet changing demands for water. It is also adaptable to different types of aquifers, making it a useful tool for water management in a variety of settings.

Sustainable: ASR is a sustainable water management method that helps to preserve and protect groundwater resources. By storing water underground, ASR reduces the risk of evaporation, which is a significant issue in arid and semi-arid regions.

Drawbacks of Aquifer Storage and Recovery

While there are many benefits to using ASR for water storage, there are also some drawbacks to consider. These include:

Limited Aquifer Capacity: The amount of water that can be stored in an aquifer is limited by the capacity of the aquifer. In some cases, the aquifer may not have enough storage capacity to meet the demands for water during times of scarcity.

Water Quality Concerns: The injection of water into an aquifer can introduce contaminants or change the chemical composition of the groundwater. This can create concerns about the safety and quality of the water for human consumption.

Infrastructure Requirements: ASR requires the construction of injection and extraction wells, as well as water treatment facilities. This infrastructure can be expensive to build and maintain, which can be a barrier to the adoption of ASR as a water management tool.

Technical Challenges: ASR requires careful management to ensure that the injection and extraction process is effective and does not damage the aquifer. This requires skilled technicians and ongoing monitoring and maintenance, which can be challenging to implement and maintain.

Case Studies of Aquifer Storage and Recovery

ASR has been implemented in many different settings around the world. The following case studies highlight some of the successes and challenges of ASR.

Orange County, California, USA

The Orange County Water District in California is one of the most extensive ASR systems in the world. The district stores up to 600,000 acre-feet of water annually in the groundwater basin, which is then extracted for use during times of scarcity. The ASR system has been in operation since the 1970s and has helped to provide a reliable water supply for the region.

Perth, Australia

Perth, Australia, is one of the driest cities in the world, and the region faces significant water scarcity issues. To address this, the city implemented an ASR system in the early 2000s, which has helped to supplement the city's water supply. The system stores water in the Leederville and Yarragadee aquifers and has been successful in providing a reliable water supply for the region.

Abu Dhabi, United Arab Emirates

Abu Dhabi, located in the Arabian Peninsula, faces significant water scarcity issues. To address this, the city implemented an ASR system in the 1990s, which stores water in the Sa'ad and Falaj Al Mualla aquifers. The ASR system has been successful in providing a reliable water supply for the city and has helped to reduce the demand for desalinated water.

Conclusion

Aquifer storage and recovery is a reliable and sustainable method of water storage that is becoming increasingly popular around the world. ASR provides a flexible and cost-effective solution for water management that can help to address issues related to water scarcity, groundwater depletion, and climate change. While there are some drawbacks to using ASR, such as technical challenges and water quality concerns, these can be managed with proper planning and management.

As we continue to face challenges related to water scarcity, it is essential to consider innovative and sustainable solutions like ASR. By working together to implement these solutions, we can ensure that we have access to reliable and safe sources of water for generations to come.

References

- American Water Works Association (2017). Water Supply and Distribution. 3rd ed. McGraw-Hill Education.
- Anderson, M. P., & Woessner, W. W. (2015). Applied groundwater modeling: simulation of flow and advective transport. Academic Press.
- Brown, M., & Steenhuis, T. (2019). Aquifer storage and recovery: a sustainable water supply solution. Springer.
- California State Water Resources Control Board (2016). Aquifer Storage and Recovery.
- Department of Water and Environmental Regulation, Government of Western Australia (2022). Aquifer Recharge.
- Koottatep, T., & Polprasert, C. (2009). Artificial recharge of groundwater: a sustainable solution for water supply in Thailand. Environmental Geology, 57(5), 967-976.
- Orange County Water District (2022). Aquifer Storage and Recovery.
- United States Environmental Protection Agency (2015). Aquifer Storage and Recovery.
- United States Geological Survey (2022). Aquifer Storage and Recovery.
- Water Research Foundation (2012). Aquifer Storage and Recovery: State of the Science and Practice.
- World Bank Group (2019). Enhancing Water Security in Arid and Semi-Arid Regions: Aquifer Storage and Recovery.

36 - Understanding Arsenic Contamination in Groundwater: Sources, Health Effects, Detection, and Removal Methods

Abstract

Arsenic contamination of groundwater is a global problem that affects millions of people worldwide. Arsenic occurs naturally in rocks and soils and is also released into the environment through human activities. High levels of arsenic in drinking water can lead to a range of health problems, including skin lesions, cancers, and cardiovascular disease. The detection of arsenic in groundwater is critical to protecting public health, and several methods are available for removing arsenic from water sources. This article provides an overview of the sources of arsenic, its effects on human health, methods for detecting and removing arsenic from groundwater, and the need for a collaborative effort to address this global problem.

Introduction

Groundwater is an essential source of drinking water for millions of people worldwide. However, in recent years, it has become apparent that some groundwater sources contain elevated levels of arsenic, which is a toxic element that can cause serious health problems in humans. Arsenic contamination of groundwater is a global problem, with millions of people in different parts of the world, including Bangladesh, India, Pakistan, and China, exposed to high levels of arsenic through drinking water.

This article aims to provide an overview of the problem of arsenic contamination in groundwater, including the sources of arsenic, its effects on human health, and methods for detecting and removing arsenic from groundwater.

Sources of Arsenic

Arsenic occurs naturally in rocks and soils and is released into the environment through weathering and erosion. It is also present in some minerals and ores, such as arsenopyrite, which is a common mineral in many gold deposits. In addition to natural sources, human activities such as mining, smelting, and the use of arsenic-containing pesticides and fertilizers can also lead to the release of arsenic into the environment.

Arsenic in Groundwater

Arsenic is often found in groundwater in areas where the underlying rocks and soils contain high levels of arsenic. The concentration of arsenic in groundwater can vary widely, ranging from a few parts per billion to several hundred parts per billion. Groundwater arsenic levels can be influenced by a range of factors, including the type of rock or soil, the presence of organic matter, and the pH and temperature of the water.

Health Effects of Arsenic

Exposure to high levels of arsenic through drinking water can lead to a range of health problems, including skin lesions, cancers of the bladder, kidney, and lung, and cardiovascular disease. Long-term exposure to even low levels of arsenic can also affect cognitive development and immune function, particularly in children.

Detection of Arsenic

The detection of arsenic in groundwater is typically carried out using a range of methods, including laboratory analysis of water samples, field-based test kits, and remote sensing techniques. Laboratory analysis is the most accurate method, but it is time-consuming and expensive. Field-based test kits are more convenient but may not provide the same level of accuracy as laboratory analysis.

Removal of Arsenic

There are several methods available for removing arsenic from groundwater, including chemical treatment, adsorption, ion exchange, and membrane filtration. Chemical treatment involves adding chemicals such as iron or aluminum salts to the water to precipitate the arsenic, which can then be removed by sedimentation or filtration. Adsorption involves using materials such as activated carbon or iron oxide to bind the arsenic, which can then be removed by sedimentation or filtration. Ion exchange involves using ion exchange resins to remove arsenic from the water, while membrane filtration involves passing the water through a membrane that selectively removes the arsenic.

Conclusion

Arsenic contamination of groundwater is a serious global problem that affects millions of people. The sources of arsenic are both natural and human-made, and its effects on human health are well-documented. The detection of arsenic in groundwater is critical to protecting public health, and there are several methods available for removing arsenic from water sources. Addressing the problem of arsenic in groundwater requires a collaborative effort between governments, international organizations, and local communities to develop and implement effective solutions to protect public health and ensure access to safe drinking water for all.

37 - Assessing Groundwater Vulnerability to Pollution: Methods and Applications

Abstract

Groundwater is an essential resource for human consumption, agriculture, and industrial activities. However, groundwater is vulnerable to pollution from various sources, including agricultural and industrial practices, landfill sites, and urbanization. Assessing groundwater vulnerability to pollution is critical to protecting this valuable resource. This article provides an overview of the methods and applications of assessing groundwater vulnerability to pollution, including vulnerability mapping, vulnerability indices, and numerical modelling. We also discuss the limitations and challenges associated with these methods and provide examples of their use in different contexts.

Introduction

Groundwater is a valuable resource that provides drinking water for millions of people worldwide, supports agriculture and industry, and plays a critical role in maintaining ecosystems. However, groundwater is vulnerable to pollution from various sources, including agricultural and industrial practices, landfill sites, and urbanization. Polluted groundwater can have significant negative impacts on human health, the environment, and economic activities that rely on clean water. Therefore, it is essential to assess groundwater vulnerability to pollution and take appropriate measures to protect this vital resource.

Assessing groundwater vulnerability to pollution is a complex task that requires a multidisciplinary approach. This article provides an overview of the methods and applications of assessing groundwater vulnerability to pollution, including vulnerability mapping, vulnerability indices, and numerical modelling. We also discuss the limitations and challenges associated with these methods and provide examples of their use in different contexts.

Vulnerability Mapping

Vulnerability mapping is a method of assessing groundwater vulnerability to pollution by identifying areas that are most susceptible to contamination. Vulnerability maps are created by integrating different data sets, such as geological, hydrological, and land-use information, to determine the likelihood and magnitude of contamination.

The vulnerability mapping approach typically involves the following steps:

- Identifying the parameters that contribute to groundwater vulnerability, such as aquifer characteristics, groundwater recharge rates, and land-use practices.

- Assigning weights to each parameter based on its relative importance in determining groundwater vulnerability.

- Combining the weighted parameters to create a vulnerability index that ranges from low to high vulnerability.

- Overlaying the vulnerability index on a geographic information system (GIS) map to identify areas that are most vulnerable to contamination.

Vulnerability mapping is a useful method for identifying areas that require increased protection from contamination. For example, vulnerability maps can be used to inform land-use planning and zoning decisions, prioritize monitoring and remediation efforts, and identify potential sources of contamination.

Vulnerability Indices

Vulnerability indices are another method of assessing groundwater vulnerability to pollution. A vulnerability index is a quantitative measure of groundwater vulnerability that considers multiple parameters, such as aquifer properties, land-use practices, and contaminant sources, and combines them into a single score or ranking.

The vulnerability index approach typically involves the following steps:

- Selecting the parameters that contribute to groundwater vulnerability, such as aquifer vulnerability, contaminant sources, and land-use practices.

- Assigning weights to each parameter based on its relative importance in determining groundwater vulnerability.

- Combining the weighted parameters to create a vulnerability index that ranges from low to high vulnerability.

- Using the vulnerability index to prioritize areas for monitoring and remediation efforts.

Vulnerability indices are useful for comparing groundwater vulnerability between different locations or over time. For example, vulnerability indices can be used to assess changes in vulnerability due to changes in land use or environmental regulations. However, vulnerability indices have some limitations, such as the subjectivity of assigning weights to parameters and the difficulty of incorporating complex processes, such as groundwater flow and contaminant transport.

Numerical Modelling

Numerical modelling is a method of assessing groundwater vulnerability to pollution that uses computer models to simulate groundwater flow and contaminant transport. Numerical models can provide detailed information about the fate and transport of contaminants in groundwater systems, including the movement of pollutants through aquifers, the rate of contaminant migration, and the impact of natural and human-made factors on groundwater quality.

The numerical modelling approach typically involves the following steps:

- Developing a conceptual model of the groundwater system, including aquifer properties, recharge and discharge rates, and sources of contamination.
- Developing a numerical model that simulates groundwater flow and contaminant transport in the conceptual model.
- Calibrating the numerical model by comparing simulated results to measured data from the groundwater system.
- Using the calibrated numerical model to simulate different scenarios, such as changes in land use or the introduction of remediation measures, to assess their impact on groundwater quality.

Numerical modelling can provide detailed information about groundwater vulnerability to pollution and can be used to evaluate the effectiveness of different management strategies. However, numerical modelling can be complex and time-consuming, and requires substantial data and expertise.

Applications

Assessing groundwater vulnerability to pollution has numerous applications in environmental management, including:

Land-use planning and zoning: Vulnerability mapping and vulnerability indices can be used to identify areas that require increased protection from contamination and inform land-use planning and zoning decisions.

Prioritizing monitoring and remediation efforts: Vulnerability mapping, vulnerability indices, and numerical modelling can be used to prioritize monitoring and remediation efforts in areas that are most vulnerable to contamination.

Evaluating the effectiveness of management strategies: Numerical modelling can be used to evaluate the effectiveness of different management strategies, such as land-use changes, remediation measures, or environmental regulations, on groundwater quality.

Environmental risk assessment: Assessing groundwater vulnerability to pollution can be used to identify potential sources of contamination and evaluate the risk of exposure to human health and the environment.

Limitations and Challenges

Assessing groundwater vulnerability to pollution has several limitations and challenges, including:

Data availability and quality: Assessing groundwater vulnerability to pollution requires access to high-quality data on aquifer properties, contaminant sources, and land-use practices. However, data availability and quality can be a significant challenge in many regions, particularly in developing countries.

Complexity: Assessing groundwater vulnerability to pollution is a complex task that requires a multidisciplinary approach and can involve complex processes, such as groundwater flow and contaminant transport.

Uncertainty: Assessing groundwater vulnerability to pollution is subject to significant uncertainty, particularly when using vulnerability indices and numerical modelling, which rely on assumptions and simplifications.

Cost and time: Assessing groundwater vulnerability to pollution can be costly and time-consuming, particularly when using numerical modelling, which requires substantial data and expertise.

Conclusion

Assessing groundwater vulnerability to pollution is critical to protecting this valuable resource from contamination. Vulnerability mapping, vulnerability indices, and numerical modelling are useful methods for assessing groundwater vulnerability to pollution, each with their advantages and limitations. The choice of method depends on the specific context and the available data and resources. However, assessing groundwater vulnerability to pollution is subject to significant uncertainty and requires a multidisciplinary approach that considers aquifer properties, land-use practices, contaminant sources, and groundwater flow and transport.

38 - Assessing Groundwater Vulnerability using DRASTIC: A Review

Abstract

Groundwater is an essential resource for human consumption, irrigation, and industrial uses. However, it is also a finite resource that requires careful management and protection. Groundwater vulnerability assessment is an effective tool for identifying areas at risk of contamination and guiding appropriate management strategies. DRASTIC is one of the widely used methods for assessing groundwater vulnerability. DRASTIC stands for Depth to water, Recharge, Aquifer media, Soil media, Topography, Impact of vadose zone, and Conductivity. This article provides an in-depth review of the DRASTIC method, including its principles, strengths, limitations, and applications. The article also examines the various modifications and improvements made to the method over the years. The review aims to provide groundwater managers and researchers with a comprehensive understanding of DRASTIC and its role in groundwater vulnerability assessment.

Introduction

Groundwater is a vital resource for human consumption, irrigation, and industrial uses. However, it is a finite resource that requires careful management and protection. Groundwater vulnerability assessment is an effective tool for identifying areas at risk of contamination and guiding appropriate management strategies. Several methods are available for assessing groundwater vulnerability, including the DRASTIC method.

The DRASTIC method was developed in the 1980s and has since been widely used for assessing groundwater vulnerability. DRASTIC stands for Depth to water, Recharge, Aquifer media, Soil media, Topography, Impact of vadose zone, and Conductivity. The method uses a combination of seven parameters to determine the vulnerability of groundwater to contamination.

This article provides an in-depth review of the DRASTIC method, including its principles, strengths, limitations, and applications. The article also examines the various modifications and improvements made to the method over the years. The review aims to provide groundwater managers and researchers with a comprehensive understanding of DRASTIC and its role in groundwater vulnerability assessment.

Principles of DRASTIC

The DRASTIC method is based on the principle that the vulnerability of groundwater to contamination is a function of the characteristics of the aquifer and the overlying soil and land use. The method uses seven parameters to assess groundwater vulnerability, each of which is given a weightage based on its importance.

Depth to water: The depth to the water table is an essential parameter in assessing groundwater vulnerability. The closer the water table is to the surface, the more vulnerable the groundwater is to contamination. The depth to water parameter is given a weightage of 5 in the DRASTIC method.

Recharge: The amount and rate of recharge are critical factors in groundwater vulnerability assessment. The more significant the recharge, the greater the potential for contamination. The recharge parameter is given a weightage of 5 in the DRASTIC method.

Aquifer media: The type of material that makes up the aquifer is an essential parameter in groundwater vulnerability assessment. The more permeable the aquifer, the more vulnerable the groundwater is to contamination. The aquifer media parameter is given a weightage of 5 in the DRASTIC method.

Soil media: The type of soil that overlies the aquifer is also an important parameter in groundwater vulnerability assessment. The more permeable the soil, the more vulnerable the groundwater is to contamination. The soil media parameter is given a weightage of 4 in the DRASTIC method.

Topography: The topography of the area is an important parameter in groundwater vulnerability assessment. Areas with steep slopes are more vulnerable to contamination as contaminants can quickly move downhill. The topography parameter is given a weightage of 3 in the DRASTIC method.

Impact of vadose zone: The vadose zone is the unsaturated zone above the water table. The impact of the vadose zone on groundwater vulnerability is assessed based on factors such as soil type, depth to the water table, and the presence of fractures. The impact of vadose zone parameter is given a weightage of 3 in the DRASTIC method.

Conductivity: The conductivity of the aquifer is a measure of its ability to transmit water. The higher the conductivity, the more vulnerable the groundwater is to contamination. The conductivity parameter is given a weightage of 2 in the DRASTIC method.

The weightage of each parameter is determined based on its importance in groundwater vulnerability assessment. The weightage values are multiplied by the parameter values to obtain a vulnerability index (VI) for each parameter. The VI values are then added together to obtain a final vulnerability index for the area under assessment.

Strengths of DRASTIC

The DRASTIC method has several strengths that make it a popular choice for assessing groundwater vulnerability. Some of these strengths are:

Easy to use: The DRASTIC method is relatively easy to use and requires only basic data such as the depth to water table, soil type, and land use. The simplicity of the method makes it accessible to groundwater managers and researchers with limited resources.

Comprehensive: The DRASTIC method assesses several parameters that influence groundwater vulnerability, including the depth to water, recharge rate, aquifer and soil media, topography, vadose zone impact, and conductivity. The method provides a comprehensive assessment of groundwater vulnerability that accounts for various factors that affect groundwater quality.

Flexibility: The DRASTIC method is flexible and can be adapted to different hydrogeological settings. The method can be modified to include additional parameters or to adjust the weightage values based on the specific conditions of the area under assessment.

Limitations of DRASTIC

The DRASTIC method also has some limitations that need to be considered when using it for groundwater vulnerability assessment. Some of these limitations are:

Lack of specificity: The DRASTIC method provides a general assessment of groundwater vulnerability and does not account for the specific contaminants or sources of contamination in the area under assessment. The method assumes that all contaminants are equally likely to occur in the area, which may not be the case.

Limited data availability: The DRASTIC method requires several data inputs, including depth to water table, soil type, and land use. In some cases, the data may be limited or unavailable, making it difficult to use the method accurately.

Over-reliance on expert judgment: The weightage values assigned to each parameter in the DRASTIC method are based on expert judgment and may not always reflect the actual importance of the parameter in the area under assessment.

Applications of DRASTIC

The DRASTIC method has several applications in groundwater vulnerability assessment, including:

Identification of areas at risk of contamination: The DRASTIC method can be used to identify areas that are most vulnerable to contamination, which can guide the development of appropriate management strategies.

Prioritization of groundwater protection efforts: The DRASTIC method can be used to prioritize groundwater protection efforts, focusing resources on areas that are most vulnerable to contamination.

Evaluation of management strategies: The DRASTIC method can be used to evaluate the effectiveness of management strategies in reducing groundwater vulnerability. The method can be used to monitor changes in vulnerability over time and assess the impact of management practices on groundwater quality.

Modifications and Improvements to DRASTIC

Over the years, several modifications and improvements have been made to the DRASTIC method to address its limitations and improve its accuracy. Some of these modifications and improvements include:

Integration of GIS: Geographic Information Systems (GIS) can be used to integrate the various data inputs required by the DRASTIC method. This can improve the accuracy of the method and make it easier to use.

Incorporation of site-specific parameters: The DRASTIC method can be modified to include site-specific parameters that are not accounted for in the original method. This can improve the accuracy of the method and make it more relevant to the specific conditions of the area under assessment.

Use of machine learning algorithms: Machine learning algorithms can be used to improve the accuracy of the DRASTIC method by identifying patterns and relationships in the data that may not be apparent to human experts.

Development of new vulnerability indices: Several new vulnerability indices have been developed that are based on the DRASTIC method but include additional parameters or modify the weightage values assigned to each parameter.

Conclusion

The DRASTIC method is a widely used approach for assessing groundwater vulnerability to contamination. The method is easy to use, comprehensive, and flexible, making it accessible to groundwater managers and researchers with limited resources. However, the method also has some limitations that need to be considered when using it for groundwater vulnerability assessment. Modifications and improvements to the method have been made over the years to address these limitations and improve its accuracy. The DRASTIC method has several applications in groundwater vulnerability assessment and can be used to identify areas at risk of contamination, prioritize groundwater protection efforts, and evaluate management strategies.

39 - Effective Management of Groundwater Resources: Challenges, Principles, and Strategies

Abstract

The article highlights the challenges faced in groundwater management, including overexploitation, pollution, climate change, and a lack of data. It also discusses the principles of groundwater management, including sustainable use, protection of the environment, integrated management, and stakeholder participation. The article also provides strategies for managing groundwater resources, such as groundwater monitoring, recharge, water conservation, land-use management, groundwater remediation, and groundwater governance. Groundwater monitoring is critical to assessing the status of groundwater resources, while groundwater recharge involves the deliberate enhancement of groundwater recharge to increase groundwater availability. Water conservation measures can reduce the demand for groundwater, and land-use management can regulate land use to protect groundwater resources. Groundwater remediation involves the treatment of contaminated groundwater to restore its quality, and groundwater governance involves the development and implementation of policies and regulations for the management of groundwater resources. Effective management of groundwater resources requires a holistic approach that balances the competing demands of different users, considers the impacts of groundwater use on the environment, and involves the participation of all stakeholders.

Introduction

Groundwater is an essential natural resource that supports various human activities, including agriculture, industry, and domestic use. Groundwater is also critical for sustaining ecosystems and biodiversity. However, groundwater resources are increasingly facing challenges due to various factors, including overexploitation, pollution, and climate change. Effective management of groundwater resources is critical for ensuring their sustainable use and protection. This article discusses the management of groundwater resources, including the challenges faced in groundwater management, the principles of groundwater management, and the strategies for managing groundwater resources.

Challenges in Groundwater Management

Groundwater management faces various challenges, including:

Overexploitation: Overexploitation occurs when the rate of groundwater extraction exceeds the rate of recharge, leading to groundwater depletion. Overexploitation can lead to land subsidence, saltwater intrusion, and other negative impacts on the environment.

Pollution: Groundwater pollution occurs when contaminants enter the groundwater system through human activities, such as industrial and agricultural practices, improper waste disposal, and oil spills. Groundwater pollution can have severe health impacts on humans and wildlife.

Climate change: Climate change can impact groundwater resources by altering the amount and distribution of rainfall, increasing evaporation, and reducing groundwater recharge. Climate change can also exacerbate other challenges in groundwater management, such as overexploitation and pollution.

Lack of data: Effective groundwater management requires accurate and up-to-date data on groundwater resources, including groundwater levels, recharge rates, and quality. However, many regions lack adequate data on groundwater resources, making it difficult to manage groundwater effectively.

Principles of Groundwater Management

Effective management of groundwater resources requires the application of several principles, including:

Sustainable use: Groundwater resources should be used in a manner that ensures their sustainability for future generations. This requires managing groundwater resources to balance the competing demands of different users, including agriculture, industry, and domestic use.

Protection of the environment: Groundwater management should consider the impacts of groundwater use on the environment, including groundwater-dependent ecosystems, surface water resources, and soil quality.

Integrated management: Groundwater management should be integrated with other water resources management activities, including surface water management, wastewater management, and water conservation.

Stakeholder participation: Effective groundwater management requires the involvement of all stakeholders, including groundwater users, community organizations, and government agencies.

Strategies for Managing Groundwater Resources

Effective management of groundwater resources requires the adoption of various strategies, including:

Groundwater monitoring: Groundwater monitoring involves the regular measurement of groundwater levels, recharge rates, and quality to assess the status of groundwater resources. Groundwater monitoring helps to identify trends in groundwater resources and inform groundwater management decisions.

Groundwater recharge: Groundwater recharge involves the deliberate enhancement of groundwater recharge to increase groundwater availability. Groundwater recharge can be achieved through various methods, including infiltration ponds, spreading basins, and injection wells.

Water conservation: Water conservation involves the adoption of measures to reduce water use and wastage. Water conservation can help to reduce the demand for groundwater and alleviate pressure on groundwater resources.

Land-use management: Land-use management involves the regulation of land use to protect groundwater resources. Land-use management measures can include the establishment of buffer zones around groundwater recharge areas, the regulation of agricultural practices, and the prohibition of certain activities, such as waste disposal, near groundwater resources.

Groundwater remediation: Groundwater remediation involves the treatment of contaminated groundwater to restore its quality. Groundwater remediation can involve various techniques, including pump-and-treat systems, in-situ treatment, and phytoremediation.

Groundwater governance: Groundwater governance involves the development and implementation of policies and regulations for the management of groundwater resources. Groundwater governance should be based on the principles of sustainable use, protection of the environment, integrated management, and stakeholder participation.

Groundwater Monitoring

Groundwater monitoring is a critical component of effective groundwater management. Groundwater monitoring involves the regular measurement of groundwater levels, recharge rates, and quality to assess the status of groundwater resources. Groundwater monitoring helps to identify trends in groundwater resources and inform groundwater management decisions.

Groundwater monitoring should be conducted using standard methods and protocols to ensure consistency and accuracy of data. Groundwater monitoring networks should be designed to provide adequate coverage of groundwater resources, including areas of high groundwater use and recharge areas.

Groundwater Recharge

Groundwater recharge involves the deliberate enhancement of groundwater recharge to increase groundwater availability. Groundwater recharge can be achieved through various methods, including infiltration ponds, spreading basins, and injection wells.

Infiltration ponds are shallow basins that are designed to capture and infiltrate stormwater runoff into the ground. Infiltration ponds can be designed to remove pollutants from stormwater runoff before it infiltrates into the ground.

Spreading basins are large shallow depressions that are used to spread water over the land surface to allow it to infiltrate into the ground. Spreading basins can be used to recharge groundwater, control floods, and enhance water quality.

Injection wells are boreholes that are used to inject water directly into the ground. Injection wells can be used to recharge groundwater and control saltwater intrusion.

Water Conservation

Water conservation involves the adoption of measures to reduce water use and wastage. Water conservation can help to reduce the demand for groundwater and alleviate pressure on groundwater resources.

Water conservation measures can include the adoption of efficient irrigation practices, such as drip irrigation, the installation of water-efficient fixtures, and the use of recycled water for non-potable uses.

Land-Use Management

Land-use management involves the regulation of land use to protect groundwater resources. Land-use management measures can include the establishment of buffer zones around groundwater recharge areas, the regulation of agricultural practices, and the prohibition of certain activities, such as waste disposal, near groundwater resources.

Buffer zones are areas around groundwater recharge areas that are protected from development and other activities that may impact groundwater quality. Buffer zones can be used to protect groundwater recharge areas and reduce the risk of pollution.

Agricultural practices, such as the use of fertilizers and pesticides, can impact groundwater quality. Regulations can be put in place to limit the use of these chemicals near groundwater resources and require best management practices to reduce the impact of agriculture on groundwater resources.

Groundwater Remediation

Groundwater remediation involves the treatment of contaminated groundwater to restore its quality. Groundwater remediation can involve various techniques, including pump-and-treat systems, in-situ treatment, and phytoremediation.

Pump-and-treat systems involve the extraction of contaminated groundwater and treatment of the extracted water before it is discharged to the environment or recharged back into the groundwater system.

In-situ treatment involves the treatment of contaminated groundwater in place without extracting the water. In-situ treatment techniques can include chemical oxidation, bioremediation, and phytoremediation.

Phytoremediation involves the use of plants to remove contaminants from the groundwater. Phytoremediation can be used to treat contaminated groundwater in place without the need for excavation or extraction.

Groundwater Governance

Groundwater governance involves the development and implementation of policies and regulations for the management of groundwater resources. Groundwater governance should be based on the principles of sustainable use, protection of the environment, integrated management, and stakeholder participation.

Groundwater governance should include the development of groundwater management plans that outline the objectives, strategies, and actions for managing groundwater resources. Groundwater governance should also include the establishment of groundwater user rights and allocation mechanisms to ensure the equitable and efficient use of groundwater resources.

Conclusion

Groundwater resources are critical for sustaining various human activities, including agriculture, industry, and drinking water supply. However, the increasing demand for groundwater resources and the impacts of climate change pose significant challenges to the sustainable management of groundwater resources.

Effective groundwater management requires a comprehensive approach that includes groundwater monitoring, groundwater recharge, water conservation, land-use management, groundwater remediation, and groundwater governance. Groundwater management should be based on the principles of sustainable use, protection of the environment, integrated management, and stakeholder participation.

Groundwater management should be adaptive and flexible, as the conditions and demands for groundwater resources can change over time. The use of advanced technologies, such as remote sensing and modeling, can improve the effectiveness of groundwater management.

In conclusion, the sustainable management of groundwater resources is essential for meeting the current and future needs of society while protecting the environment. Effective groundwater management requires the collaboration of various stakeholders, including water managers, policymakers, scientists, and the public. By adopting a comprehensive and integrated approach to groundwater management, we can ensure the long-term sustainability of groundwater resources for generations to come.

40 - The Impact of Climate Change on Groundwater Resources and Adaptation Measures for Sustainable Management

Abstract

This article discusses the impact of climate change on groundwater resources and the adaptation measures that can be taken for sustainable management. Climate change affects groundwater recharge, storage, and quality through changes in rainfall patterns, rising temperatures, sea-level rise, extreme weather events, and changes in land use. Changes in water quality due to saltwater intrusion, groundwater contamination, increased water temperature, and changes in water chemistry are also addressed. The article suggests measures such as water conservation, groundwater recharge, monitoring and management, integrated water resources management, and managed aquifer recharge to ensure sustainable groundwater management.

Introduction

Climate change refers to long-term changes in the Earth's climate, particularly changes that have occurred since the beginning of the Industrial Revolution in the mid- to late-19th century. These changes are primarily driven by human activities that have led to the release of large amounts of greenhouse gases, particularly carbon dioxide, into the atmosphere. This increase in greenhouse gases has caused the Earth's temperature to rise, leading to a range of effects, including melting glaciers, rising sea levels, more frequent and intense heatwaves, droughts and floods, and changes in precipitation patterns. Climate change is considered one of the most significant and urgent challenges facing humanity today, and there is a growing need for global action to mitigate its impacts and adapt to its effects.

Climate change is one of the biggest challenges faced by humanity today. Its impact on the environment and various natural resources has been severe, and groundwater resources have not been spared either. Groundwater is a crucial resource that is used for drinking, irrigation, and various other purposes. However, with climate change, groundwater resources are being affected, and it is important to understand the impact of climate change on groundwater resources.

What is groundwater?

Groundwater is water that is found beneath the earth's surface in the pores and fractures of rocks and soil. It is the primary source of water for many rural and urban communities around the world. Groundwater is replenished through rainfall and surface water that seeps into the ground. It is also recharged through natural processes such as infiltration, percolation, and evapotranspiration.

How does climate change impact groundwater resources?

Climate change has a direct impact on groundwater resources. Changes in rainfall patterns, temperature, and other climate-related factors have a significant impact on groundwater recharge, storage, and quality. The following are some of the ways in which climate change is affecting groundwater resources.

Changes in rainfall patterns

Climate change has led to changes in rainfall patterns. In many parts of the world, rainfall has become less predictable, and extreme weather events such as floods and droughts have become more frequent. This has a direct impact on groundwater recharge. In areas with less rainfall, groundwater recharge is reduced, which leads to a decline in the groundwater table. Conversely, in areas with heavy rainfall, excess water can lead to groundwater contamination.

Rising temperatures

Climate change has led to rising temperatures, which have a direct impact on groundwater recharge. As temperatures rise, evapotranspiration rates increase, which means that more water is lost to the atmosphere. This reduces the amount of water available for groundwater recharge.

Sea level rise

Sea level rise is another impact of climate change that affects groundwater resources. As sea levels rise, saline water infiltrates coastal aquifers, which makes the groundwater saline and unsuitable for human consumption and agriculture.

Extreme weather events

Extreme weather events such as floods and droughts are becoming more frequent due to climate change. Floods can lead to groundwater contamination, while droughts can reduce groundwater recharge, leading to a decline in the groundwater table.

Changes in land use

Climate change can also lead to changes in land use, which can have a significant impact on groundwater resources. For example, deforestation can lead to reduced groundwater recharge, while urbanization can lead to increased runoff and reduced infiltration, which can also reduce groundwater recharge.

Impact of climate change on groundwater quality

Climate change not only affects groundwater quantity but also quality. The following are some of the ways in which climate change is impacting groundwater quality.

Saltwater intrusion

As sea levels rise, saline water infiltrates coastal aquifers, which makes the groundwater saline and unsuitable for human consumption and agriculture.

Groundwater contamination

Extreme weather events such as floods can lead to groundwater contamination. Floodwaters can carry pollutants from agricultural and industrial sources, which can infiltrate the groundwater.

Increased water temperature

Rising temperatures can also increase the temperature of groundwater. This can lead to the growth of microorganisms, which can contaminate the groundwater and make it unsuitable for human consumption.

Changes in water chemistry

Changes in rainfall patterns and temperature can also lead to changes in water chemistry. For example, increased rainfall can lead to an increase in dissolved organic matter, which can lead to the growth of microorganisms and the production of disinfection by-products.

Adaptation measures for groundwater management

Given the impact of climate change on groundwater resources, it is important to adopt adaptation measures for groundwater management. The following are some of the measures that can be adopted.

Water conservation

Water conservation is an essential measure that can help to reduce the demand for groundwater. This can be achieved through the adoption of water-efficient technologies, the use of drought-tolerant crops, and the promotion of water conservation practices.

Groundwater recharge

Groundwater recharge can help to increase the availability of groundwater. This can be achieved through the use of artificial recharge techniques such as injection wells and recharge basins. Natural recharge methods such as vegetative measures and land use changes can also be used to increase groundwater recharge.

Monitoring and management

Effective monitoring and management of groundwater resources are critical for sustainable groundwater management. This includes the measurement of groundwater levels and quality, the development of groundwater models, and the establishment of groundwater management plans.

Integrated water resources management

Integrated water resources management (IWRM) is an approach that promotes the coordinated development and management of water, land, and related resources. This approach can help to ensure the sustainable management of groundwater resources by considering the interconnected nature of water resources.

Managed aquifer recharge

Managed aquifer recharge (MAR) is a technique that involves the intentional recharge of groundwater to enhance the availability and quality of groundwater resources. This can be achieved through the use of infiltration basins, injection wells, and other techniques.

Climate-resilient water supply infrastructure

Water supply infrastructure that is resilient to climate change is essential for ensuring the sustainable management of groundwater resources. This includes the development of climate-resilient water supply systems, the establishment of water quality monitoring systems, and the promotion of efficient water use.

Conclusion

Climate change has a significant impact on groundwater resources, and it is important to adopt adaptation measures to ensure the sustainable management of groundwater resources. The adoption of water conservation measures, groundwater recharge techniques, and effective monitoring and management can help to ensure the sustainable management of groundwater resources. The promotion of integrated water resources management and the development of climate-resilient water supply infrastructure are also critical for ensuring the sustainable management of groundwater resources in the face of climate change.

41 - Effective Water Conservation Measures for a Sustainable Future

Abstract

Water is a precious resource, and its conservation is crucial for the sustainability of our planet. In this article, we discuss ten effective water conservation measures that can help to reduce water consumption and ensure the availability of freshwater resources for future generations. These measures include efficient irrigation techniques, greywater recycling, rainwater harvesting, installation of efficient fixtures and appliances, water metering, pricing and incentives, water conservation education, policies and regulations, sustainable agriculture practices, and reforestation and watershed management. By adopting these measures, individuals, communities, industries, and governments can work together to ensure the sustainable use of water resources and preserve this precious resource for future generations.

Introduction

Water is an essential resource for all living beings on earth. It is required for various purposes such as drinking, cooking, cleaning, and agriculture, among others. With the growing population and industrialization, the demand for water has increased significantly. However, the availability of freshwater resources is limited, and the situation is getting worse due to climate change and overexploitation. Therefore, it has become crucial to conserve and manage water resources efficiently. In this article, we will discuss various water conservation measures that can help to reduce water consumption and preserve this precious resource.

1. Efficient Irrigation Techniques

Agriculture is the largest consumer of freshwater globally, accounting for around 70% of the total water consumption. However, much of this water is wasted due to inefficient irrigation techniques. Traditional flood irrigation and sprinkler systems waste a significant amount of water due to evaporation, runoff, and deep percolation. Therefore, modern irrigation techniques such as drip irrigation, subsurface irrigation, and micro-sprinklers should be promoted.

Drip irrigation is a water-efficient technique that delivers water directly to the plant roots, minimizing evaporation and runoff. Subsurface irrigation involves placing water underground, which reduces water loss due to evaporation and also reduces weed growth. Micro-sprinklers are similar to traditional sprinklers but are designed to deliver water more efficiently by spraying it in a fine mist, reducing water loss due to evaporation.

2. Greywater Recycling

Greywater is wastewater generated from domestic activities such as washing machines, dishwashers, and showers. This water can be recycled and reused for non-potable purposes such as toilet flushing, garden irrigation, and car washing, among others. Greywater recycling can save a significant amount of freshwater resources and reduce the amount of wastewater discharged into the environment. However, it is important to use appropriate filtration and disinfection systems to ensure that the recycled water is safe for reuse.

3. Rainwater Harvesting

Rainwater harvesting is a technique that involves collecting and storing rainwater for later use. This technique can be used for domestic as well as agricultural purposes. Rainwater harvesting systems can range from simple rooftop catchment systems to more complex systems that involve storage tanks and filtration systems. The collected rainwater can be used for various non-potable purposes such as irrigation, car washing, and toilet flushing, among others. Rainwater harvesting can help to reduce the demand for freshwater resources and also reduce the risk of flooding.

4. Efficient Fixtures and Appliances

Efficient fixtures and appliances such as low-flow toilets, showerheads, and faucets, and energy-efficient washing machines and dishwashers can significantly reduce water consumption. Low-flow fixtures and appliances reduce water consumption by using less water per use, while energy-efficient appliances reduce water consumption by using less water for washing and rinsing. Installing efficient fixtures and appliances can help to reduce the demand for freshwater resources and also lower water bills.

5. Water Metering

Water metering is a technique that involves measuring the amount of water consumed by households, industries, and agriculture. Water metering can help to identify leaks and wastage, and also encourage users to reduce water consumption. Water metering can also help water utilities to manage water resources efficiently by identifying high consumption areas and planning for future demand. In some countries, water metering is mandatory for all households, and users are charged based on their consumption.

6. Water Pricing and Incentives

Water pricing and incentives can be used to encourage users to reduce water consumption. Higher prices for water can encourage users to use water more efficiently and reduce wastage. Incentives such as rebates and subsidies for efficient fixtures and appliances can also encourage users to invest in water-efficient technologies. Water pricing and incentives can also help to fund water conservation programs and infrastructure.

7. Water Conservation Education

Water conservation education is essential to raise awareness about the importance of water conservation and the need to use water efficiently. Education can be provided through various mediums such as schools, media, and community programs. The education should cover topics such as the water cycle, water conservation techniques, and the impact of climate change on water resources. By educating the public, we can encourage them to adopt water conservation measures in their daily lives and reduce water consumption.

8. Water Conservation Policies and Regulations

Water conservation policies and regulations can be implemented at the local, regional, and national levels to encourage water conservation and efficient water use. Policies can include mandatory water metering, water pricing and incentives, and regulations on water use for industries and agriculture. Water conservation policies and regulations can help to manage water resources efficiently and ensure equitable distribution of water resources.

9. Sustainable Agriculture Practices

Sustainable agriculture practices can help to reduce water consumption in agriculture. Practices such as crop rotation, conservation tillage, and precision farming can help to reduce water consumption and improve crop yields. Precision farming involves using technology to optimize crop production and minimize water use. Sustainable agriculture practices can help to reduce the demand for freshwater resources and ensure food security.

10. Reforestation and Watershed Management

Reforestation and watershed management can help to conserve and manage water resources efficiently. Forests act as natural water catchments and help to maintain the water cycle. Watershed management involves protecting and managing the land surrounding rivers and lakes to ensure water quality and quantity. Reforestation and watershed management can help to reduce soil erosion, maintain water quality, and reduce the risk of floods and droughts.

Conclusion

Water conservation measures are crucial to ensure the availability of freshwater resources for future generations. By adopting efficient irrigation techniques, recycling greywater, harvesting rainwater, installing efficient fixtures and appliances, water metering, pricing, and incentives, providing education, implementing policies and regulations, promoting sustainable agriculture practices, and managing watersheds, we can reduce water consumption and preserve this precious resource. It is important to note that water conservation is not only the responsibility of individuals but also of governments, industries, and communities. Therefore, collective efforts are required to ensure the sustainable use of water resources.

42 - Mapping the Potential for Rainwater Harvesting: A Comprehensive Review

Abstract

As water scarcity continues to be a pressing issue globally, rainwater harvesting (RWH) has emerged as a promising solution to mitigate water shortage. However, the success of RWH implementation largely depends on the availability of a reliable map of the potential for RWH. Mapping the potential for RWH involves identifying and analyzing various factors such as rainfall patterns, topography, land use, and soil type, among others. This comprehensive review article aims to provide an overview of the existing methods and approaches for mapping the potential for RWH. It examines the various factors that influence the potential for RWH and discusses the available tools and techniques for mapping RWH potential at different spatial scales. The article also highlights the challenges and limitations associated with RWH mapping and provides recommendations for future research in this field.

Introduction

Water scarcity is a growing concern worldwide, and it is expected to worsen in the coming years due to population growth, urbanization, and climate change. The availability of fresh water is limited, and the demand for water is increasing at an alarming rate. One of the solutions to mitigate water shortage is rainwater harvesting (RWH), which involves collecting and storing rainwater for various uses such as irrigation, domestic use, and industrial purposes. RWH has the potential to reduce water demand from traditional sources and provide a reliable and sustainable source of water.

The success of RWH implementation largely depends on the availability of a reliable map of the potential for RWH. Mapping the potential for RWH involves identifying and analyzing various factors such as rainfall patterns, topography, land use, and soil type, among others. A comprehensive understanding of these factors is essential to determine the suitability and feasibility of RWH in a given location. Therefore, the objective of this review article is to provide an overview of the existing methods and approaches for mapping the potential for RWH. The article will examine the various factors that influence the potential for RWH and discuss the available tools and techniques for mapping RWH potential at different spatial scales. The article will also highlight the challenges and limitations associated with RWH mapping and provide recommendations for future research in this field.

Factors Affecting Rainwater Harvesting Potential

The potential for RWH varies depending on several factors, including rainfall patterns, topography, land use, and soil type, among others. Understanding these factors is crucial in mapping the potential for RWH.

Rainfall Patterns

Rainfall is a fundamental factor in determining the potential for RWH. The amount, intensity, and distribution of rainfall in a given location influence the amount of water that can be harvested. Areas with high rainfall are generally considered suitable for RWH. However, the temporal and spatial variability of rainfall patterns must also be considered. The temporal variability refers to the variation in rainfall patterns over time, while the spatial variability refers to the variation in rainfall patterns across space. Areas with high temporal variability may experience periods of drought or low rainfall, which may affect the availability of water for harvesting. On the other hand, areas with high spatial variability may have localized rainfall patterns, making it difficult to capture and harvest the water effectively.

Topography

Topography refers to the shape and elevation of the land surface. It influences the flow of water and the potential for RWH. Areas with steep slopes and rugged terrain are generally not suitable for RWH due to the high runoff rates, which make it difficult to capture and store the water effectively. On the other hand, areas with gentle slopes and flat terrain are ideal for RWH as they allow for the collection and storage of water.

Land Use

Land use refers to the activities that take place on the land surface, such as agriculture, urbanization, and forestry. Land use affects the potential for RWH as it affects the surface permeability, vegetation cover, and soil characteristics. Areas with high levels of urbanization and impervious surfaces such as concrete and asphalt are not suitable for RWH as they prevent the infiltration of water into the soil. On the other hand, areas with vegetation cover such as forests and agricultural lands are ideal for RWH as they promote the infiltration of water into the soil.

Soil Type

Soil type refers to the physical and chemical characteristics of the soil, such as texture, structure, and porosity. Soil characteristics affect the infiltration capacity and water-holding capacity of the soil, which are critical in determining the potential for RWH. Areas with sandy or rocky soils may have low water-holding capacity, while areas with heavy clay soils may have low infiltration capacity, making them unsuitable for RWH. On the other hand, areas with loamy soils are ideal for RWH as they have high water-holding capacity and infiltration capacity.

Mapping Rainwater Harvesting Potential

Mapping the potential for RWH involves identifying and analyzing the various factors that influence the potential for RWH. Several methods and approaches have been developed for mapping RWH potential at different spatial scales, including global, regional, and local scales.

Global Scale

At the global scale, satellite-based remote sensing techniques are used to estimate the potential for RWH. These techniques use satellite data to estimate the amount of rainfall and the land surface characteristics that influence the potential for RWH, such as land cover and topography. The data are analyzed using geographic information system (GIS) software to produce maps of the potential for RWH. These maps provide a broad overview of the potential for RWH at the global scale, but they may not be accurate enough for local planning purposes.

Regional Scale

At the regional scale, statistical models are used to estimate the potential for RWH based on historical rainfall data and land surface characteristics. These models use regression analysis to identify the relationship between rainfall and the land surface characteristics that influence the potential for RWH, such as topography and soil type. The models are calibrated using historical rainfall data and validated using independent data sets. The output of these models is a map of the potential for RWH at the regional scale.

Local Scale

At the local scale, site-specific data such as rainfall data, topography data, and soil data are used to estimate the potential for RWH. These data are analyzed using GIS software to produce maps of the potential for RWH at the local scale. The maps provide a detailed and accurate assessment of the potential for RWH at the site-specific level, which is essential for planning and implementation purposes.

Mapping the potential for rainwater harvesting (RWH) at a local scale involves the use of various techniques and tools to identify suitable areas for RWH systems. These techniques can be broadly categorized into remote sensing, statistical modelling, and site-specific data analysis.

Remote sensing techniques involve the use of satellite imagery and aerial photographs to obtain information on rainfall, land use, topography, and soil characteristics. This data can then be used to develop maps that show the potential for RWH in a given area. For example, satellite data can be used to estimate the amount of rainfall in a particular area and to identify areas with high and low rainfall. This information can be combined with other data such as soil type and topography to produce maps that show the potential for RWH in different parts of the area.

Statistical modelling techniques involve the use of statistical methods to analyze data and develop models that predict the potential for RWH. These models can be based on historical rainfall data, soil type, topography, and other factors that influence the potential for RWH. For example, a regression analysis can be used to identify the relationship between rainfall and the potential for RWH in a given area. The resulting model can then be used to predict the potential for RWH in different parts of the area.

Site-specific data analysis involves the collection and analysis of data on soil type, topography, and other factors that influence the potential for RWH in a particular location. This data can then be used to develop maps that show the potential for RWH in that location. For example, a site-specific analysis can be used to identify areas with good soil infiltration rates and areas with high water tables, which are suitable for different types of RWH systems.

The choice of technique depends on the availability and quality of data, as well as the scale of analysis. Remote sensing techniques are useful for large-scale analysis, while site-specific data analysis is suitable for small-scale analysis. Statistical modelling can be used for both small and large-scale analysis, depending on the data available. These techniques can help to identify suitable areas for RWH systems and to design systems that are appropriate for the local conditions.

Challenges and Limitations

Mapping the potential for RWH is a complex process that involves several challenges and limitations. One of the main challenges is the lack of accurate and reliable data, particularly in developing countries. The availability of data such as rainfall data, topography data, and soil data is limited in many regions, making it difficult to estimate the potential for RWH accurately.

Another challenge is the complexity of the factors that influence the potential for RWH. The interaction between factors such as rainfall patterns, topography, land use, and soil type is complex, and it may be difficult to capture this complexity accurately in a map.

In addition, the available tools and techniques for mapping the potential for RWH are often limited by their spatial resolution and accuracy. Satellite-based remote sensing techniques provide a broad overview of the potential for RWH but may not be accurate enough for local planning purposes. On the other hand, site-specific data may provide a detailed and accurate assessment of the potential for RWH but may not be feasible to collect and analyze at a large scale.

Conclusion and Recommendations

Mapping the potential for RWH is a critical step in planning and implementing RWH systems. It helps to identify areas that are suitable for RWH and to design systems that are appropriate for the local conditions. However, mapping the potential for RWH is a complex process that requires accurate and reliable data and the use of appropriate tools and techniques.

To overcome the challenges and limitations of mapping the potential for RWH, it is recommended to:

Improve data collection and management: Efforts should be made to improve the collection and management of data such as rainfall data, topography data, and soil data. This can be achieved through the use of advanced technologies such as remote sensing and GIS.

Develop integrated models: Integrated models that consider the interaction between factors that influence the potential for RWH should be developed. These models should be able to capture the complexity of the factors and provide accurate estimates of the potential for RWH.

Combine different techniques: Different techniques such as remote sensing, statistical modelling, and site-specific data analysis should be combined to produce more accurate and reliable maps of the potential for RWH.

Involve local communities: Local communities should be involved in the mapping process to ensure that their knowledge and expertise are taken into account. This can help to identify areas that are suitable for RWH and to design systems that are appropriate for the local conditions.

Monitor and evaluate: RWH systems should be monitored and evaluated regularly to assess their performance and to identify areas for improvement. This can help to ensure that the systems are functioning as intended and that they are providing the expected benefits.

In conclusion, mapping the potential for RWH is a critical step in planning and implementing RWH systems. It helps to identify areas that are suitable for RWH and to design systems that are appropriate for the local conditions. However, mapping the potential for RWH is a complex process that requires accurate and reliable data and the use of appropriate tools and techniques. Efforts should be made to improve data collection and management, develop integrated models, combine different techniques, involve local communities, and monitor and evaluate RWH systems to ensure their effectiveness.

43 - Stream-Aquifer Interaction: Understanding the Complexities and Strategies for Sustainable Management

Abstract

This article explores the complex process of Stream-Aquifer Interaction (SAI) and its critical role in the hydrology of the landscape. SAI is influenced by various factors such as land use, human activities, and climate change, making its sustainable management challenging. The article discusses the latest monitoring technologies, improved hydrologic models, and management strategies being developed to manage SAI sustainably, such as managed aquifer recharge, integrated water management, land use management, and climate change adaptation. Implementing these strategies can ensure the sustainable use of water resources, protect the health of both streams and aquifers, and contribute to a more resilient water management system.

Introduction

Streams and aquifers are interdependent and closely related. The water exchange between these two systems is called Stream Aquifer Interaction (SAI). SAI plays an important role in maintaining streamflow, groundwater recharge, and water quality. Understanding the dynamics of SAI is crucial for sustainable water management.

This article will discuss the concept of SAI, its significance, factors influencing SAI, and methods to study SAI. It will also highlight the challenges associated with managing SAI and the current research trends in this field.

Concept of SAI

Streams and aquifers are interconnected systems, where water flows through the surface and subsurface. The surface water in streams and rivers originates from precipitation, which then flows through the landscape and into the stream channel. In contrast, groundwater originates from the infiltration of precipitation into the subsurface, where it recharges the aquifers.

SAI refers to the movement of water between streams and aquifers, which occurs through two main processes: gaining and losing streams. A gaining stream is one that receives water from the underlying aquifer, whereas a losing stream is one that loses water to the underlying aquifer.

In gaining streams, the water table intersects the streambed, resulting in water flowing from the aquifer into the stream. The stream water is then recharged into the aquifer, and the cycle continues. In contrast, losing streams have a water table below the streambed, and the stream water infiltrates into the underlying aquifer.

Significance of SAI

SAI plays a crucial role in maintaining the ecological and hydrological balance of streams and aquifers. It influences the quantity and quality of water available for human use and supports aquatic and riparian ecosystems. SAI also contributes to the regulation of water temperature, nutrient cycling, and sediment transport in streams.

The exchange of water between streams and aquifers affects the water quality of both systems. For example, groundwater can be a source of nutrients and pollutants that can contaminate the stream water. Similarly, stream water can recharge the aquifer with dissolved oxygen, organic matter, and other nutrients.

SAI is also significant for groundwater recharge. In arid and semi-arid regions, streams can be the primary source of recharge for the underlying aquifers. Conversely, in humid regions, aquifers can act as a storage reservoir for excess streamflow during wet periods, which can then be used during dry periods.

Factors influencing SAI

Several factors influence the dynamics of SAI. These factors include geology, hydrology, climate, land use, and human activities.

Geology

The geology of the landscape determines the permeability and porosity of the subsurface, which affects the rate and direction of groundwater flow. In permeable geologic formations, such as sand and gravel, groundwater flows more easily and quickly, resulting in more rapid SAI. Conversely, in impermeable formations, such as clay and shale, groundwater flow is slower, resulting in less rapid SAI.

Hydrology

The hydrology of the landscape determines the flow of surface water and groundwater. The amount and timing of precipitation, evapotranspiration, and runoff affect the flow of water in streams and aquifers. In wetland areas, where the water table is near the surface, SAI is more frequent and extensive. Conversely, in dryland areas, where the water table is deeper, SAI is less frequent and limited.

Climate

Climate affects the amount and timing of precipitation, which in turn affects the flow of water in streams and aquifers. In arid and semi-arid regions, where precipitation is limited, SAI is crucial for maintaining streamflow and groundwater recharge. Conversely, in humid regions, where precipitation is abundant, SAI is less important.

Land use

Land use affects the flow of water in streams and aquifers by altering the surface and subsurface hydrology of the landscape. Land use practices, such as agriculture, urbanization, and forestry, can increase the runoff and reduce the infiltration of precipitation, which can decrease the amount of water available for SAI. Land use can also increase the amount of pollutants and nutrients in the water, which can affect the water quality of both systems.

Human activities

Human activities, such as groundwater pumping, dam construction, and water diversion, can also affect SAI. Groundwater pumping can lower the water table and reduce the flow of water from aquifers to streams, resulting in less SAI. Dam construction can alter the flow of water in streams, reducing the amount of water available for recharge of the underlying aquifers. Water diversion can also reduce the flow of water in streams, affecting the availability of water for SAI.

Methods to Study SAI

Several methods are used to study SAI, including hydrologic modelling, groundwater monitoring, streamflow measurements, and tracer tests.

Hydrologic modelling

Hydrologic models are used to simulate the movement of water in streams and aquifers. These models use data on precipitation, evapotranspiration, soil moisture, and other factors to simulate the flow of water in the landscape. Hydrologic models can be used to predict the impacts of changes in land use, climate, and human activities on SAI.

Groundwater monitoring

Groundwater monitoring involves measuring the water level and quality in wells and boreholes. Groundwater monitoring can provide data on the direction and rate of groundwater flow, which can help to understand the dynamics of SAI.

Streamflow measurements

Streamflow measurements involve measuring the amount of water flowing in streams and rivers. Streamflow measurements can provide data on the direction and rate of SAI, which can help to understand the impact of changes in land use, climate, and human activities on streamflow and groundwater recharge.

Tracer tests

Tracer tests involve introducing a known amount of a tracer substance into the stream or aquifer and measuring its movement. Tracer tests can provide data on the direction and rate of SAI, as well as information on the characteristics of the subsurface flow paths.

Challenges associated with Managing SAI

Managing SAI is challenging due to the complex interactions between streams and aquifers and the multiple factors that influence SAI. Some of the main challenges associated with managing SAI include:

Limited data availability: Data on SAI is often limited, particularly in regions with limited resources or in areas where SAI is difficult to measure.

Conflicting demands: The demands for water from both streams and aquifers can be in conflict. Managing SAI requires balancing these conflicting demands and ensuring sustainable use of water resources.

Land use change: Land use change can have a significant impact on SAI. Managing SAI requires considering the potential impacts of land use change and implementing strategies to minimize these impacts.

Climate change: Climate change can alter the flow of water in streams and aquifers, affecting SAI. Managing SAI requires considering the potential impacts of climate change and implementing strategies to adapt to these impacts.

Current Research Trends

Current research in SAI is focused on improving our understanding of the complex interactions between streams and aquifers and developing strategies to manage SAI sustainably. Some of the current research trends in SAI include:

Developing new monitoring technologies: New monitoring technologies, such as remote sensing and geophysical methods, are being developed to improve our ability to measure SAI.

Improving hydrologic models: Hydrologic models are being improved to better simulate SAI and to predict the impacts of changes in land use, climate, and human activities on SAI.

Developing new management strategies: New management strategies, such as managed aquifer recharge and integrated water management, are being developed to manage SAI sustainably.

Studying the impacts of climate change: Research is being conducted to study the impacts of climate change on SAI and to develop strategies to adapt to these impacts.

Understanding the impacts of land use change: Research is being conducted to understand the impacts of land use change on SAI and to develop strategies to minimize these impacts.

Identifying the role of biota in SAI: Research is being conducted to identify the role of biota, such as plants and microbes, in SAI and to develop strategies to enhance their role in SAI.

Management Strategies

Several management strategies can be used to manage SAI sustainably. These strategies include:

Managed aquifer recharge

Managed aquifer recharge involves intentionally recharging aquifers with surface water, either through infiltration or injection. Managed aquifer recharge can help to increase the amount of water available for SAI and to improve the quality of groundwater.

Integrated water management

Integrated water management involves managing both surface water and groundwater as part of an integrated system. Integrated water management can help to balance the competing demands for water from both streams and aquifers and to ensure sustainable use of water resources.

Land use management

Land use management involves implementing strategies to minimize the impacts of land use change on SAI. These strategies may include reducing the amount of impervious surfaces, increasing infiltration rates, and implementing agricultural practices that promote soil health and reduce runoff.

Climate change adaptation

Climate change adaptation involves developing strategies to adapt to the impacts of climate change on SAI. These strategies may include implementing water conservation measures, developing new water supply sources, and increasing the resilience of water infrastructure.

Conclusion

Stream-aquifer interaction is a complex process that plays a critical role in the hydrology of the landscape. SAI is influenced by a variety of factors, including land use, human activities, and climate. Managing SAI sustainably is challenging due to the complex interactions between streams and aquifers and the multiple factors that influence SAI. However, new monitoring technologies, improved hydrologic models, and new management strategies are being developed to manage SAI sustainably. These strategies include managed aquifer recharge, integrated water management, land use management, and climate change adaptation. By implementing these strategies, we can ensure sustainable use of water resources and protect the health of both streams and aquifers.

44 - Conjunctive Use of Surface Water and Groundwater for Sustainable Water Management

Abstract

The conjunctive use of surface water and groundwater is a water management strategy that can help in ensuring a reliable and sustainable water supply. This approach involves the coordinated use of both surface water and groundwater resources to meet water demands. This article explores the best practices for implementing conjunctive use and the challenges associated with it. We also examine case studies from India, Australia, and the United States that demonstrate the benefits of conjunctive use, such as increased water availability, improved water quality, and improved agricultural productivity. To ensure the successful implementation of conjunctive use, an integrated and coordinated approach to water resources management is required, involving all stakeholders. The use of innovative technologies, such as remote sensing, geographic information systems (GIS), and artificial intelligence (AI), can also help in the effective management of conjunctive use. By adopting these best practices, countries can ensure the sustainability of conjunctive use and the effective management of water resources in the long run.

Introduction

Water is the most important natural resource for sustaining life on earth. Water is a scarce resource, and its availability and distribution are not uniform across different regions. The availability of water depends on various factors like rainfall, geography, and climatic conditions. Surface water and groundwater are the two main sources of water that are used for various purposes like drinking, irrigation, industrial use, and many more. In this article, we will discuss the conjunctive use of surface water and groundwater, its benefits, and challenges.

What is Conjunctive Use of Surface Water and Groundwater?

Conjunctive use of surface water and groundwater refers to the use of both surface water and groundwater resources in a coordinated and integrated manner to meet the water demand. It involves the simultaneous use of both sources of water to maximize the benefits and minimize the limitations of each source. Conjunctive use is a management strategy that aims to optimize the use of water resources by integrating surface water and groundwater management.

Benefits of Conjunctive Use of Surface Water and Groundwater

Increased Water Availability: Conjunctive use of surface water and groundwater increases the overall availability of water for various uses. It helps in ensuring a reliable and sustainable water supply by utilizing both sources of water.

Improved Water Quality: Surface water and groundwater have different qualities and characteristics. By using both sources of water, the overall quality of water can be improved. For example, the use of groundwater can dilute the high salinity levels in surface water, thereby improving the water quality.

Increased Agricultural Productivity: Conjunctive use of surface water and groundwater can significantly increase agricultural productivity. Surface water can be used for irrigation during the wet season, while groundwater can be used during the dry season when surface water is scarce. This can help in ensuring a steady water supply for crops and increasing their yield.

Reduced Dependence on a Single Source: By using both surface water and groundwater, the dependence on a single source of water is reduced. This helps in reducing the risks associated with water scarcity or quality issues in a particular source.

Improved Groundwater Recharge: The use of surface water can help in recharging the groundwater aquifers. During the wet season, excess surface water can be allowed to percolate into the ground, thereby recharging the groundwater aquifers. This can help in replenishing the depleted groundwater resources.

Challenges of Conjunctive Use of Surface Water and Groundwater

Lack of Coordination: Conjunctive use of surface water and groundwater requires a high level of coordination and integration between different water management agencies. In many cases, there is a lack of coordination between different agencies, which can hinder the effective implementation of conjunctive use.

Regulatory Challenges: The regulatory framework for conjunctive use is often complex and fragmented. There may be conflicting regulations and policies at the state and federal levels, which can make it difficult to implement conjunctive use.

Technical Challenges: Conjunctive use of surface water and groundwater requires a high level of technical expertise. There are technical challenges associated with the management and operation of conjunctive use systems, such as the optimization of pumping rates, monitoring of groundwater levels, and management of water quality.

Environmental Concerns: Conjunctive use can have environmental impacts, such as the depletion of surface water resources and the overexploitation of groundwater resources. There may also be concerns about the impact of pumping on surface water ecosystems and the quality of water resources.

Cost: Conjunctive use of surface water and groundwater can be expensive. The cost of infrastructure, such as pumps, pipelines, and monitoring equipment, can be significant. There may also be operational costs associated with the management and maintenance of the conjunctive use system.

To overcome the challenges associated with the conjunctive use of surface water and groundwater, several best practices have been developed. These best practices can help in the effective implementation of conjunctive use and ensure the sustainable use of water resources.

Integrated Water Resources Management (IWRM): IWRM is a management approach that aims to coordinate and integrate the management of water resources across different sectors and stakeholders. IWRM can help in the effective implementation of conjunctive use by ensuring the coordination between different water management agencies and stakeholders.

Stakeholder Participation: Stakeholder participation is crucial for the successful implementation of conjunctive use. The participation of local communities, water user associations, and other stakeholders can help in ensuring the sustainability of conjunctive use and the effective management of water resources.

Monitoring and Evaluation: Monitoring and evaluation are essential for the effective management of conjunctive use. Regular monitoring of groundwater levels, surface water flows, and water quality can help in identifying the challenges and opportunities associated with conjunctive use. Evaluation can help in determining the effectiveness of conjunctive use and identifying areas for improvement.

Adaptation to Climate Change: Climate change can significantly impact the availability and distribution of water resources. The conjunctive use of surface water and groundwater can help in adapting to climate change by ensuring a reliable and sustainable water supply. Adaptation strategies, such as the use of climate-resilient crops and water-efficient irrigation techniques, can also help in the effective implementation of conjunctive use.

Innovative Technologies: Innovative technologies, such as remote sensing, geographic information systems (GIS), and artificial intelligence (AI), can help in the effective management of conjunctive use. These technologies can help in monitoring and predicting water availability, identifying areas for conjunctive use, and optimizing the management of water resources.

Case Studies

Several countries have implemented the conjunctive use of surface water and groundwater. The following are some case studies that demonstrate the benefits and challenges associated with conjunctive use.

India: India has implemented conjunctive use of surface water and groundwater in several regions to ensure a reliable and sustainable water supply. For example, in the state of Gujarat, the Sujalam Sufalam Water Conservation Project aims to recharge the groundwater aquifers by using excess surface water during the wet season. The project has helped in improving the availability and quality of water for irrigation and domestic use.

Australia: In Australia, conjunctive use of surface water and groundwater has been implemented in several regions to ensure a sustainable water supply. For example, the Murray-Darling Basin Plan aims to manage the surface water and groundwater resources in the Murray-Darling Basin in an integrated and coordinated manner. The plan has helped in improving the efficiency of water use and reducing the risks associated with water scarcity and quality issues.

United States: In the United States, conjunctive use of surface water and groundwater has been implemented in several regions to ensure a reliable and sustainable water supply. For example, the Central Valley Project in California uses both surface water and groundwater to meet the water demand for agriculture, urban, and industrial uses. The project has helped in improving the availability and quality of water and reducing the risks associated with water scarcity and quality issues.

Conclusion

The conjunctive use of surface water and groundwater is a management strategy that can help in ensuring a reliable and sustainable water supply. Conjunctive use can increase water availability, improve water quality, increase agricultural productivity, reduce dependence on a single source, and improve groundwater recharge. However, there are several challenges associated with conjunctive use, such as the lack of coordination, regulatory challenges, technical challenges, environmental concerns, and cost. To overcome these challenges, several best practices have been developed, such as integrated and coordinated approach to water resources management, stakeholder participation, and the use of innovative technologies. By adopting these best practices, countries can ensure the sustainability of conjunctive use and the effective management of water resources in the long run.

45 - River Bank Filtration: Harnessing the Natural Power of Riverbanks for Sustainable Water Treatment

Abstract

This article provides an in-depth overview of river bank filtration (RBF), a sustainable method of water treatment that utilizes the natural filtering capacity of riverbanks to purify surface water. RBF has been used successfully in many parts of the world and is a cost-effective and environmentally friendly alternative to traditional water treatment methods. The article discusses the principles of RBF, the benefits and limitations of the method, and its applications in various parts of the world. Ultimately, the article highlights the potential of RBF as a viable solution for communities that lack access to clean water or are looking for a more sustainable method of water treatment.

Introduction

Water is a vital resource for all living beings, and it is essential to ensure the quality and safety of water for human consumption. One of the techniques used for water treatment is river bank filtration (RBF). RBF is an environmentally friendly, cost-effective, and sustainable water treatment technique that utilizes the natural filtering capacity of river banks to purify surface water. In this article, we will discuss RBF, its history, the process of RBF, the benefits and limitations of RBF, and its application in different parts of the world.

History of RBF

The use of RBF dates back to ancient times. People have been using this technique to access clean water for centuries. However, it was not until the early 20th century that the scientific community recognized the potential of RBF as a water treatment method. In the early 1900s, researchers in Germany began to study the effects of riverbank filtration on the quality of water. They found that RBF could remove bacteria and other pathogens from surface water, making it safe for human consumption.

During World War II, RBF was used extensively by the German army to provide clean water to soldiers. After the war, RBF gained popularity in Europe and the United States as a cost-effective and sustainable method of water treatment. Today, RBF is used in many countries around the world, including Germany, the United States, India, and China.

Process of RBF

The process of RBF involves pumping surface water from a river or lake through the porous soil and gravel layers of the riverbank. The soil and gravel layers act as a natural filter, removing impurities from the water. The filtered water is then collected from wells located on the other side of the riverbank. The depth and length of the soil and gravel layers, as well as the distance between the wells and the river, can vary depending on the location and the characteristics of the riverbank.

There are two types of RBF: direct and indirect. In direct RBF, water is pumped directly from the river or lake into the wells located on the other side of the riverbank. In indirect RBF, the water is first pumped into a surface water collection system, such as a pond or reservoir, before being pumped through the riverbank and into the wells. Indirect RBF is often used when the quality of the surface water is poor, and additional treatment is needed.

Benefits of RBF

RBF has several benefits over traditional water treatment methods. First, RBF is environmentally friendly. It does not require the use of chemicals, such as chlorine or fluoride, which can be harmful to the environment. RBF also does not require the construction of large treatment plants, which can be costly and can have negative environmental impacts.

Second, RBF is a cost-effective method of water treatment. The process of RBF is simple and requires minimal maintenance. Once the wells are drilled and the collection system is in place, the only ongoing costs are for pumping the water and monitoring the quality of the water. This makes RBF an attractive option for communities that cannot afford the cost of traditional water treatment methods.

Third, RBF produces high-quality water. The natural filtering capacity of the soil and gravel layers of the riverbank removes impurities from the water, including bacteria, viruses, and other pathogens. The water produced by RBF is often of higher quality than water treated with traditional water treatment methods, such as chlorination.

Fourth, RBF is a sustainable method of water treatment. It utilizes a natural resource, the river or lake, and does not require the use of non-renewable resources, such as fossil fuels. RBF also does not produce any waste or byproducts that can harm the environment.

Limitations of RBF

Despite its many benefits, RBF does have some limitations. One limitation is that the quality of the water produced by RBF can be affected by the quality of the surface water. If the surface water is highly contaminated, the soil and gravel layers of the riverbank may not be able to remove all of the impurities from the water. In such cases, additional treatment may be required before the water is safe for human consumption.

Another limitation of RBF is that it is not suitable for all locations. RBF requires a suitable riverbank with the right soil and gravel layers to act as a natural filter. In some locations, the riverbank may not be suitable for RBF, which can limit its applicability.

Finally, RBF may not be able to remove all types of contaminants from the water. For example, RBF may not be effective at removing certain chemicals or heavy metals from the water. In such cases, additional treatment may be required to remove these contaminants.

Applications of RBF

RBF has been used successfully in many parts of the world, including Germany, the United States, India, and China. In Germany, RBF has been used for over a century to provide clean drinking water to cities and towns. The city of Berlin, for example, uses RBF to provide up to 70% of its drinking water. RBF is also used extensively in the United States, particularly in the Midwest, where it is used to treat water for drinking and irrigation.

In India, RBF has been used to provide clean drinking water to rural communities. The state of Gujarat, for example, has implemented RBF to provide safe drinking water to over 500 villages. RBF has also been used in China to provide clean water to communities along the Yellow River.

Conclusion

River bank filtration is an environmentally friendly, cost-effective, and sustainable method of water treatment that utilizes the natural filtering capacity of riverbanks to purify surface water. RBF has been used successfully in many parts of the world, including Germany, the United States, India, and China. RBF produces high-quality water, is cost-effective, and has minimal environmental impact. However, RBF does have some limitations, including the quality of the surface water, the suitability of the riverbank, and the ability to remove certain types of contaminants from the water. Despite these limitations, RBF remains an attractive option for communities that cannot afford the cost of traditional water treatment methods or are looking for a more sustainable method of water treatment.

46 - Exploring the Hydrology of Lakes: Understanding the Water Balance and Management Strategies

Abstract

This article explores the hydrology of lakes, including the processes that govern their water balance and the factors that influence their water quality. Lakes play an important role in the global water cycle, serving as important sources of water for human use and habitats for aquatic organisms. However, human activities such as land use changes, climate change, and pollution can lead to water quality issues such as eutrophication, acidification, and contamination. Effective lake management strategies must address these issues and consider the complex interactions between natural processes and human activities. This article discusses various management strategies that can be used to address water quality issues in lakes, including nutrient management, watershed management, dam management, and contamination management. By protecting and managing our lakes, we can ensure their continued availability as important water resources and habitats for aquatic organisms.

Introduction

Lakes are one of the most important water resources on our planet. They are often formed by natural processes such as tectonic activity, glaciation, and volcanic activity. Lakes play a vital role in the hydrological cycle by storing and releasing water over time. They also provide numerous ecological services, such as habitat for fish, birds, and other wildlife, recreational opportunities, and water supply for irrigation, drinking, and industrial use.

Hydrology is the scientific study of water in the environment, including its properties, distribution, and movement. In this article, we will discuss the hydrology of lakes, including their formation, water budget, and the factors that influence their water balance.

Formation of Lakes

Lakes can form in a variety of ways, including tectonic activity, glaciation, volcanic activity, and erosion. Tectonic lakes are formed when the earth's crust moves and creates a depression that fills with water. Glacial lakes are formed when glaciers carve out a basin that fills with meltwater. Volcanic lakes are formed when volcanic activity creates a depression that fills with water. Erosion lakes are formed when water erodes a depression in the earth's surface.

The size and shape of a lake are determined by the geology of the surrounding area. For example, a lake that is located in a steep valley will be narrow and deep, while a lake that is located in a flat area will be wide and shallow. The depth of a lake is also influenced by the amount of precipitation it receives, as well as the amount of evaporation that occurs.

Water Budget of Lakes

The water budget of a lake is the balance between the water that enters the lake and the water that leaves the lake. The water budget of a lake is affected by several factors, including precipitation, evaporation, runoff, and groundwater inflow and outflow.

Precipitation

Precipitation is the most important source of water for lakes. Rainfall, snowmelt, and other forms of precipitation can all contribute to the water budget of a lake. The amount of precipitation that falls on a lake is influenced by several factors, including the climate of the area, the topography of the surrounding area, and the prevailing winds.

Evaporation

Evaporation is the process by which water is converted from a liquid to a gas and enters the atmosphere. The rate of evaporation is influenced by several factors, including temperature, humidity, wind speed, and the surface area of the lake. Lakes with a larger surface area will generally experience higher rates of evaporation than lakes with a smaller surface area.

Runoff

Runoff is the water that flows over the surface of the land and enters a lake. Runoff can be influenced by several factors, including the amount of precipitation that falls, the type of soil in the area, and the topography of the surrounding area.

Groundwater Inflow and Outflow

Groundwater inflow and outflow can also affect the water budget of a lake. Groundwater is water that is stored in underground aquifers. If the water table is above the level of the lake, groundwater will flow into the lake, increasing its water budget. If the water table is below the level of the lake, groundwater will flow out of the lake, decreasing its water budget.

Factors that Affect the Water Balance of Lakes

Several factors can affect the water balance of lakes, including climate, land use, and human activities.

Climate

The climate of an area can have a significant impact on the water balance of a lake. Areas with high levels of precipitation and low levels of evaporation will generally have higher water budgets than areas with low levels of precipitation and high levels of evaporation. Climate change can also have a significant impact on the water balance of lakes, as changes in temperature and precipitation patterns can alter the amount of water available for lakes.

Land Use

Land use can also affect the water balance of lakes. Urbanization, agricultural activities, and deforestation can all increase runoff and decrease groundwater recharge, which can decrease the water budget of lakes. Additionally, the use of fertilizers and pesticides in agricultural activities can lead to increased nutrient runoff, which can lead to eutrophication and other water quality issues in lakes.

Human Activities

Human activities such as damming, water diversion, and water withdrawals can also affect the water balance of lakes. Dams can alter the natural flow of water into and out of lakes, which can have a significant impact on their water budgets. Water diversions can also reduce the amount of water available for lakes, while water withdrawals can lower lake levels and decrease the water budget.

Hydrological Processes in Lakes

Lakes are dynamic systems that are constantly changing as a result of several hydrological processes, including mixing, stratification, and turnover.

Mixing

Mixing is the process by which the water in a lake is circulated and distributed throughout the lake. Mixing can be caused by several factors, including wind, temperature differences, and the density of the water. In lakes that are thermally stratified, mixing can occur during the spring and fall when the temperature of the water in the lake is more uniform.

Stratification

Stratification is the process by which the water in a lake separates into distinct layers based on differences in temperature and density. In lakes that are thermally stratified, the warmer, less dense water is found at the surface, while the colder, denser water is found at the bottom. This stratification can create unique habitats for fish and other aquatic organisms.

Turnover

Turnover is the process by which the water in a lake is mixed from top to bottom. Turnover can occur in response to changes in temperature or wind patterns. During turnover, nutrients and other materials that have accumulated at the bottom of the lake are mixed back into the water column, which can have important implications for the ecology of the lake.

Water Quality Issues in Lakes

Lakes are vulnerable to a variety of water quality issues, including eutrophication, acidification, and contamination.

Eutrophication

Eutrophication is the process by which lakes become nutrient-rich and support excessive plant growth, which can lead to a depletion of oxygen in the water and the death of fish and other aquatic organisms. Eutrophication can be caused by a variety of factors, including runoff from agricultural activities, wastewater discharges, and atmospheric deposition.

Acidification

Acidification is the process by which lakes become more acidic due to the deposition of acidic compounds from the atmosphere. Acidification can have significant impacts on the ecology of lakes, including the death of fish and other aquatic organisms, as well as changes in the types of plants that can grow in the lake.

Contamination

Lakes can also be contaminated by a variety of pollutants, including chemicals, heavy metals, and pathogens. Contamination can be caused by a variety of human activities, including industrial activities, sewage discharges, and agricultural activities. Contamination can have significant impacts on the ecology of lakes, as well as on human health and the economy.

Lake Management strategies

There are several management strategies that can be used to address water quality issues in lakes, including:

Nutrient Management: Nutrient management strategies can be used to reduce the amount of nutrients that enter lakes, which can help to prevent eutrophication. These strategies can include the use of best management practices in agriculture, such as reducing fertilizer use and promoting cover crops, and the use of advanced wastewater treatment technologies to remove nutrients from wastewater before discharge.

Watershed Management: Watershed management strategies can be used to reduce the amount of runoff and pollution that enters lakes from surrounding land areas. These strategies can include the use of vegetated buffer strips, wetland restoration, and stormwater management practices.

Dam Management: Dam management strategies can be used to manage the water levels in lakes and maintain suitable habitat conditions for aquatic organisms. These strategies can include the use of selective withdrawal systems, which allow for the selective removal of water from different depths in the lake, and the use of flow-through devices to maintain suitable water levels.

Contamination Management: Contamination management strategies can be used to address contamination issues in lakes. These strategies can include the use of pollutant source controls, such as regulations on industrial discharges, as well as the use of advanced treatment technologies to remove pollutants from wastewater before discharge.

Effective lake management strategies must be tailored to the specific needs and characteristics of each lake. The development of successful management strategies requires a comprehensive understanding of the hydrology of the lake, as well as the social and economic factors that influence lake management decisions.

Conclusion

Lakes are important water resources that play a vital role in the hydrological cycle. The hydrology of lakes is influenced by several factors, including precipitation, evaporation, runoff, and groundwater inflow and outflow. Climate, land use, and human activities can all affect the water balance of lakes and lead to water quality issues. Hydrological processes such as mixing, stratification, and turnover are important for the ecology of lakes. It is important to protect and manage our lakes to ensure their continued availability as a source of water and as important habitats for aquatic organisms. Effective lake management strategies must address the complex interactions between natural processes and human activities, while also considering the economic and social implications of lake management decisions.

The hydrology of lakes is a complex and dynamic system that is influenced by several factors, including precipitation, evaporation, runoff, and groundwater inflow and outflow. The water balance of lakes can be affected by climate, land use, and human activities, which can lead to water quality issues such as eutrophication, acidification, and contamination. Effective lake management strategies must address these issues and consider the complex interactions between natural processes and human activities. By protecting and managing our lakes, we can ensure their continued availability as important water resources and habitats for aquatic organisms.

47 - Rejuvenating Ponds and Lakes: Techniques and Best Practices

Abstract

Ponds and lakes are important natural resources that support a wide range of plant and animal life, and provide numerous benefits to humans. However, due to factors such as pollution, climate change, and human activities, many of these water bodies have become degraded and are in need of rejuvenation. This article provides an overview of the techniques and best practices for rejuvenating ponds and lakes. It discusses the causes of degradation, the benefits of rejuvenation, and the various methods available for improving water quality, restoring ecological balance, and enhancing recreational value. The article also highlights the importance of community involvement and collaboration in successful rejuvenation projects.

Introduction

Ponds and lakes are vital components of our ecosystems and provide numerous benefits, such as water supply, flood control, recreation, and habitat for plant and animal species. They are also important cultural and historical features, serving as focal points for communities and providing opportunities for cultural and recreational activities. However, these water bodies are under threat from a range of factors, including pollution, overfishing, climate change, and human activities such as urbanization, agriculture, and industry. As a result, many ponds and lakes have become degraded, with poor water quality, loss of biodiversity, and reduced recreational value. Rejuvenation of these water bodies is essential for their preservation and for the benefits they provide to both the environment and society.

Causes of Pond and Lake Degradation

The degradation of ponds and lakes can be attributed to a range of factors, including:

Pollution: The discharge of untreated sewage and industrial waste into water bodies leads to the accumulation of nutrients such as nitrogen and phosphorus, which can cause eutrophication. This leads to the growth of algae and other aquatic plants, which can deplete oxygen levels and harm aquatic life.

Climate change: Global warming and changes in precipitation patterns can affect water quality, alter aquatic habitats, and increase the frequency and severity of algal blooms and harmful algal blooms (HABs).

Overfishing: Overfishing can reduce fish populations and disrupt the ecological balance of ponds and lakes, leading to the proliferation of invasive species and a decline in water quality.

Habitat loss: Human activities such as urbanization, agriculture, and industrial development can lead to the destruction of wetlands and other important habitats, reducing the biodiversity and ecological functioning of ponds and lakes.

Poor land use practices: Improper land use practices, such as excessive use of fertilizers and pesticides, can lead to runoff and the accumulation of nutrients in water bodies, contributing to eutrophication.

Benefits of Rejuvenating Ponds and Lakes

Rejuvenation of ponds and lakes can bring a range of benefits, such as:

Improved water quality: Rejuvenation techniques can help to reduce nutrient levels and other pollutants, leading to clearer water and improved aquatic habitats.

Restoration of ecological balance: By reducing the impacts of pollution, overfishing, and habitat loss, rejuvenation can help to restore the natural balance of ponds and lakes, improving the biodiversity and functioning of these ecosystems.

Enhanced recreational value: Rejuvenation can make ponds and lakes more attractive for recreational activities such as fishing, boating, and swimming, improving the quality of life for local communities.

Protection of water resources: Rejuvenation can help to protect the water resources that are essential for human needs such as drinking water, irrigation, and industrial uses.

Techniques for Rejuvenating Ponds and Lakes

There are a range of techniques available for rejuvenating ponds and lakes, depending on the specific causes and extent of degradation. Some of the most common techniques are discussed below.

Aeration: Aeration involves the addition of oxygen to the water body, which can improve water quality and support aquatic life. This can be achieved through the use of diffusers or fountains, which create turbulence and increase oxygen exchange between the water and air. Aeration can also help to prevent the buildup of harmful gases such as hydrogen sulfide and methane, which can be toxic to aquatic life.

Nutrient management: Reducing nutrient levels in ponds and lakes can help to control eutrophication and promote the growth of beneficial aquatic plants. This can be achieved through the use of chemical treatments such as alum or ferric chloride, which can bind with nutrients and prevent their release into the water. Nutrient management can also involve the use of biological treatments such as the addition of beneficial bacteria or the introduction of filter feeders such as mussels, which can help to control algal blooms.

Fish management: Proper fish management can help to control populations and maintain the ecological balance of ponds and lakes. This can involve the removal of invasive species, the stocking of native fish, and the implementation of fishing regulations to prevent overfishing.

Habitat restoration: Restoring wetlands and other important habitats can help to improve the biodiversity and ecological functioning of ponds and lakes. This can involve the removal of invasive species and the planting of native vegetation, as well as the creation of nesting sites and other habitat features for aquatic species.

Sediment management: Sediment can accumulate in ponds and lakes over time, leading to reduced water depth, degraded habitats, and increased nutrient levels. Sediment management techniques such as dredging or sediment removal can help to improve water quality and restore the ecological balance of the water body.

Best Practices for Rejuvenating Ponds and Lakes

Rejuvenation of ponds and lakes requires careful planning and implementation, and the following best practices can help to ensure successful outcomes:

Conduct a thorough assessment: Before implementing rejuvenation techniques, it is important to conduct a comprehensive assessment of the water body to identify the specific causes of degradation and determine the most appropriate course of action.

Involve the community: Successful rejuvenation projects require the involvement and support of local communities, stakeholders, and organizations. Engaging with these groups can help to build awareness and support for the project, as well as provide valuable input and resources.

Use an integrated approach: Rejuvenation techniques should be integrated and tailored to the specific needs and conditions of the water body. This may involve the use of multiple techniques, such as nutrient management and fish management, to achieve the desired outcomes.

Monitor progress: Regular monitoring and evaluation of the rejuvenation project can help to track progress, identify any issues or challenges, and make necessary adjustments to the approach.

Maintain ongoing management: Rejuvenation is an ongoing process, and ongoing management is required to ensure the long-term success and sustainability of the project. This may involve regular maintenance of equipment and infrastructure, as well as ongoing monitoring and management of the water body.

Conclusion

Rejuvenation of ponds and lakes is essential for the preservation of these important natural resources and the benefits they provide to both the environment and society. By understanding the causes of degradation and implementing appropriate rejuvenation techniques, we can restore the ecological balance, improve water quality, and enhance the recreational value of these valuable water bodies. Effective rejuvenation requires community involvement, careful planning, and ongoing management, and by following best practices, we can ensure the long-term success and sustainability of these important ecosystems.

48 - Urban Hydrology: Understanding the Water Cycle in Urban Areas

Abstract

Urban hydrology is a subfield of hydrology that studies the water cycle in urban areas. Urbanization has significantly altered the hydrology of urban areas, leading to increased flooding, water pollution, and reduced water availability. This article discusses the impact of urbanization on the water cycle, the importance of urban hydrology, the various hydrological processes in urban areas, and the challenges associated with urban hydrology. The article highlights the need for accurate and up-to-date data, a coordinated and integrated approach to manage the impacts of urbanization, infrastructure design and maintenance, proactive measures to manage the impacts of climate change, and strong governance and policy frameworks based on scientific evidence.

Introduction

Urban hydrology is a subfield of hydrology that studies the water cycle in urban areas. It deals with the study of various hydrological processes like precipitation, infiltration, runoff, and water quality in urban environments. Urbanization has significantly altered the hydrology of urban areas, leading to increased flooding, water pollution, and reduced water availability. Urban hydrology aims to understand these changes and develop strategies to manage water resources in urban areas.

In this article, we will discuss various aspects of urban hydrology, including the impact of urbanization on the water cycle, the importance of urban hydrology, the various hydrological processes in urban areas, and the challenges associated with urban hydrology.

Impact of Urbanization on the Water Cycle

Urbanization has a significant impact on the water cycle in urban areas. The natural hydrological cycle is disrupted due to the construction of buildings, pavements, and other impervious surfaces, which reduce infiltration and increase runoff. Urbanization also increases the demand for water, leading to the abstraction of groundwater and surface water. This, in turn, reduces the amount of water available for other uses.

Urbanization has also led to increased water pollution due to the discharge of untreated or poorly treated sewage, industrial effluents, and stormwater runoff. This has resulted in the degradation of water quality and increased risks to public health.

Importance of Urban Hydrology

Urban hydrology is essential for the sustainable management of water resources in urban areas. It helps to identify the various hydrological processes in urban areas and understand their interactions. This knowledge is crucial for the development of effective strategies to manage water resources in urban areas.

Urban hydrology also helps to mitigate the adverse effects of urbanization on the water cycle. It provides insights into the impacts of urbanization on water resources and helps to develop sustainable solutions to address these challenges.

Hydrological Processes in Urban Areas

The various hydrological processes in urban areas include precipitation, infiltration, evapotranspiration, surface runoff, and groundwater recharge. Each of these processes plays a critical role in the urban water cycle.

Precipitation

Precipitation is the primary source of water in the urban water cycle. It can be in the form of rain, snow, or sleet. The amount and intensity of precipitation in urban areas are affected by various factors, including urbanization, topography, and climate change.

Infiltration

Infiltration is the process by which water enters the soil. In urban areas, infiltration is reduced due to the presence of impervious surfaces like buildings, pavements, and roads. This leads to an increase in surface runoff and a decrease in groundwater recharge.

Evapotranspiration

Evapotranspiration is the process by which water is lost to the atmosphere through evaporation from the soil and transpiration from plants. In urban areas, evapotranspiration is reduced due to the absence of vegetation and the presence of impervious surfaces. This leads to an increase in surface runoff and a decrease in groundwater recharge.

Surface Runoff

Surface runoff is the flow of water over the surface of the land. In urban areas, surface runoff is increased due to the presence of impervious surfaces, which prevent infiltration. This can lead to flooding and erosion, which can cause damage to infrastructure and property.

Groundwater Recharge

Groundwater recharge is the process by which water infiltrates the soil and recharges the groundwater table. In urban areas, groundwater recharge is reduced due to the presence of impervious surfaces, which prevent infiltration. This can lead to a decrease in the availability of groundwater, which can impact the sustainability of water resources in urban areas.

Challenges Associated with Urban Hydrology

Urban hydrology is a complex field that is faced with several challenges. These challenges include:

Data availability and accuracy: Urban hydrology requires accurate and up-to-date data on hydrological processes in urban areas. However, collecting and maintaining this data can be challenging due to the fragmented nature of urban areas, the presence of impervious surfaces, and the complex interactions between hydrological processes. This can lead to errors and uncertainty in hydrological models and predictions.

Urbanization: Urbanization has significantly altered the hydrology of urban areas, leading to increased flooding, water pollution, and reduced water availability. Managing these impacts requires a coordinated and integrated approach that takes into account the social, economic, and environmental factors that drive urbanization.

Infrastructure: The built infrastructure in urban areas, such as stormwater drainage systems, canals, and dams, can significantly impact the hydrology of urban areas. The design and maintenance of this infrastructure can impact the quality and quantity of water available in urban areas.

Climate Change: Climate change is expected to increase the frequency and intensity of extreme weather events such as floods and droughts, which can have significant impacts on the hydrology of urban areas. Managing these impacts requires a comprehensive and proactive approach that takes into account the potential impacts of climate change on urban hydrology.

Governance and Policy: Effective management of urban hydrology requires strong governance and policy frameworks that are based on scientific evidence and take into account the needs and aspirations of local communities. However, the fragmented nature of urban governance can make it challenging to coordinate and implement policies across different sectors and jurisdictions.

Conclusion

Urban hydrology plays a critical role in the sustainable management of water resources in urban areas. The impact of urbanization on the water cycle has led to increased flooding, water pollution, and reduced water availability. Urban hydrology provides essential insights into the various hydrological processes in urban areas, including precipitation, infiltration, evapotranspiration, surface runoff, and groundwater recharge. Despite the challenges associated with data availability and accuracy, urbanization, infrastructure, climate change, governance, and policy, urban hydrology remains a vital field that helps to develop sustainable solutions to address the challenges facing urban water resources. Effective management of urban hydrology requires a coordinated and integrated approach that takes into account the social, economic, and environmental factors that drive urbanization, as well as the potential impacts of climate change on urban hydrology.

49 - Exploring the Complex Hydrology of Hard Rock Regions: Implications for Water Management and Quality

Abstract

Hydrology is the scientific study of water in the Earth's systems, including the movement, distribution, and quality of water. Hard rock regions, also known as crystalline rock regions, are areas characterized by rocky terrain and low permeability soils. The hydrology of these regions differs from that of other regions, and understanding it is crucial for effective water management in such areas. In this article, we will explore the hydrology of hard rock regions in detail, including the various aspects that contribute to it.

Introduction

Hard rock regions are primarily made up of igneous and metamorphic rocks such as granite, gneiss, and schist. These rocks are usually impermeable, meaning they do not allow water to pass through easily. As a result, groundwater in these regions is stored in fractures, joints, and fissures in the rocks. The hydrology of hard rock regions is, therefore, characterized by a complex network of underground water channels.

The study of the hydrology of hard rock regions is essential for several reasons. Firstly, it is necessary for managing the water resources in these regions. Secondly, it is essential for understanding the impact of human activities such as mining and construction on water resources. Lastly, it is critical for predicting the behavior of water systems in these regions during extreme weather events such as floods and droughts.

Surface Water and Groundwater Interactions

The hydrology of hard rock regions is closely linked to the interactions between surface water and groundwater. In areas with high precipitation, surface water runoff can seep into the fractures and fissures in the rocks and recharge the groundwater system. This process is known as infiltration, and it is essential for maintaining groundwater levels in hard rock regions.

The movement of groundwater in hard rock regions is also influenced by the topography of the area. In areas with a high slope, groundwater flows through the fractures and fissures in the rocks downhill towards the streams and rivers. In contrast, in areas with a low slope, groundwater flows laterally towards the nearest stream or river.

The interaction between surface water and groundwater in hard rock regions is also affected by the seasonal variation in precipitation. During the wet season, the increased surface water runoff can lead to an increase in groundwater recharge. In contrast, during the dry season, the lack of precipitation can result in a decrease in groundwater recharge. As a result, water availability in hard rock regions is highly dependent on the seasonal variation in precipitation.

Fracture and Joint System

One of the most significant features of the hydrology of hard rock regions is the fracture and joint system in the rocks. These features are essential for storing and transmitting groundwater in hard rock regions. Fractures and joints are formed in the rocks due to tectonic activity and weathering. These features can vary in size from a few millimeters to several meters and can form complex networks that can extend for several kilometers.

The permeability of hard rock regions is primarily determined by the fracture and joint system. The larger and more connected the fracture and joint system, the more permeable the rock is, and the easier it is for water to flow through it. Conversely, the smaller and less connected the fracture and joint system, the less permeable the rock is, and the more difficult it is for water to flow through it.

The fracture and joint system in hard rock regions can also be influenced by human activities such as mining and construction. These activities can result in the formation of new fractures and joints in the rock, which can alter the groundwater flow patterns and recharge rates. Therefore, it is essential to consider the impact of human activities on the fracture and joint system in hard rock regions when managing water resources.

Karst Systems

Karst systems are a type of hydrological system that is common in limestone and dolomite rock formations. These systems are characterized by the dissolution of the rock by groundwater, resulting in the formation of underground caves, sinkholes, and other features. While karst systems are not typically found in hard rock regions, they can occur in areas where hard rock formations are overlain by carbonate rocks.

In these situations, the overlying carbonate rocks can dissolve, resulting in the formation of karst features. The interaction between groundwater and the overlying carbonate rocks can also affect the groundwater flow patterns in the hard rock formations below.

Water Quality

The hydrology of hard rock regions can also impact water quality. Since the water in these regions is stored in fractures and fissures in the rocks, it is not filtered by soil layers like it is in other regions. As a result, the water in hard rock regions can contain higher levels of dissolved minerals and contaminants such as heavy metals.

The water quality in hard rock regions can also be affected by human activities such as mining and agriculture. These activities can introduce contaminants into the groundwater system, which can impact water quality. Therefore, it is essential to monitor water quality in hard rock regions and take steps to mitigate any negative impacts on water quality.

Water Management

Effective water management in hard rock regions requires an understanding of the hydrology of the area. One important aspect of water management in these regions is the management of groundwater resources. Since groundwater is the primary source of water in hard rock regions, it is essential to monitor groundwater levels and recharge rates to ensure sustainable use.

One way to manage groundwater resources in hard rock regions is through artificial recharge. Artificial recharge involves the intentional recharge of groundwater through surface infiltration or injection wells. This process can help maintain groundwater levels in hard rock regions and provide a reliable source of water during dry periods.

Another important aspect of water management in hard rock regions is the protection of water sources from contamination. This can be achieved through the implementation of regulations and best practices for activities such as mining and agriculture, which can introduce contaminants into the groundwater system.

Conclusion

The hydrology of hard rock regions is characterized by a complex network of underground water channels, primarily stored in fractures, joints, and fissures in the rocks. Understanding the hydrology of hard rock regions is essential for effective water management in these areas. The interaction between surface water and groundwater, the fracture and joint system, karst systems, water quality, and water management are all important aspects of the hydrology of hard rock regions. By taking these factors into account, we can ensure the sustainable use and protection of water resources in hard rock regions.

50 - Hydrology of Mountainous Regions: Understanding the Unique Water Cycle and Mitigating the Impacts of Climate Change

Abstract

Mountainous regions are important components of the global water cycle, containing a range of water storage systems that provide water to downstream communities and ecosystems. However, the impacts of climate change, including increased temperatures, changes in precipitation patterns, and more frequent extreme weather events, are having a significant impact on hydrology in mountainous regions. In this article, we explore the unique hydrologic environment of mountainous regions and the impacts of climate change on water resources. We also discuss mitigation strategies, such as improving water storage and management systems, managing land use and vegetation, promoting sustainable water use practices, addressing pollution and water quality issues, and adapting to changing hydrologic conditions, to ensure a sustainable future for mountainous hydrology.

Introduction

Hydrology is the study of the movement, distribution, and quality of water on Earth. Mountainous regions are unique hydrologic environments due to their steep topography, diverse geology, and high precipitation rates. These environments play a critical role in the water cycle, as they are often the source of freshwater for downstream communities and ecosystems. In this article, we will explore the hydrology of mountainous regions, including the factors that influence water availability, the types of water storage systems, and the impact of climate change on these systems.

Factors Influencing Water Availability in Mountainous Regions

Mountainous regions receive high levels of precipitation due to orographic lifting, a process in which air is forced to rise over mountains, cooling and causing moisture to condense and fall as precipitation. The amount and intensity of precipitation are influenced by various factors, including the location, elevation, and topography of the mountain range, as well as regional climate patterns.

Temperature is another critical factor that affects water availability in mountainous regions. High-elevation regions typically experience cooler temperatures, which reduce evaporation rates and increase the amount of water stored in the soil and snowpack. Additionally, temperature affects the timing of snowmelt and the availability of water throughout the year. Warmer temperatures can cause earlier snowmelt and reduce the amount of water available in the summer months.

The geology of mountainous regions also plays a role in water availability. Some geologic formations, such as impermeable rocks, can prevent water from infiltrating into the ground and reduce groundwater recharge. Other formations, such as karst, can create large underground reservoirs that store and release water over time.

Types of Water Storage Systems in Mountainous Regions

Mountainous regions contain a variety of water storage systems, including snowpack, glaciers, rivers, lakes, and groundwater. Each of these systems plays a critical role in providing water to downstream communities and ecosystems.

Snowpack is a vital water storage system in mountainous regions, as it provides a slow release of water during the spring and summer months. The snowpack acts as a natural reservoir, holding water that can be released slowly into streams and rivers as the snow melts. Snowpack also helps to reduce the risk of flooding during the spring runoff.

Glaciers are another essential water storage system in mountainous regions, as they can store water for long periods of time. Glaciers form when snow accumulates and compresses into ice, and they can release water into streams and rivers throughout the year. Glaciers are particularly important in regions with limited precipitation, as they provide a consistent source of freshwater.

Rivers and lakes are the most visible water storage systems in mountainous regions, and they provide water for drinking, irrigation, and other uses. Rivers are formed when water flows downhill and collects in streams, eventually forming a larger river. Lakes are formed when water is trapped in a depression or valley, and they can provide a significant source of water storage.

Groundwater is also an important water storage system in mountainous regions, as it provides a source of water for wells and springs. Groundwater is stored in underground reservoirs, such as aquifers, and can be replenished by snowmelt and precipitation.

Impact of Climate Change on Hydrology in Mountainous Regions

Climate change is having a significant impact on hydrology in mountainous regions. As temperatures increase, snowpack and glaciers are melting at a faster rate, which can lead to reduced water availability during the summer months. Additionally, changes in precipitation patterns can alter the timing and amount of water that is available for storage in snowpack and groundwater systems.

One of the most significant impacts of climate change on mountainous regions is the increase in extreme weather events, such as floods and droughts. Extreme weather events can have a significant impact on water availability, as floods can lead to erosion and the loss of soil nutrients, while droughts can cause water shortages and negatively impact agriculture and ecosystems.

Another impact of climate change on mountainous hydrology is the increased frequency and severity of wildfires. Wildfires can alter the hydrology of a region by reducing the amount of vegetation, which can increase the likelihood of erosion and flooding. Additionally, wildfires can release pollutants into water sources, which can impact the quality of freshwater.

Climate change also has the potential to impact the quality of freshwater in mountainous regions. Warmer temperatures can lead to increased water temperatures, which can impact aquatic ecosystems and lead to the growth of harmful algal blooms. Additionally, changes in precipitation patterns can alter the amount and quality of pollutants that enter water sources, leading to degraded water quality.

Mitigation Strategies for Climate Change Impacts on Mountainous Hydrology

To mitigate the impacts of climate change on mountainous hydrology, a range of strategies can be employed. These strategies include:

Improving water storage and management systems. This can include the construction of new reservoirs, the improvement of existing water storage infrastructure, and the implementation of water conservation practices.

Managing land use and vegetation. This can involve reducing the risk of wildfire through vegetation management practices, such as prescribed burning, and reducing the amount of impermeable surfaces in urban areas to increase groundwater recharge.

Promoting sustainable water use practices. This can involve promoting the use of water-efficient technologies, such as drip irrigation, and implementing water reuse and recycling programs.

Addressing pollution and water quality issues. This can involve implementing best management practices for agriculture and industry to reduce the amount of pollutants that enter water sources.

Adapting to changing hydrologic conditions. This can include the development of drought and flood management plans, as well as the implementation of measures to protect ecosystems and communities from the impacts of extreme weather events.

Conclusion

Mountainous regions are unique hydrologic environments that play a critical role in the water cycle. These regions contain a range of water storage systems, including snowpack, glaciers, rivers, lakes, and groundwater, which provide water to downstream communities and ecosystems. However, the impacts of climate change, including increased temperatures, changes in precipitation patterns, and more frequent extreme weather events, are having a significant impact on hydrology in mountainous regions. Mitigation strategies, such as improving water storage and management systems, managing land use and vegetation, promoting sustainable water use practices, addressing pollution and water quality issues, and adapting to changing hydrologic conditions, can help to mitigate the impacts of climate change and ensure a sustainable future for mountainous hydrology.

51 - Hydrology and Sustainable Management of Springs: Understanding the Complexities and Importance

Abstract

Springs are an important source of freshwater and play a critical role in the hydrological cycle. The hydrology of springs is complex and influenced by geological, climatic, and ecological factors. This article provides an overview of the hydrology of springs, including their classification based on discharge, geological formation, and surrounding vegetation. The sustainable management of springs requires the protection of their recharge areas, regular monitoring of their health, and the implementation of restoration and water allocation strategies that prioritize the needs of different user groups. The article emphasizes the importance of protecting and managing springs to ensure the sustainability of freshwater resources and the ecological health of aquatic and terrestrial habitats.

Introduction

Springs are natural outlets of groundwater that emerge on the surface of the earth. They play a critical role in the hydrological cycle and contribute significantly to the availability of water resources. Springs are found in various geological formations and climatic conditions and are an essential source of freshwater for human consumption, agriculture, and livestock. This article provides a comprehensive review of the hydrology of springs, including their formation, characteristics, classification, and management.

Formation of Springs

Springs are formed due to the percolation of water through the soil and rocks into the underground aquifers. The aquifers are porous and permeable geological formations that can store and transmit groundwater. The formation of springs depends on the topography, geology, and climatic conditions of the region. The two primary mechanisms that contribute to the formation of springs are:

Recharge Area: The recharge area is the zone where water infiltrates the ground and recharges the aquifers. This process occurs in areas with high precipitation and permeable soils and rocks. The water moves through the subsurface and collects in the aquifers. The pressure created by the weight of the water in the aquifers pushes the water towards the surface, and when the water table intersects with the ground surface, a spring is formed.

Artesian Aquifers: Artesian aquifers are confined aquifers that are sandwiched between impermeable layers of rocks. When the recharge area of the aquifer is at a higher elevation than the discharge area, the water in the aquifer is under pressure due to the weight of the overlying rocks. The pressure in the aquifer can force the water to the surface, forming an artesian spring.

Characteristics of Springs

Springs have unique hydrological and ecological characteristics that distinguish them from other surface water sources. Some of the primary characteristics of springs are:

Flow Rate: The flow rate of springs varies from a few liters per day to several cubic meters per second, depending on the size and geological formation of the aquifer. The flow rate of springs is highly dependent on the precipitation and recharge rates of the aquifer.

Water Quality: The quality of spring water is generally high, as it is filtered through the soil and rocks before it emerges on the surface. The water from springs is generally free from contaminants and pathogens, making it suitable for human consumption.

Temperature: The temperature of spring water varies depending on the geological formation and the depth of the aquifer. The temperature of the water in the aquifer is generally constant, and the water emerging from the spring can be warm or cold, depending on the location and depth of the spring.

Ecological Significance: Springs are critical habitats for aquatic and terrestrial plants and animals. The constant flow of water and the unique ecological conditions provided by springs create a diverse range of habitats for aquatic and terrestrial species.

Classification of Springs

Springs can be classified based on their morphology, flow rate, and geological formation. Some of the primary classifications of springs are:

Morphological Classification

Springs can be classified based on their morphology, including their shape and size. Some of the common morphological classifications of springs are:

Vauclusian Springs: Vauclusian springs are large, deep springs that emerge from karstic aquifers. These springs are characterized by their high flow rates and the presence of subterranean channels and caverns.

Plunge Pools: Plunge pools are small springs that emerge from narrow openings in the ground. These springs are characterized by their steep, cascading flows and are commonly found in mountainous regions.

Terrace Springs: Terrace springs are flat, shallow springs that emerge from the side of hills or terraces. These springs are characterized by their low flow rates and the presence of vegetation around the spring.

Flow Rate Classification

Springs can also be classified based on their flow rate, which is typically categorized into three classes:

Low-Flow Springs: Low-flow springs have a flow rate of less than 10 liters per minute and are usually found in arid and semi-arid regions.

Medium-Flow Springs: Medium-flow springs have a flow rate between 10 and 100 liters per minute and are typically found in areas with moderate precipitation.

High-Flow Springs: High-flow springs have a flow rate of more than 100 liters per minute and are usually found in areas with high precipitation and permeable geology.

Geological Formation Classification

Springs can also be classified based on the geological formation that gives rise to them. Some of the common geological formations that give rise to springs are:

Karstic Springs: Karstic springs are formed in areas with soluble rocks such as limestone, dolomite, and gypsum. These springs are characterized by their high flow rates and the presence of subterranean channels and caverns.

Fracture Springs: Fracture springs are formed in areas with hard, impermeable rocks such as granite or basalt. These springs emerge from fractures and fissures in the rocks and are typically low-flow springs.

Fault Springs: Fault springs are formed in areas where the earth's crust is under tectonic stress. These springs emerge from fractures and faults in the earth's crust and are typically medium to high-flow springs.

Recharge Area of Spring

The recharge area of a spring is the location where water infiltrates into the ground and percolates through the soil and rock formations before emerging as a spring. Determining the recharge area of a spring is important for identifying the sources of groundwater that feed the spring and for managing the land use practices that can impact water quality and quantity. Here are some methods for finding the recharge area of a spring:

Tracing Studies: Tracing studies involve adding a tracer substance to the groundwater near the spring and tracking its movement to identify the recharge area. Common tracer substances include dyes, chemicals, and isotopes.

Geologic Mapping: Geologic mapping involves mapping the rock formations and geologic features in the area surrounding the spring to identify the potential recharge areas.

Hydrologic Modelling: Hydrologic modelling involves using computer models to simulate the movement of groundwater and surface water in the area surrounding the spring to identify the recharge area.

Determining the recharge area of a spring can help to identify the sources of groundwater that feed the spring and manage land use practices that impact water quality and quantity.

Spring Rejuvenation

Spring rejuvenation involves restoring the flow and quality of a spring that has become degraded or damaged due to human activities. Here are some methods for rejuvenating a spring:

Protection of Recharge Area: Protecting the recharge area of the spring from human activities such as deforestation, agriculture, and mining is essential for restoring the flow and quality of the spring.

Reforestation: Planting trees and other vegetation in the recharge area can help to increase infiltration rates, reduce soil erosion, and improve water quality.

Erosion Control: Implementing erosion control measures such as terracing, check dams, and vegetative barriers can help to reduce soil erosion and sedimentation in the spring.

Habitat Restoration: Restoring the habitat surrounding the spring by removing invasive species, planting native vegetation, and restoring riparian zones can help to improve water quality and create a healthy ecosystem.

Implementing protection, reforestation, erosion control, and habitat restoration measures can help to restore the health of a spring and ensure the sustainability of freshwater resources.

Management of Springs

Springs are a valuable resource and require careful management to ensure their sustainability and availability. Some of the primary management strategies for springs are:

Protection: Springs and their recharge areas should be protected from human activities such as deforestation, mining, and agricultural practices that can cause soil erosion and contamination of groundwater. The protection of springs and their recharge areas can be achieved through land use regulations and zoning laws.

Monitoring: Regular monitoring of the flow rate, water quality, and ecological conditions of springs is essential to assess their health and sustainability. This information can be used to develop management strategies that ensure the long-term sustainability of springs.

Restoration: Restoration of degraded or damaged springs can be achieved through ecological restoration techniques such as reforestation, erosion control, and habitat restoration.

Water Allocation: The allocation of water from springs should be based on sustainable water management practices that ensure the equitable distribution of water resources among different user groups. This can be achieved through the development of water allocation plans that prioritize the needs of different user groups.

Conclusion

Springs are a critical source of freshwater, playing a significant role in the hydrological cycle and contributing to the availability of water resources. Springs are formed due to the percolation of water through the soil and rocks into the underground aquifers. They have unique hydrological and ecological characteristics that distinguish them from other surface water sources, including their flow rate, water quality, temperature, and ecological significance. Springs can be classified based on their morphology, flow rate, and geological formation, and finding the recharge area of a spring and rejuvenating it are important aspects of sustainable water management. Overall, understanding the hydrology of springs is essential for the sustainable management of freshwater resources.

52 - Understanding Snow Hydrology: Processes, Impacts, and Applications

Abstract

Snow hydrology is an essential component of the global water cycle, influencing water availability, water quality, and ecosystem health. This article aims to provide a comprehensive review of snow hydrology, including the physical processes of snow formation, accumulation, and melt, the impacts of snowmelt on hydrological systems and human activities, and the applications of snow hydrology in various fields. We also discuss current research trends and future directions in snow hydrology, highlighting the need for improved monitoring and modelling of snowmelt dynamics and their effects on water resources.

Introduction

Snow is a critical component of the global water cycle, accounting for about 12% of the Earth's land surface and storing a substantial amount of freshwater. Snowmelt contributes to streamflow, groundwater recharge, and irrigation, affecting the availability and quality of water resources for human consumption, agriculture, and industry. However, snowmelt can also cause floods, landslides, and erosion, posing significant challenges to water management and infrastructure development. Therefore, understanding the physical processes of snow formation, accumulation, and melt, as well as their impacts on hydrological systems and human activities, is essential for sustainable water resources management.

This article provides a comprehensive overview of snow hydrology, covering the following topics:

- Physical processes of snow formation, accumulation, and melt

- Impacts of snowmelt on hydrological systems and human activities
- Applications of snow hydrology in various fields
- Current research trends and future directions in snow hydrology

Physical Processes of Snow Formation, Accumulation, and Melt

Snow formation is a complex process that involves the deposition of water vapor onto preexisting ice crystals or other solid particles in the atmosphere, such as dust or pollen. As these ice crystals grow, they become snowflakes, which can take on a variety of shapes and sizes depending on temperature, humidity, and other atmospheric conditions. Snowflakes can also undergo physical transformations during their descent to the ground, such as aggregation and riming, which can affect their properties, such as density and size.

Once snow reaches the ground, it can accumulate and form a snowpack, which is a layer of snow that varies in depth, density, and composition. Snowpack properties are influenced by a range of factors, including temperature, precipitation, wind, and solar radiation. For example, warmer temperatures can lead to snowmelt and compaction, while colder temperatures can promote sublimation and preservation of snow. Additionally, wind can redistribute snow and create snowdrifts, while solar radiation can induce meltwater production and snow metamorphism.

Snowmelt is the process by which snowpacks transform into liquid water and flow downstream. The rate and timing of snowmelt depend on various factors, including temperature, precipitation, solar radiation, wind, and topography. As snowpacks melt, they release water into the ground and streams, influencing water availability, water quality, and ecosystem health. The snowmelt process can also cause hazards such as floods, landslides, and debris flows, which can affect human safety and infrastructure.

Impacts of Snowmelt on Hydrological Systems and Human Activities

Snowmelt has significant impacts on hydrological systems and human activities. In regions where snow is a primary source of water, snowmelt can affect water availability and quality for drinking, irrigation, and industrial use. For example, snowmelt contributes to the annual water supply of many regions in North America, including the western United States and Canada. Changes in snowmelt dynamics due to climate change can alter the timing and magnitude of streamflow, affecting water management and planning.

Snowmelt can also affect water quality by transporting pollutants, such as sediment and nutrients. Furthermore, snowmelt can also affect ecosystems and biodiversity by altering water availability and temperature regimes. Many aquatic species, including fish and amphibians, depend on seasonal snowmelt for their life cycles, and changes in snowmelt timing and magnitude can affect their reproductive success and survival. Snowmelt can also influence vegetation growth and productivity, as well as soil moisture and nutrient availability.

However, snowmelt can also pose hazards to human activities and infrastructure, particularly in mountainous regions. Rapid snowmelt can cause floods and landslides, damaging property and disrupting transportation networks. In somc cases, snowmelt can also lead to drought conditions in areas where the snowpack is a primary source of water, as the meltwater may be insufficient to meet demand during the dry season.

Applications of Snow Hydrology in Various Fields

Snow hydrology has numerous applications in various fields, including water resources management, climate change mitigation, and hazard assessment. Snowmelt modelling and forecasting are critical tools for water managers, providing them with information on the timing and magnitude of streamflow and groundwater recharge. Snowmelt modelling can also inform drought planning and emergency response, as well as support irrigation and hydropower operations.

Furthermore, snow hydrology can inform climate change mitigation strategies by assessing the impacts of changes in snowmelt dynamics on water resources and ecosystems. Understanding the effects of climate change on snowmelt can also inform land-use planning and management, as well as support the development of adaptive strategies.

Snow hydrology is also critical for hazard assessment and management. Snowmelt can cause floods, landslides, and other hazards, and understanding the dynamics of snowmelt can inform hazard mapping and warning systems. Snowmelt monitoring can also support avalanche forecasting and mitigation, reducing the risk of accidents and fatalities in mountainous regions.

Current Research Trends and Future Directions in Snow Hydrology

Current research in snow hydrology is focused on improving our understanding of snowmelt dynamics and their impacts on water resources and ecosystems. Advances in remote sensing and modelling techniques have enabled researchers to better monitor and predict snowmelt processes, as well as assess the impacts of climate change on snowmelt.

One of the major challenges in snow hydrology is the lack of observational data, particularly in remote and mountainous regions. Improving snow monitoring networks and data sharing can help to address this challenge, providing researchers and water managers with more accurate and reliable information on snowmelt dynamics.

Another area of research in snow hydrology is the development of new modelling approaches that incorporate the complex physical and biological processes involved in snow formation, accumulation, and melt. These models can simulate the interactions between snow, vegetation, and soil, as well as the effects of climate change on snowmelt dynamics.

Finally, there is a growing recognition of the importance of integrating indigenous knowledge and perspectives into snow hydrology research and management. Indigenous communities have unique knowledge and practices related to snow and water management, and their input can inform more effective and sustainable water management strategies.

Conclusion

Snow hydrology is a critical component of the global water cycle, influencing water availability, water quality, and ecosystem health. Understanding the physical processes of snow formation, accumulation, and melt, as well as their impacts on hydrological systems and human activities, is essential for sustainable water resources management. Advances in remote sensing and modelling techniques, as well as increased collaboration with indigenous communities, are expected to enhance our understanding of snowmelt dynamics and inform more effective and sustainable water management strategies.

53 - Hydrology of Desert Areas: Understanding Water in Arid Regions

Abstract

Deserts are known for their extreme aridity, making water one of the scarcest and most valuable resources in these areas. Hydrology, the study of water and its movement, is crucial in understanding the distribution, availability, and management of water in desert areas. This article provides an in-depth exploration of the hydrology of desert areas, covering topics such as precipitation, evapotranspiration, groundwater recharge, and surface water flow. The article also discusses the impacts of human activities, such as land use changes and climate change, on the hydrology of desert areas. By improving our understanding of the hydrology of desert areas, we can better manage water resources and ensure sustainable development in these vulnerable regions.

Introduction

Deserts cover around one-third of the Earth's land surface and are characterized by their extreme aridity, with low and unpredictable precipitation levels, high evapotranspiration rates, and limited surface water resources. Water is one of the scarcest and most valuable resources in desert areas, and understanding the hydrology of these regions is crucial for sustainable development and the management of water resources.

Hydrology is the study of water and its movement in the environment. In desert areas, hydrology plays a critical role in understanding the distribution, availability, and management of water resources. In this article, we provide an in-depth exploration of the hydrology of desert areas, covering topics such as precipitation, evapotranspiration, groundwater recharge, and surface water flow. We also discuss the impacts of human activities, such as land use changes and climate change, on the hydrology of desert areas.

Precipitation

Precipitation is the primary source of water in desert areas, and the amount and intensity of precipitation largely determine the availability of water resources in these regions. However, precipitation in desert areas is highly variable, both spatially and temporally. Spatially, precipitation patterns vary depending on the topography, latitude, and prevailing winds in a region. For example, areas close to mountain ranges or coasts may receive more precipitation than areas in the interior of a desert. Temporally, precipitation in desert areas can be highly irregular, with long periods of drought interrupted by short, intense rainfall events.

The variability of precipitation in desert areas presents significant challenges for water management. In some desert areas, such as the Sahara and the Arabian Peninsula, annual precipitation levels are extremely low, with less than 100 mm per year in some areas. In these regions, water scarcity is a significant challenge, and communities often rely on groundwater resources or imported water to meet their needs. In other desert areas, such as the Sonoran Desert in the southwestern United States and northern Mexico, precipitation levels can be relatively high, with up to 500 mm per year in some areas. However, even in these regions, water availability can be highly variable, with periods of drought and water scarcity.

Evapotranspiration

Evapotranspiration is the process by which water is transferred from the land surface to the atmosphere through evaporation and transpiration by plants. In desert areas, high temperatures and low humidity levels result in high evapotranspiration rates, which can exceed precipitation levels. This means that much of the precipitation that falls in desert areas is lost to the atmosphere through evapotranspiration rather than contributing to groundwater recharge or surface water flow.

The high evapotranspiration rates in desert areas also have significant implications for vegetation and ecosystem health. In areas where water is scarce, vegetation may be limited or nonexistent, leading to desertification and land degradation. In some cases, human activities such as agriculture and grazing can exacerbate these processes, leading to further loss of vegetation and soil erosion.

Groundwater Recharge

Groundwater is an essential source of water in desert areas, particularly in regions where surface water resources are limited or nonexistent. Groundwater recharge refers to the process by which water infiltrates into the ground and replenishes underground aquifers. The amount and rate of groundwater recharge in desert areas are influenced by a range of factors, including precipitation levels, soil types, vegetation cover, and land use.

In regions with low precipitation levels, groundwater recharge can be slow and sporadic, with recharge occurring only during rare rainfall events. In contrast, in areas with higher precipitation levels or in areas with more permeable soils, groundwater recharge can be more frequent and significant. However, even in areas with relatively high groundwater recharge rates, over-pumping of aquifers or changes in land use can deplete groundwater resources over time, leading to water scarcity and other environmental impacts.

Surface Water Flow

Surface water resources in desert areas are limited, with most rivers and streams either seasonal or ephemeral, meaning they only flow during and immediately after rainfall events. The irregularity of surface water flow in desert areas presents significant challenges for water management and can lead to conflicts over water resources.

Human activities such as urbanization, agriculture, and mining can also impact surface water resources in desert areas. For example, the construction of dams and reservoirs can alter the natural flow of rivers and streams, while irrigation can divert water away from natural watercourses, impacting downstream users and ecosystems. The impacts of climate change, including changes in precipitation patterns and increased evapotranspiration rates, can also impact surface water resources in desert areas, leading to further water scarcity and environmental degradation.

Impacts of Human Activities

Human activities such as land use changes, urbanization, and agriculture can have significant impacts on the hydrology of desert areas. For example, the conversion of natural vegetation to agriculture or urban areas can reduce groundwater recharge rates, increase evapotranspiration rates, and alter the natural flow of rivers and streams.

In some cases, human activities can also lead to environmental degradation and land degradation, further exacerbating water scarcity issues in desert areas. Mining activities, for example, can lead to the depletion of groundwater resources and the contamination of surface water resources through the release of pollutants.

Climate change is also expected to have significant impacts on the hydrology of desert areas. Changes in precipitation patterns, increased temperatures, and increased evapotranspiration rates are likely to exacerbate water scarcity issues in many desert regions, leading to further environmental degradation and social impacts.

Conclusion

The hydrology of desert areas is complex and highly variable, with water scarcity and limited water resources presenting significant challenges for sustainable development and the management of water resources. Human activities, including land use changes and climate change, are likely to exacerbate these challenges, leading to further environmental degradation and social impacts.

Improving our understanding of the hydrology of desert areas is crucial for sustainable development and the effective management of water resources. This includes the development of innovative water management strategies, such as the use of recycled water and desalination, as well as the conservation and restoration of natural ecosystems and water resources.

By taking a holistic approach to water management in desert areas, we can ensure the sustainable use of water resources and promote the long-term health and resilience of ecosystems and communities in these vulnerable regions.

54 - Understanding the Hydrology of Arid Regions: Challenges and Opportunities for Managing Water Resources

Abstract

Arid regions, which are home to about one-third of the global population, face a significant water scarcity problem. Hydrology plays a crucial role in managing water resources in these regions. This article explores the characteristics of arid climates, the water cycle in arid regions, and the challenges and opportunities for managing water resources in these areas. Arid climates are characterized by low precipitation levels, high evaporation rates, and infertile soil. The water cycle in arid regions is marked by low precipitation levels, high evaporation rates, and limited surface water resources. The challenges of managing water resources in arid regions include water scarcity, groundwater depletion, and water quality issues. The opportunities for managing water resources in arid regions include water conservation, desalination, and water harvesting.

Introduction

Arid regions, which are characterized by low precipitation levels and high evaporation rates, are home to approximately one-third of the world's population. Water scarcity is a major problem in these regions, and managing this scarce resource is crucial for sustainable development. Hydrology, which is the study of water in the Earth's system, plays a critical role in understanding the water cycle and managing water resources. In this article, we will explore the hydrology of arid regions, including the characteristics of arid climates, the water cycle in arid regions, and the challenges and opportunities for managing water resources in these areas.

Characteristics of Arid Climates

Arid climates are characterized by low precipitation levels and high evaporation rates. Precipitation in arid regions is generally less than 250 millimeters per year, and in some areas, it can be less than 50 millimeters per year. The high evaporation rates in arid regions are due to the high temperatures and low humidity levels. The combination of low precipitation and high evaporation rates leads to a low water availability in arid regions.

The temperature in arid regions is highly variable, with hot temperatures during the day and cold temperatures at night. This temperature variability is due to the lack of cloud cover, which allows for the rapid cooling of the Earth's surface at night. The lack of cloud cover also leads to high levels of solar radiation, which can cause water to evaporate quickly.

The soil in arid regions is often dry and infertile, with low organic matter content. The lack of vegetation in these regions is due to the lack of water and the high temperatures, which make it difficult for plants to survive. The lack of vegetation also leads to soil erosion, as there are no roots to hold the soil in place.

The Water Cycle in Arid Regions

The water cycle in arid regions is characterized by low precipitation levels, high evaporation rates, and limited surface water resources. The main sources of water in these regions are groundwater and rainfall. The water cycle in arid regions can be divided into four main components: precipitation, evaporation, infiltration, and runoff.

Precipitation: Precipitation in arid regions is often infrequent and unpredictable. When it does occur, it usually takes the form of short, intense bursts of rain. The amount of precipitation that falls in arid regions is highly variable and can range from less than 50 millimeters per year to over 250 millimeters per year.

Evaporation: Evaporation rates in arid regions are high due to the high temperatures and low humidity levels. The high evaporation rates can lead to a rapid loss of surface water and can make it difficult to maintain surface water resources.

Infiltration: Infiltration is the process by which water penetrates the soil surface and enters the groundwater system. In arid regions, infiltration rates are generally low due to the dry and compacted soil.

Runoff: Runoff is the process by which water flows over the surface of the ground and into streams and rivers. In arid regions, runoff is often limited due to the lack of precipitation and the high evaporation rates. When precipitation does occur, the water can quickly evaporate or be absorbed by the dry soil.

Challenges and Opportunities for Managing Water Resources in Arid Regions

Managing water resources in arid regions is a challenging task due to the limited availability of water and the high demand for water by humans and ecosystems. In this section, we will discuss some of the challenges and opportunities for managing water resources in arid regions.

Challenges

Water Scarcity: Water scarcity is a major challenge in arid regions. With limited precipitation and high evaporation rates, the available water resources are often insufficient to meet the demands of human populations and ecosystems. Climate change is also exacerbating this problem, as it is causing more frequent and intense droughts in arid regions.

Groundwater Depletion: Groundwater is the main source of water in many arid regions, but overuse and mismanagement have led to groundwater depletion. In some areas, the groundwater levels have dropped to the point where wells have gone dry, and the land has subsided due to the collapse of the aquifer.

Water Quality: Water quality is another challenge in arid regions, as the limited water resources are often contaminated with pollutants. Industrial activities, agricultural practices, and human settlements all contribute to water pollution in arid regions, which can have serious health and environmental consequences.

Opportunities

Water Conservation: Water conservation is a key strategy for managing water resources in arid regions. This can include the use of more efficient irrigation systems in agriculture, the promotion of water-saving technologies in households and businesses, and the implementation of water reuse and recycling programs.

Desalination: Desalination is the process of removing salt and other minerals from seawater or brackish water to produce freshwater. In arid regions where freshwater is scarce, desalination can be a valuable source of water. However, desalination is energy-intensive and can be costly, making it a less viable option in some areas.

Water Harvesting: Water harvesting is the collection and storage of rainwater for later use. In arid regions, where precipitation is infrequent but often intense, water harvesting can be a valuable source of water. This can include the use of rainwater harvesting systems on buildings, the construction of small-scale dams and reservoirs, and the development of underground water storage systems.

Conclusion

Managing water resources in arid regions is a critical challenge that requires innovative solutions and careful management. Arid regions are characterized by low precipitation levels, high evaporation rates, and limited surface water resources. The hydrology of these regions is complex, and understanding the water cycle is essential for managing water resources. Despite the challenges associated with water scarcity, groundwater depletion, and water quality, there are also opportunities for managing water resources in arid regions. Water conservation, desalination, and water harvesting are just a few examples of strategies that can be used to address the water scarcity problem. With careful planning and management, it is possible to ensure a sustainable water supply for the populations and ecosystems that depend on it in arid regions.

55 - Hydrology of Coastal Regions: Key Processes, Influencing Factors, and Impacts on Ecosystems and Human Communities

Abstract

Coastal regions are dynamic environments where the interaction between water and land is constantly changing. The hydrology of coastal regions plays a crucial role in shaping the landscape and the diverse ecosystems that inhabit these areas. This article provides an overview of the key hydrological processes that occur in coastal regions, including tides, waves, coastal currents, and groundwater. The article also discusses the factors that influence these processes, such as topography, climate, geology, vegetation, and human activities. The impacts of hydrological changes on coastal ecosystems, including changes in salinity, habitat loss, nutrient imbalances, and erosion, are also explored. Additionally, the article highlights the impacts of hydrological changes on human communities, such as infrastructure damage, reduced freshwater availability, and impacts on human health.

Introduction

Hydrology is the study of water in the earth's system, including the movement, distribution, and quality of water. Coastal regions are areas where the land meets the sea, and water interacts with the land. Hydrology of coastal regions is, therefore, concerned with the study of the hydrological processes that occur in coastal areas. These processes are influenced by a variety of factors, including tides, waves, currents, and the salinity of the water.

Coastal regions are also home to a diverse range of ecosystems that are sensitive to changes in the hydrological regime. For example, wetlands, estuaries, and mangrove forests are important habitats for a variety of flora and fauna. Changes in the hydrological regime can affect the productivity and biodiversity of these ecosystems, as well as the availability of freshwater for human use.

In this article, we will explore the hydrology of coastal regions, including the key hydrological processes, the factors that influence these processes, and the impacts of hydrological changes on coastal ecosystems and human communities.

Key Hydrological Processes

The hydrology of coastal regions is influenced by a variety of physical and biological processes. Some of the key hydrological processes in coastal regions are:

Tides: Tides are the periodic rise and fall of sea level caused by the gravitational forces of the moon and the sun. Tidal cycles are typically 12.4 hours long, with two high tides and two low tides occurring each day. The height of the tides can vary significantly depending on the location and the phase of the moon. Tides have a significant impact on the hydrology of coastal regions. They can cause fluctuations in water levels, which can influence the movement of water and the distribution of sediment. Tidal cycles can also affect the salinity of coastal waters, particularly in estuaries, where freshwater and seawater mix.

Waves: Waves are caused by the movement of wind over the surface of the ocean. They can have a significant impact on the hydrology of coastal regions, particularly in areas where the shoreline is exposed to high-energy wave action. Waves can cause erosion of the shoreline, as well as the transport of sediment and nutrients.

Coastal currents: Coastal currents are the movements of water along the coastline. They are influenced by a variety of factors, including tides, waves, and wind. Coastal currents can transport sediment and nutrients, as well as affect the distribution of water temperature and salinity.

Groundwater: Groundwater is water that is stored beneath the surface of the earth. In coastal regions, groundwater can interact with surface water, particularly in areas where the water table is close to the surface. Groundwater can discharge into coastal waters, affecting the salinity and nutrient content of the water.

Factors Influencing Hydrological Processes

The hydrology of coastal regions is influenced by a variety of factors. Some of the key factors that influence hydrological processes in coastal regions are:

Topography: The topography of coastal regions can influence the movement of water and the distribution of sediment. Areas with steep slopes may experience more erosion and sediment transport than areas with gentle slopes. Topography can also affect the distribution of groundwater and the discharge of groundwater into coastal waters.

Climate: Climate can influence the hydrological regime of coastal regions. Regions with high levels of precipitation may experience more runoff and freshwater input to coastal waters. Climate can also affect the intensity of storms and wave action, which can have a significant impact on the hydrology of coastal regions.

Geology: The geology of coastal regions can influence the movement of water and the distribution of sediment. Areas with porous and permeable rocks may allow for the movement of water through the subsurface, affecting groundwater discharge into coastal waters. The composition of the rocks can also affect the chemical composition of the water.

Vegetation: Vegetation plays an important role in the hydrology of coastal regions. Plants can influence the movement of water and sediment, as well as the distribution of nutrients. Coastal vegetation, such as mangrove forests and salt marshes, can also provide important habitats for a variety of flora and fauna.

Human activities: Human activities, such as land use changes, development, and water withdrawals, can have a significant impact on the hydrology of coastal regions. Land use changes can affect the amount of runoff and sediment transport, while development can alter the topography and drainage patterns of coastal regions. Water withdrawals can reduce the amount of freshwater input to coastal waters, affecting the salinity and nutrient content of the water.

Impacts on Coastal Ecosystems

Changes in the hydrological regime of coastal regions can have significant impacts on coastal ecosystems. Some of the key impacts include:

Changes in salinity: Changes in the salinity of coastal waters can have a significant impact on the productivity and biodiversity of coastal ecosystems. Some species are adapted to specific salinity ranges, and changes in salinity can affect their growth, reproduction, and survival. Seawater intrusion, which is the movement of saltwater into freshwater aquifers, can lead to increased salinity levels in coastal groundwater and surface water. This can affect the availability of freshwater for plants and animals that are adapted to lower salinity levels.

Habitat loss: Changes in the hydrological regime of coastal regions can lead to the loss of important habitats, such as wetlands and estuaries. Seawater intrusion can also lead to the loss of freshwater wetlands, which can have negative impacts on the plants and animals that depend on these habitats. Wetlands provide important ecological functions, including water filtration, nutrient cycling, and habitat provision.

Nutrient imbalances: Changes in the hydrological regime of coastal regions can also lead to imbalances in nutrient levels in coastal waters. For example, seawater intrusion can increase the levels of nutrients, such as nitrogen and phosphorus, in coastal groundwater and surface water. This can lead to eutrophication, which is the excessive growth of algae and other aquatic plants. Eutrophication can lead to oxygen depletion in coastal waters, which can have negative impacts on fish and other aquatic animals.

Erosion: Changes in the hydrological regime of coastal regions can also lead to increased erosion of the shoreline. Seawater intrusion can lead to the loss of coastal vegetation, which can provide important protection against wave action and erosion. Erosion can lead to the loss of valuable land and infrastructure, as well as negative impacts on coastal ecosystems.

It is important to manage the hydrology of coastal regions in a sustainable way to minimize negative impacts on both coastal ecosystems and human communities. This can involve strategies such as reducing water withdrawals, protecting and restoring coastal habitats, and promoting sustainable land use practices.

Impacts on Human Communities

Changes in the hydrology of coastal regions can also have significant impacts on human communities. Some of the key impacts include:

Water availability: Changes in the hydrological regime of coastal regions can affect the availability of freshwater for human use. This can be particularly important in regions where freshwater resources are limited.

Infrastructure damage: Changes in the hydrological regime of coastal regions can lead to damage to infrastructure, such as roads, buildings, and utilities. This can have significant economic impacts on communities.

Increased flooding: Changes in the hydrological regime of coastal regions can lead to increased flooding, which can have significant impacts on human communities. Flooding can damage homes and businesses, as well as affect transportation and emergency services.

Public health: Changes in the hydrological regime of coastal regions can also have impacts on public health. For example, increased nutrient inputs can lead to harmful algal blooms, which can affect the safety of seafood and recreational waters.

Conclusion

The hydrology of coastal regions is a complex and dynamic system that is influenced by a variety of physical, biological, and human factors. Changes in the hydrological regime of coastal regions can have significant impacts on the diverse ecosystems that inhabit these areas, as well as the human communities that rely on them. As the global climate continues to change, understanding the hydrology of coastal regions and its interactions with other environmental systems is crucial for effective management and conservation of these valuable ecosystems.

56 - Seawater Intrusion in Coastal Aquifers: Causes, Effects, and Solutions

Abstract

This article provides an overview of seawater intrusion in coastal aquifers, including its causes, effects, and solutions. Seawater intrusion occurs when saltwater infiltrates freshwater aquifers, leading to a decrease in freshwater availability and quality. This can have significant environmental, economic, and societal effects, including a decrease in biodiversity, a negative impact on agricultural productivity and tourism, and conflicts over water resources. Sustainable groundwater management, land-use planning, seawater intrusion barriers, and desalination are all potential solutions to address seawater intrusion. Overall, addressing seawater intrusion is essential to ensuring the long-term availability of freshwater resources and protecting the environment, economy, and society.

Introduction

The availability of freshwater is crucial for the survival and development of human society. Groundwater has been the primary source of freshwater for human consumption and agriculture worldwide. Coastal regions rely on groundwater extracted from aquifers near the shorelines, as they are often free from pollution and saline intrusion. However, seawater intrusion is a significant threat to these coastal aquifers, as it leads to a decrease in water quality and quantity. Seawater intrusion is a natural process that occurs when saltwater infiltrates into freshwater aquifers due to various factors. In this article, we will discuss the causes, effects, and solutions to seawater intrusion in coastal aquifers.

What is Seawater Intrusion?

Seawater intrusion is a phenomenon where saline water infiltrates into freshwater aquifers, leading to an increase in the salt concentration of groundwater. This process occurs due to the natural gradient between saltwater and freshwater, which is caused by the difference in density between the two types of water. The pressure of freshwater is different than that of saltwater, which creates a gradient between the two. This gradient allows seawater to infiltrate into freshwater aquifers when the pressure of saltwater exceeds that of freshwater.

Causes of Seawater Intrusion

Seawater intrusion can be caused by various factors, including natural and human activities. Some of the primary causes of seawater intrusion are discussed below.

Natural Causes

The natural causes of seawater intrusion include geological processes and sea-level rise. The geological processes that contribute to seawater intrusion include coastal erosion, subsidence, and faults. Coastal erosion occurs when waves and tides remove sediment from the coastline, resulting in the exposure of underlying geological formations. This process can create conduits that allow seawater to infiltrate into freshwater aquifers. Subsidence occurs when the land sinks due to natural geological processes, such as tectonic activity or compaction of sediments. This process can cause the freshwater aquifers to become shallower, allowing seawater to infiltrate more easily. Faults are geological structures that can act as pathways for seawater to infiltrate into freshwater aquifers.

Sea-level rise is another natural factor that contributes to seawater intrusion. As sea levels rise, the freshwater aquifers near the coast become shallower, making them more susceptible to seawater intrusion.

Human Activities

Human activities such as groundwater pumping, land-use changes, and saltwater intrusion barriers can also contribute to seawater intrusion.

Groundwater pumping is a major contributor to seawater intrusion. When groundwater is extracted from aquifers faster than it can be replenished, the pressure of freshwater decreases, allowing seawater to infiltrate into the aquifer. This process is exacerbated in coastal regions where the aquifers are already shallow and the natural gradient between freshwater and seawater is high.

Land-use changes such as urbanization and agriculture can also contribute to seawater intrusion. When land-use changes occur, the amount of rainfall that infiltrates into the ground decreases, leading to a decrease in freshwater recharge. This decrease in freshwater recharge can cause the pressure of freshwater to decrease, allowing seawater to infiltrate into the aquifer.

Saltwater intrusion barriers are structures designed to prevent seawater from infiltrating into freshwater aquifers. While these structures can be effective in certain cases, they can also create unintended consequences. For example, if the barrier is too close to the coastline, it can cause the freshwater aquifer to become stagnant, leading to a decrease in water quality.

Effects of Seawater Intrusion

Seawater intrusion can have significant effects on the environment, economy, and society. Some of the primary effects of seawater intrusion are discussed below.

Environmental Effects

Seawater intrusion can have a significant impact on the environment, particularly on coastal ecosystems. The increase in salt concentration can affect the vegetation and wildlife that depend on freshwater, leading to a decrease in biodiversity. The intrusion of seawater can also lead to an increase in soil salinity, which can negatively impact agricultural productivity and limit the types of crops that can be grown in the affected areas.

Economic Effects

Seawater intrusion can have a significant economic impact, particularly in coastal regions where agriculture and tourism are important industries. The decrease in freshwater availability can lead to a decrease in agricultural productivity, which can have a ripple effect on the local economy. Similarly, the decrease in water quality can negatively impact the tourism industry, as visitors may be deterred by the poor water quality.

Societal Effects

Seawater intrusion can have significant societal effects, particularly in areas where access to freshwater is already limited. The decrease in freshwater availability can lead to conflicts over water resources and exacerbate existing social and economic inequalities. Additionally, the decrease in water quality can lead to health problems for individuals who rely on groundwater as their primary source of drinking water.

Solutions to Seawater Intrusion

Seawater intrusion is a complex issue that requires a multifaceted approach to address. Some of the primary solutions to seawater intrusion are discussed below.

Sustainable Groundwater Management

Sustainable groundwater management is essential to addressing seawater intrusion. This involves managing groundwater resources in a way that ensures their long-term availability while also preventing seawater intrusion. This can be achieved through a combination of measures, such as reducing groundwater pumping rates, increasing the use of alternative water sources, and implementing groundwater recharge projects.

Land-Use Planning

Land-use planning is another important solution to seawater intrusion. This involves managing land use in a way that reduces the impact on groundwater resources. For example, limiting urbanization in areas with shallow aquifers can reduce the demand for groundwater and prevent seawater intrusion. Similarly, implementing sustainable agricultural practices can reduce water use and limit the impact of land-use changes on groundwater resources.

Seawater Intrusion Barriers

Seawater intrusion barriers can be an effective solution to prevent seawater intrusion. These structures can be designed to prevent seawater from infiltrating into freshwater aquifers while also allowing freshwater to flow freely. However, it is important to ensure that these structures are designed and implemented in a way that does not create unintended consequences, such as stagnant water or a decrease in water quality.

Desalination

Desalination is a technology that can be used to treat seawater and make it suitable for human consumption and agricultural use. While desalination can be expensive, it can be an important solution in areas where freshwater resources are limited and seawater intrusion is a significant problem.

Conclusion

Seawater intrusion is a significant threat to coastal aquifers worldwide. It is a complex issue that requires a multifaceted approach to address, including sustainable groundwater management, land-use planning, seawater intrusion barriers, and desalination. Addressing seawater intrusion is essential to ensuring the long-term availability of freshwater resources and protecting the environment, economy, and society.

57 - Understanding the Hydrology of Deltaic Regions: Processes, Impacts, and Management Strategies

Abstract

This article provides an overview of the hydrology of deltaic regions, which are complex and dynamic systems that are shaped by a range of physical and ecological processes. The article describes the factors that influence water flow and sediment transport in deltaic systems, including tides, river discharge, and ocean currents, and discusses the critical role of sediment transport in shaping the structure and morphology of these systems over time. The article also highlights the impacts of human activity on deltaic hydrology, including alterations to land use patterns, which can increase sediment and nutrient inputs, and the construction of dams and levees, which can alter natural flow patterns and lead to increased erosion. Finally, the article outlines a range of management strategies that can be used to mitigate these impacts, including the restoration of natural flow patterns, the management of sediment transport, and the implementation of regulations and voluntary measures to reduce human impacts on these systems.

Introduction

Hydrology is the study of water on Earth, including its distribution, circulation, and properties. Deltaic regions, which are formed at the mouth of a river, are one of the most dynamic and complex areas in terms of hydrology. The combination of river water, tides, and ocean currents create a complex network of water flow and transport that affects both natural and human systems. In this article, we will explore the hydrology of deltaic regions, including the factors that influence water flow, the role of sediment transport, and the impacts of human activity on these systems.

Factors Influencing Water Flow in Deltaic Regions

The flow of water in deltaic regions is influenced by a variety of factors, including tides, river discharge, and ocean currents.

Tides are a major driver of water flow in these regions, as they create a cyclic rise and fall of water levels that can significantly affect the flow of river water. When the tide is high, ocean water is pushed into the delta, creating a backflow of river water that can slow or even reverse the direction of water flow. Conversely, when the tide is low, river water can flow more freely into the ocean, leading to increased discharge and sediment transport.

River discharge is another major factor influencing water flow in deltaic regions. The amount of water flowing through a river varies depending on a range of factors, including precipitation, temperature, and land use. In deltaic regions, high river discharge can lead to increased sediment transport, as more sediment is carried downstream by the fast-moving water. This can have significant impacts on the shape and structure of the delta, as well as on the distribution of plant and animal communities.

Finally, ocean currents also play a role in the hydrology of deltaic regions. These currents can create eddies and other flow patterns that influence the movement of water in the delta. In some cases, ocean currents can even reverse the direction of water flow, leading to unusual or unexpected hydrological conditions.

Sediment Transport in Deltaic Regions

Sediment transport is a critical component of the hydrology of deltaic regions, as it shapes the structure and morphology of these systems over time. Sediment transport can occur in a variety of ways, including through river flow, tidal currents, and wave action. In deltaic regions, sediment transport is often dominated by river flow, which can carry large quantities of sediment downstream during periods of high discharge.

The sediment carried by river water is typically divided into three categories: suspended sediment, bedload sediment, and washload sediment. Suspended sediment refers to fine-grained particles that are carried by the water column, while bedload sediment refers to larger particles that move along the riverbed. Washload sediment refers to particles that are too small to settle on the riverbed, but that are still carried by the water column.

In deltaic regions, sediment transport can be influenced by a range of factors, including the shape of the river channel, the gradient of the river bed, and the presence of vegetation. For example, in areas with a steep gradient, sediment transport may be dominated by bedload sediment, while in areas with a flatter gradient, suspended sediment may be more important. Similarly, the presence of vegetation can reduce the speed of water flow, leading to increased deposition of sediment in certain areas.

Human Impacts on Deltaic Hydrology

Human activities can have a significant impact on the hydrology of deltaic regions, as they can alter the flow of water and sediment transport patterns. One of the most significant impacts of human activity on deltaic hydrology is the construction of dams and levees. Dams can reduce river discharge, leading to decreased sediment transport and altered flow patterns downstream. Levees can also alter the flow of water, as they can restrict the movement of water and sediment and lead to increased erosion in other areas.

Another significant impact of human activity on deltaic hydrology is the alteration of land use patterns. Agricultural activities, for example, can increase the amount of sediment and nutrients carried by river water, leading to increased eutrophication and altered ecological communities. Urban development can also have significant impacts, as it can increase the amount of impervious surface area and lead to increased runoff, which can alter the flow of water and increase the risk of flooding.

Climate change is also having a significant impact on the hydrology of deltaic regions. Rising sea levels and changes in precipitation patterns can alter the flow of water and sediment transport patterns in these systems, leading to increased erosion and altered ecological communities. In addition, increased temperatures can lead to more frequent and severe droughts, which can reduce river discharge and alter the hydrology of the delta.

Management strategies for deltaic hydrology

Effective management of deltaic hydrology requires a holistic approach that takes into account the complex interactions between water flow, sediment transport, and ecological systems. One important strategy is the restoration of natural flow patterns, which can help to restore ecological systems and reduce the risk of flooding. This can be achieved through the removal of dams and levees, the restoration of wetlands and other natural areas, and the implementation of more natural flow regimes.

Another key management strategy is the management of sediment transport. This can be achieved through a range of methods, including the strategic placement of sediment, the construction of sediment traps and dikes, and the use of vegetation to stabilize riverbanks and prevent erosion. In addition, the management of land use patterns can help to reduce sediment and nutrient inputs, which can improve water quality and reduce the risk of eutrophication.

Finally, the management of human impacts on deltaic hydrology requires a combination of regulatory and voluntary measures. This can include the implementation of regulations to limit the construction of levees and dams, the promotion of sustainable land use practices, and the development of education and outreach programs to increase public awareness of the importance of deltaic systems.

Conclusion

Deltaic regions are complex and dynamic systems that are shaped by a range of hydrological processes. The flow of water and sediment in these systems is influenced by a variety of factors, including tides, river discharge, and ocean currents. Sediment transport is a critical component of deltaic hydrology, shaping the structure and morphology of these systems over time. Human activities can have a significant impact on deltaic hydrology, altering the flow of water and sediment transport patterns and leading to increased erosion and altered ecological communities. Effective management of deltaic hydrology requires a holistic approach that takes into account the complex interactions between water flow, sediment transport, and ecological systems. This can include the restoration of natural flow patterns, the management of sediment transport, and the implementation of regulations and voluntary measures to reduce human impacts on these systems.

58 - Techniques for Effective Water Resources Management: Balancing Competing Demands for Sustainability and Resilience

Abstract

Effective water resources management is essential for sustainable development and environmental protection. This article discusses ten key techniques for managing water resources effectively, including integrated water resources management, water conservation, water harvesting, water pricing, water rights and allocation, water quality management, water use efficiency, stakeholder participation, climate change adaptation, and ecosystem-based approaches. These techniques can help to balance the competing demands for water, reduce waste and promote sustainability, and ensure the availability of water resources for future generations. By implementing these techniques, we can ensure that our water resources are managed in a way that is sustainable, equitable, and resilient.

Introduction

Water resources management is a crucial issue in today's world as the demand for water is constantly increasing. With the rapid population growth, urbanization, and climate change, it is becoming more important than ever to manage our water resources effectively. Water resources management refers to the process of planning, developing, and managing water resources to meet the demands of various sectors while ensuring sustainability and environmental protection. Effective water resources management is crucial for the economic, social, and environmental development of a country. In this article, we will discuss some techniques for effective water resources management.

1. Integrated Water Resources Management (IWRM)

Integrated Water Resources Management (IWRM) is a comprehensive approach to water resources management that takes into account the complex interactions between various water users and the environment. It involves the integration of various disciplines, including hydrology, ecology, economics, and social sciences, to develop a holistic approach to water resources management. IWRM focuses on balancing the competing demands for water, including agriculture, industry, domestic use, and environmental needs, while ensuring the sustainability of the water resources and the ecosystems that depend on them. The key principles of IWRM include participation, integration, adaptive management, and sustainability.

2. Water Conservation

Water conservation is the practice of using water efficiently to reduce waste and promote sustainability. Water conservation techniques can be used in various sectors, including agriculture, industry, and domestic use. In agriculture, water conservation techniques include drip irrigation, precision farming, and crop rotation. In industry, water conservation techniques include water recycling, wastewater treatment, and water-efficient technologies. In domestic use, water conservation techniques include fixing leaks, using water-efficient appliances, and reducing water use in landscaping.

3. Water Harvesting

Water harvesting is the process of collecting and storing rainwater for future use. Water harvesting techniques include rooftop rainwater harvesting, surface runoff harvesting, and groundwater recharge. Water harvesting can be used in both rural and urban areas to supplement water supplies and reduce the demand for groundwater and surface water. Water harvesting can also help to reduce the impact of floods and droughts, improve soil moisture, and promote vegetation growth.

4. Water Pricing

Water pricing is an economic tool that can be used to manage water resources effectively. Water pricing involves setting a price for water that reflects its true value, including the cost of production, treatment, distribution, and environmental costs. Water pricing can encourage water users to use water more efficiently and can help to allocate water resources to the most economically valuable uses. Water pricing can also generate revenue that can be used to invest in water infrastructure and conservation measures.

5. Water Rights and Allocation

Water rights and allocation are legal tools that can be used to manage water resources effectively. Water rights are legal entitlements to use water, and they can be allocated to different users based on various criteria, including priority, need, and efficiency. Water allocation involves the distribution of water resources among various users, and it can be done through various mechanisms, including market-based systems, administrative allocation, and negotiated agreements. Effective water rights and allocation systems can ensure that water resources are used efficiently and fairly and can help to resolve conflicts between different water users.

6. Water Quality Management

Water quality management involves the protection and improvement of water quality to ensure that it is suitable for various uses, including drinking, irrigation, and recreation. Water quality management techniques include source water protection, pollution prevention, and treatment technologies. Effective water quality management can help to reduce the impact of water pollution on human health and the environment, and can help to ensure that water resources are available for sustainable use.

7. Water Use Efficiency

Water use efficiency is the measure of the amount of water used to produce a unit of output or service. Water use efficiency techniques include water-efficient technologies, improved management practices, and behavioral changes. Improving water use efficiency can help to reduce the demand for water and increase the productivity of water-intensive industries, such as agriculture and manufacturing. It can also help to reduce the impact of water use on the environment and increase the resilience of water resources in the face of climate change.

8. Stakeholder Participation

Stakeholder participation is the involvement of various stakeholders, including water users, communities, and government agencies, in the planning and management of water resources. Effective stakeholder participation can help to ensure that water resources are managed in a way that reflects the needs and interests of all stakeholders, including those who are traditionally marginalized. Stakeholder participation can also help to build trust and foster collaboration among different water users, which can be crucial in resolving conflicts and achieving sustainable water management outcomes.

9. Climate Change Adaptation

Climate change adaptation involves the development of strategies and measures to manage the impacts of climate change on water resources. Climate change is expected to have significant impacts on water resources, including changes in precipitation patterns, increased frequency and intensity of floods and droughts, and changes in the quality and quantity of water resources. Climate change adaptation measures can include water storage and distribution systems, water efficiency technologies, and ecosystem-based approaches, such as wetland restoration and riverbank stabilization.

10. Ecosystem-Based Approaches

Ecosystem-based approaches involve the integration of ecological processes and functions into water resources management. Ecosystem-based approaches can help to promote the resilience and sustainability of water resources by maintaining and restoring the natural processes that support them. Ecosystem-based approaches can include wetland restoration, riverbank stabilization, and reforestation. Ecosystem-based approaches can also provide additional benefits, such as carbon sequestration, habitat conservation, and recreational opportunities.

Conclusion

Effective water resources management is crucial for the sustainable development of our societies and the protection of our environment. Water resources management techniques can help to balance the competing demands for water, reduce waste and promote sustainability, and ensure the availability of water resources for future generations. The techniques discussed in this article, including integrated water resources management, water conservation, water harvesting, water pricing, water rights and allocation, water quality management, water use efficiency, stakeholder participation, climate change adaptation, and ecosystem-based approaches, are all important tools for effective water resources management. By implementing these techniques, we can ensure that our water resources are managed in a way that is sustainable, equitable, and resilient.

59 - Integrated Water Resources Management (IWRM): A Holistic Approach to Sustainable Water Management

Abstract

This article provides an in-depth overview of Integrated Water Resources Management (IWRM), a holistic approach to managing water resources that takes into account the social, economic, and environmental dimensions of water management. The article explains the principles of IWRM, its benefits, and its challenges. IWRM is based on the principle of integrated planning and management, which involves the integration of all aspects of water management, including water supply, water quality, wastewater management, and water demand management. The article highlights the benefits of IWRM, such as improved water resource management, enhanced stakeholder participation, increased resilience to climate change, improved water quality, improved ecosystem health, and improved social and economic outcomes. It also discusses the challenges that IWRM faces, including lack of political will, lack of resources, institutional fragmentation, limited stakeholder participation, and limited capacity. Overall, the article emphasizes the importance of adopting an integrated and holistic approach to water management to ensure that water resources are managed sustainably and equitably, and contribute to broader social, economic, and environmental goals.

Introduction

Integrated Water Resources Management (IWRM) is a holistic approach to managing water resources, taking into account the social, economic, and environmental dimensions of water management. It is an approach that recognizes the complex and interconnected nature of water resources, and seeks to balance competing demands for water in a way that is sustainable and equitable. IWRM is based on the principle of integrated planning and management, which involves the integration of all aspects of water management, including water supply, water quality, wastewater management, and water demand management.

IWRM is increasingly being recognized as a key approach to addressing the challenges of water management in the 21st century, as water resources become increasingly scarce and the demand for water continues to grow. In this article, we will explore the concept of IWRM in more detail, examining its principles, benefits, and challenges.

The Principles of IWRM

IWRM is based on a number of key principles, which are designed to ensure that water resources are managed in a sustainable and equitable way. These principles include:

Integrated planning and management: IWRM involves the integration of all aspects of water management, including water supply, water quality, wastewater management, and water demand management. This requires a holistic and interdisciplinary approach to water management.

Stakeholder participation: IWRM recognizes the importance of stakeholder participation in water management. This means involving all relevant stakeholders in the planning, implementation, and monitoring of water management activities, including government agencies, local communities, NGOs, and the private sector.

Integrated water management at the river basin level: IWRM is based on the principle of managing water resources at the river basin level, rather than at the national or local level. This involves considering the entire river basin as a unit of management, and taking into account the interrelationships between water resources, land use, and ecosystems.

Balancing competing demands for water: IWRM seeks to balance the competing demands for water, including the needs of agriculture, industry, domestic users, and the environment. This requires a comprehensive understanding of water demand and supply, as well as the social, economic, and environmental impacts of water management decisions.

Adapting to climate change: IWRM recognizes the importance of adapting to the impacts of climate change on water resources. This means developing strategies to cope with changes in water availability, floods, droughts, and sea level rise.

The Benefits of IWRM

IWRM has a number of benefits, both in terms of improving the sustainability and equity of water management, and in terms of contributing to broader social, economic, and environmental goals. Some of the key benefits of IWRM include:

Improved water resource management: IWRM provides a framework for improving the management of water resources, by taking into account the social, economic, and environmental dimensions of water management. This can help to reduce conflicts over water resources, and ensure that water is used in a sustainable and equitable way.

Enhanced stakeholder participation: IWRM recognizes the importance of stakeholder participation in water management. By involving all relevant stakeholders in water management activities, IWRM can help to build trust and cooperation, and ensure that the needs and concerns of all stakeholders are taken into account.

Increased resilience to climate change: IWRM provides a framework for adapting to the impacts of climate change on water resources. By developing strategies to cope with changes in water availability, floods, droughts, and sea level rise, IWRM can help to build resilience to the impacts of climate change.

Improved water quality: IWRM can help to improve water quality by taking a holistic and interdisciplinary approach to water management. By considering the entire river basin as a unit of management, IWRM can help to identify and address sources of pollution and other water quality issues.

Improved ecosystem health: IWRM can help to improve the health of ecosystems by taking into account the interrelationships between water resources, land use, and ecosystems. By managing water resources in a way that is compatible with ecosystem needs, IWRM can help to protect and restore ecosystems.

Improved social and economic outcomes: IWRM can contribute to broader social and economic goals by improving water management. This can include improving access to water for domestic users, supporting agricultural productivity, and promoting economic development.

Challenges of IWRM

While IWRM has many benefits, it also faces a number of challenges in implementation. Some of the key challenges of IWRM include:

Lack of political will: IWRM requires strong political commitment and support to be implemented effectively. In some cases, governments may be reluctant to adopt IWRM due to concerns about the costs or potential conflicts with other policy priorities.

Lack of resources: IWRM requires significant financial and technical resources to be implemented effectively. In some cases, governments may lack the resources needed to support IWRM, or may be unable to secure funding from international donors or other sources.

Institutional fragmentation: Water management is often fragmented across multiple government agencies, which can make it difficult to implement integrated and coordinated approaches to water management. IWRM requires strong institutional coordination and collaboration, which can be challenging to achieve.

Limited stakeholder participation: While stakeholder participation is a key principle of IWRM, it can be difficult to achieve in practice. Stakeholders may have different interests and priorities, which can make it difficult to reach consensus on water management decisions.

Limited capacity: Implementing IWRM requires technical expertise and capacity in a range of areas, including water management, environmental management, and stakeholder engagement. In some cases, governments may lack the capacity needed to implement IWRM effectively.

Conclusion

Integrated Water Resources Management (IWRM) is a holistic approach to managing water resources, taking into account the social, economic, and environmental dimensions of water management. IWRM is based on the principle of integrated planning and management, which involves the integration of all aspects of water management, including water supply, water quality, wastewater management, and water demand management.

IWRM has a number of benefits, including improved water resource management, enhanced stakeholder participation, increased resilience to climate change, improved water quality, improved ecosystem health, and improved social and economic outcomes. However, IWRM also faces a number of challenges, including lack of political will, lack of resources, institutional fragmentation, limited stakeholder participation, and limited capacity.

Despite these challenges, IWRM is increasingly being recognized as a key approach to addressing the challenges of water management in the 21st century. By adopting an integrated and holistic approach to water management, governments and other stakeholders can ensure that water resources are managed in a sustainable and equitable way, and contribute to broader social, economic, and environmental goals.

60 - An Overview of Surface Water Modelling: Applications, Models, Data Requirements, and Challenges

Abstract

Surface water modelling is a process that involves creating mathematical or computer models to simulate the behavior of water in a given area. This article provides an overview of surface water modelling, its applications, types of models, data requirements, and challenges. Surface water modelling has applications in flood management, irrigation, environmental studies, and water resource management. There are two main types of surface water models: physically-based models and empirically-based models. Some of the most commonly used models include HEC-RAS, SWMM, MIKE SHE, SWAT, HSPF, USLE, and SHE. Data requirements for surface water modelling include topographic data, soil properties, vegetation cover, climate data, and hydrological data. Surface water modelling presents challenges such as data availability, computational power, model complexity, calibration and validation, and uncertainty.

Introduction

Surface water modelling is the process of creating mathematical or computer models that simulate the behaviour of water in a given area. Surface water modelling is used in a variety of contexts, including flood management, irrigation, environmental studies, and water resource management.

This article will provide an in-depth overview of surface water modelling, including its applications, types of models, data requirements, and challenges. We will also discuss some of the most commonly used surface water modelling software and tools.

Applications of Surface Water Modelling

Surface water modelling has a wide range of applications, including:

Flood modelling: Surface water modelling is used to simulate the behaviour of rivers and other water bodies during heavy rain or snowmelt, helping to predict potential flood events and identify areas at risk.

Irrigation planning: Surface water modelling is used to simulate the availability of water for irrigation purposes, helping to plan and optimize water use for agriculture.

Environmental studies: Surface water modelling is used to simulate the impact of pollutants on water bodies, helping to understand the impact of human activities on the environment.

Water resource management: Surface water modelling is used to simulate the availability of water resources, helping to plan and manage water resources for drinking, irrigation, and other purposes.

Types of Surface Water Models

There are two main types of surface water models: physically-based models and empirically-based models.

Physically-based models: Physically-based models are based on the physical laws governing the behaviour of water, including the laws of conservation of mass, energy, and momentum. These models use complex equations to simulate the behaviour of water in a given area, taking into account factors such as topography, soil properties, and vegetation cover. Physically-based models are typically more accurate than empirically-based models, but they require a significant amount of data and computational power to run.

Empirically-based models: Empirically-based models are based on statistical relationships between input and output variables. These models use historical data to identify relationships between factors such as rainfall, water flow, and sediment transport. Empirically-based models are simpler and easier to use than physically-based models, but they are generally less accurate and cannot be used to predict behaviour outside the range of historical data.

Surface Water Models

There are several surface water models available, both physically-based and empirically-based. Some of the most commonly used models include:

HEC-RAS: The Hydrologic Engineering Center's River Analysis System (HEC-RAS) is a physically-based model that simulates water flow in rivers and streams. HEC-RAS is widely used for floodplain management, bridge and culvert design, and other applications.

SWMM: The Storm Water Management Model (SWMM) is a physically-based model that simulates the behaviour of urban drainage systems, including stormwater runoff and wastewater flow. SWMM is widely used for stormwater management and water quality studies.

MIKE SHE: The MIKE SHE model is a physically-based model that simulates the movement of water and energy through the soil and vegetation. MIKE SHE is widely used for groundwater recharge studies, water resource management, and environmental impact assessments.

SWAT: The Soil and Water Assessment Tool (SWAT) is a physically-based model that simulates the movement of water, sediment, and nutrients in large watersheds. SWAT is widely used for water resource management, environmental impact assessments, and agricultural planning.

HSPF: The Hydrological Simulation Program - FORTRAN (HSPF) is a physically-based model that simulates the movement of water, sediment, and nutrients in watersheds. HSPF is widely used for water quality studies, environmental impact assessments, and watershed management.

USLE: The Universal Soil Loss Equation (USLE) is an empirically-based model that predicts soil erosion due to water and wind. USLE is widely used for soil conservation planning and land use management.

SHE: The Spatially distributed Hydrologic modelling of Erosion and sedimentation (SHE) model is a physically-based model that simulates the movement of water, sediment, and nutrients in watersheds. SHE is widely used for water quality studies, environmental impact assessments, and soil conservation planning.

These models vary in complexity, data requirements, and the types of applications they are suited for. Choosing the appropriate model for a given application depends on factors such as the scale of the system being modelled, the available data, and the specific research questions being addressed.

Data Requirements for Surface Water Modelling

Surface water modelling requires a significant amount of data, including:

Topographic data: Topographic data is used to create a digital elevation model (DEM) of the area being modelled. This data is typically obtained through LiDAR or satellite imagery.

Soil properties: Soil properties, such as soil type and hydraulic conductivity, are used to simulate the movement of water through the soil.

Vegetation cover: Vegetation cover affects the infiltration of water into the soil and the rate of evapotranspiration.

Climate data: Climate data, such as rainfall and temperature, is used to simulate the input of water into the system.

Hydrological data: Hydrological data, such as streamflow and water levels, is used to calibrate and validate the model.

Challenges in Surface Water Modelling

Surface water modelling presents several challenges, including:

Data availability: Surface water modelling requires a significant amount of data, which may not always be available or of sufficient quality.

Computational power: Physically-based models require a significant amount of computational power to run, which can limit their use in some applications.

Model complexity: Physically-based models are complex and require significant expertise to develop and use effectively.

Calibration and validation: Surface water models must be calibrated and validated using hydrological data, which can be time-consuming and difficult.

Uncertainty: Surface water modelling involves significant uncertainty, due to factors such as data quality, model simplifications, and natural variability in the system

Conclusion

Surface water modelling is a useful tool in various fields, including flood management, irrigation planning, environmental studies, and water resource management. Choosing the appropriate model for a given application depends on factors such as the scale of the system being modelled, the available data, and the specific research questions being addressed. While surface water modelling presents several challenges, its benefits outweigh the difficulties, and with continued advancements in technology and data collection, surface water modelling will continue to play a critical role in the management and understanding of water resources.

61 - Surface Water Modelling Codes and Their Applications in Water Resources Management and Environmental Protection

Abstract

Surface water modelling codes and surface water quality modelling codes are essential tools for managing and protecting water resources. These codes simulate complex hydrological and water quality processes, and can be used to evaluate the impact of human activities and climate change on surface water resources. This article provides an overview of the most commonly used surface water modelling codes and surface water quality modelling codes, their applications in water resources management and environmental protection, and their strengths and limitations. The article emphasizes the importance of careful consideration of input data, model assumptions, and computational resources when using these codes. Despite their limitations, surface water modelling codes and surface water quality modelling codes are invaluable for developing effective water management strategies, protecting aquatic ecosystems and human health, and mitigating the impact of climate change on water resources.

Introduction

Surface water is one of the most important natural resources on earth. It provides numerous benefits to society, including water supply, transportation, recreation, and ecological services. However, the quality of surface water can be impacted by various human activities, such as urbanization, agriculture, and industry, which can lead to water pollution and degradation. In order to manage and protect surface water resources, it is important to use surface water modelling codes and surface water quality modelling codes. In this article, we will discuss these codes in detail, including their applications, strengths, and limitations.

Surface Water Modelling Codes

Surface water modelling codes are computer programs that simulate the flow of water on the surface of the earth. They use mathematical equations to predict the movement of water across the landscape, including rivers, lakes, and wetlands. These codes can be used to analyze and predict the effects of different factors on surface water flow, such as rainfall, land use changes, and climate change. They are also used to develop flood maps, water allocation plans, and water supply assessments.

There are many surface water modelling codes available, each with its own strengths and limitations. Some of the most commonly used codes are:

HEC-RAS (Hydrologic Engineering Center's River Analysis System): HEC-RAS is a widely used software program for hydraulic modelling of rivers and streams. It is used to analyze water levels, flows, and velocities in rivers and streams, as well as to predict the effects of changes in the river channel or surrounding landscape.

MIKE 11: MIKE 11 is a comprehensive hydrodynamic and hydraulic modelling software package for simulating the movement of water in rivers, estuaries, and coastal zones. It can be used to model water quality and sediment transport, as well as to analyze flood risk and water resource management.

SWMM (Storm Water Management Model): SWMM is a dynamic modelling tool for stormwater runoff and wastewater management. It can simulate the quantity and quality of stormwater runoff from urban areas, and can be used to evaluate the effectiveness of different stormwater management practices.

LISFLOOD-FP: LISFLOOD-FP is a hydrodynamic model that simulates the flow of water over floodplains. It can be used to simulate flooding events, as well as to analyze the effects of land use changes and climate change on flood risk.

Delft3D: Delft3D is a three-dimensional modelling tool for simulating the movement of water, sediment, and waves in estuaries, coastal areas, and rivers. It can be used to analyze the effects of human activities on the marine environment, such as dredging and port construction.

Surface Water Quality Modelling Codes

Surface water quality modelling codes are computer programs that simulate the fate and transport of pollutants in surface water. They use mathematical equations to predict how pollutants move and degrade in water bodies, as well as to evaluate the effectiveness of different pollution control strategies. These codes are essential for assessing the impact of human activities on water quality, and for developing strategies to manage and protect water resources.

There are many surface water quality modelling codes available, each with its own strengths and limitations. Some of the most commonly used codes are:

QUAL2K: QUAL2K is a water quality modelling tool that simulates the fate and transport of pollutants in rivers and streams. It can be used to analyze the impact of point and non-point source pollution on water quality, as well as to evaluate the effectiveness of pollution control measures.

WASP (Water Quality Analysis Simulation Program): WASP is a dynamic modelling tool for simulating the transport and fate of pollutants in surface water. It can be used to evaluate the impact of various pollutants on water quality, as well as to develop management strategies to protect water resources.

CE-QUAL-W2: CE-QUAL-W2 is a water quality modelling code that simulates the physical, chemical, and biological processes in lakes and reservoirs. It can be used to evaluate the impact of nutrient loading, algae growth, and sedimentation on water quality, as well as to develop management strategies for improving water quality.

QUAL2E: QUAL2E is a water quality modelling code that simulates the fate and transport of pollutants in estuaries and coastal waters. It can be used to evaluate the impact of pollutants on water quality, as well as to develop management strategies for protecting aquatic ecosystems and human health.

SWAT (Soil and Water Assessment Tool): SWAT is a comprehensive modelling tool for assessing the impact of land use and management practices on water quality. It can simulate the transport of sediment, nutrients, and pesticides in surface water, as well as to evaluate the effectiveness of different management practices in reducing pollutant loading.

Applications of Modelling Codes for Surface Water and Surface Water Quality

Surface water modelling codes and surface water quality modelling codes have many applications in water resources management and environmental protection. Some of the most important applications are:

Flood mapping and flood risk assessment: Surface water modelling codes can be used to develop flood maps and to evaluate the risk of flooding in different areas. This information is essential for developing flood warning systems, emergency response plans, and flood management strategies.

Water allocation and management: Surface water modelling codes can be used to evaluate water availability and to develop water allocation plans for different uses, such as irrigation, drinking water supply, and industrial use. These codes can also be used to simulate the impact of water management strategies, such as water conservation measures and water pricing policies.

Environmental impact assessment: Surface water quality modelling codes can be used to evaluate the impact of human activities, such as agriculture, industry, and urbanization, on water quality and aquatic ecosystems. This information is essential for developing effective pollution control strategies and for protecting aquatic ecosystems and human health.

Climate change impact assessment: Surface water modelling codes can be used to simulate the impact of climate change on surface water resources, such as changes in rainfall patterns, temperature, and snowmelt. This information is essential for developing adaptation strategies and for mitigating the impact of climate change on water resources.

Water quality monitoring and management: Surface water quality modelling codes can be used to monitor water quality and to evaluate the effectiveness of pollution control measures. This information is essential for developing water quality management plans and for protecting human health and the environment.

Strengths and Limitations of Modelling Codes for Surface Water and Surface Water Quality

Surface water modelling codes and surface water quality modelling codes have many strengths and limitations, which should be considered when selecting a code for a particular application. Some of the most important strengths and limitations are:

Strengths of surface water modelling codes:

- Can simulate complex hydrological processes and interactions between surface water and groundwater.
- Can be used to evaluate the impact of different management strategies on surface water resources.
- Can be used to develop flood maps and to evaluate flood risk in different areas.
- Can be used to simulate the impact of climate change on surface water resources.

Limitations of surface water modelling codes:

- Require detailed input data and calibration to accurately simulate surface water processes.
- May be limited by computational resources and time required to run simulations.
- May be subject to uncertainty and error, particularly in predicting extreme events and rare phenomena.
- May require specialized expertise to use and interpret results.

Strengths of surface water quality modelling codes:

- Can simulate the fate and transport of pollutants in surface water and evaluate the impact of different pollution control strategies.
- Can be used to evaluate the impact of human activities on water quality and aquatic ecosystems.
- Can be used to develop water quality management plans and to monitor the effectiveness of pollution control measures.
- Can be used to evaluate the impact of climate change on water quality.

Limitations of surface water quality modelling codes:

- Require detailed input data on pollutant sources, concentrations, and loading rates.
- May be limited by uncertainty and variability in pollutant fate and transport processes.
- May be subject to error due to incomplete or inaccurate data on pollutant sources and concentrations.
- May require specialized expertise to use and interpret results.

Conclusion

Surface water modelling codes and surface water quality modelling codes are essential tools for managing and protecting water resources. These codes can simulate complex hydrological and water quality processes, and can be used to evaluate the impact of human activities and climate change on surface water resources. However, these codes also have limitations, and require careful consideration of input data, model assumptions, and computational resources. Despite these limitations, surface water modelling codes and surface water quality modelling codes are invaluable for developing effective water management strategies, protecting aquatic ecosystems and human health, and mitigating the impact of climate change on water resources.

62 - Tips for Conducting Accurate and Reliable Surface Water Modelling Studies

Abstract

Surface water modelling studies are crucial tools for assessing surface water resources and informing water resource management decisions. To conduct accurate and reliable surface water modelling studies, it is essential to follow certain tips. These tips include defining study objectives, collecting high-quality data, selecting an appropriate model, calibrating and validating the model, and communicating results effectively. Defining clear study objectives ensures that the data collected and model selected align with the study's goals. Collecting comprehensive and representative data ensures that the model accurately represents the study area. Selecting an appropriate model and calibrating and validating it ensures the model produces reliable results. Finally, communicating results effectively ensures that stakeholders, decision-makers, and the public understand the study's implications and can make informed decisions about surface water management. By following these tips, surface water modelling studies can produce accurate and reliable results to inform water resource management decisions.

Introduction

Surface water modelling studies are important tools for assessing the quality and quantity of surface water resources. These studies provide valuable information for water resource managers, planners, and decision-makers in determining how best to manage and allocate water resources. Surface water modelling studies are essential for understanding the impact of land use change, climate change, and other factors on surface water quality and quantity.

In this article, we will discuss some tips for conducting surface water modelling studies. These tips include defining study objectives, collecting data, selecting an appropriate model, calibrating and validating the model, and communicating results effectively.

Define Study Objectives

The first step in conducting a surface water modelling study is to define the study objectives. Study objectives will guide the selection of data sources, model parameters, and the overall scope of the study. Study objectives should be specific, measurable, achievable, relevant, and time-bound (SMART).

For example, a study objective might be to model the impact of land use change on surface water quality in a specific watershed. The objective is specific because it focuses on a particular type of change, measurable because it can be quantified, achievable because it is within the scope of the study, relevant because it is related to surface water quality, and time-bound because it has a specific timeframe.

Collect Data

The second step in conducting a surface water modelling study is to collect data. Data collection is critical because the accuracy and reliability of the model depend on the quality of the data used. The data required for a surface water modelling study will depend on the study objectives and the model selected. Generally, data will be required for the following:

- Hydrology: precipitation, temperature, evapotranspiration, streamflow, etc.
- Topography: elevation, slope, aspect, etc.
- Land use/cover: vegetation, impervious surfaces, etc.
- Soils: soil texture, permeability, infiltration, etc.
- Water quality: nutrients, metals, bacteria, etc.

Data can be collected from various sources, such as satellite imagery, aerial photography, ground surveys, and existing databases. The data should be representative of the study area and as spatially and temporally comprehensive as possible.

Select an Appropriate Model

The third step in conducting a surface water modelling study is to select an appropriate model. Surface water modelling models vary in complexity, and the model selected should be appropriate for the study objectives and data available. Some commonly used surface water models include:

- SWAT (Soil and Water Assessment Tool)
- HEC-HMS (Hydrologic Engineering Center - Hydrologic Modelling System)
- MIKE SHE (Modular Integrated Hydrologic Modelling System)

Each model has its strengths and weaknesses and is designed for a specific purpose. It is essential to understand the capabilities and limitations of each model before selecting one for a particular study.

Calibrate and Validate the Model

The fourth step in conducting a surface water modelling study is to calibrate and validate the model. Calibration and validation are essential steps in ensuring the accuracy and reliability of the model.

Calibration involves adjusting the model parameters to simulate the observed hydrologic conditions accurately. Calibration typically involves comparing simulated results to observed data, and adjusting model parameters until the simulated results match the observed data.

Validation involves testing the model's ability to predict hydrologic conditions that were not used in the calibration process. Validation typically involves comparing simulated results to independent data sets.

Communicate Results Effectively

The final step in conducting a surface water modelling study is to communicate results effectively. The results of the study should be communicated in a way that is clear and understandable to stakeholders, decision-makers, and the public. Effective communication of study results involves the following:

- Use of clear and concise language
- Use of visual aids, such as graphs, charts, and maps
- Providing context and explaining technical terms
- Presenting the limitations and uncertainties of the study
- Providing recommendations and implications for decision-making

It is essential to tailor the communication of study results to the target audience. For example, technical reports may be appropriate for scientists and engineers, while summaries and infographics may be more suitable for the general public.

Conclusion

Surface water modelling studies are essential tools for assessing the quality and quantity of surface water resources. The tips outlined in this article, including defining study objectives, collecting data, selecting an appropriate model, calibrating and validating the model, and communicating results effectively, can help ensure the accuracy and reliability of these studies. By following these tips, water resource managers, planners, and decision-makers can make informed decisions about the management and allocation of surface water resources.

63 - Understanding Rainfall-Runoff Modelling: Applications, Types, and Limitations

Abstract

This article provides an overview of rainfall-runoff modelling, a crucial tool for simulating hydrological processes in catchments. The article discusses the different types of models, including conceptual, semi-distributed, and distributed models, and their applications in water resources management, flood forecasting, and water quality management. The limitations of rainfall-runoff models are also discussed, including data requirements, model complexity, calibration and validation, and uncertainty. Overall, the article highlights the importance of rainfall-runoff modelling in understanding the hydrological processes in a catchment and developing sustainable water resource management strategies.

Introduction

Rainfall-runoff modelling is the process of estimating the amount of water that runs off a catchment or watershed after a rainfall event. The process involves simulating the hydrological response of a catchment to a given rainfall event, taking into account the physical characteristics of the catchment, such as its topography, land use, and soil properties. This modelling approach is essential for the proper management of water resources, flood forecasting, and water quality management.

There are different methods of rainfall-runoff modelling, ranging from simple empirical models to complex physically-based models. In this article, we will discuss the different types of rainfall-runoff models and their applications.

Types of Rainfall-Runoff Models

Rainfall-runoff models can be broadly classified into two types: empirical models and physically-based models.

Empirical Models

Empirical models are simple and easy to use models that do not require detailed information on the physical characteristics of the catchment. They are based on statistical relationships between rainfall and runoff, and are often used when data is limited or unavailable. These models are widely used in hydrology, especially for small catchments, where the physical processes of the catchment are less important. Some examples of empirical models include the Unit Hydrograph (UH) and the Rational Method.

Unit Hydrograph (UH)

The Unit Hydrograph (UH) is a widely used empirical model for rainfall-runoff modelling. It is based on the concept that the runoff from a rainfall event can be represented by a hydrograph that is proportional to the rainfall excess. The UH assumes that the runoff from a given catchment is a function of the rainfall input and the time it takes for the water to flow through the catchment.

The UH can be derived by convoluting the rainfall excess with a response function that describes the time-varying response of the catchment. The UH is typically assumed to be a triangular function with a peak at the time of concentration, which is the time it takes for the water to flow from the most distant point in the catchment to the outlet.

The UH can be used to estimate the runoff from a rainfall event by multiplying the rainfall excess by the UH and then summing over all the time steps. The UH is often used in conjunction with the SCS-CN method for estimating the rainfall excess.

Rational Method

The Rational Method is a widely used empirical model for estimating peak flows in small urban catchments. The method is based on the assumption that the peak flow from a catchment is proportional to the rainfall intensity and the catchment area.

The method can be expressed as follows:

$$Q = CIA$$

Where Q is the peak flow rate (m^3/s), C is a runoff coefficient that takes into account the imperviousness of the catchment, I is the rainfall intensity (mm/hr), and A is the catchment area (ha).

The Rational Method is a simple and easy to use method, but it has some limitations. It assumes that the rainfall intensity is constant over the catchment, which is often not the case in real-world situations. It also does not take into account the time-varying response of the catchment.

Physically-Based Models

Physically-based models are more complex and require detailed information on the physical characteristics of the catchment. They are based on the principles of conservation of mass and energy, and simulate the physical processes that govern the hydrological response of a catchment to a rainfall event. These models are often used for larger catchments, where the physical processes are more important.

Some examples of physically-based models include the Soil and Water Assessment Tool (SWAT) and the Hydrological Simulation Program-FORTRAN (HSPF).

Soil and Water Assessment Tool (SWAT)

The Soil and Water Assessment Tool (SWAT) is a physically-based model that simulates the hydrological processes in a catchment. The model consists of different components that represent the physical processes of the catchment, such as land use, soil properties, and vegetation. The model also includes a weather generator that generates synthetic rainfall data for the catchment.

The SWAT model uses a semi-distributed approach, where the catchment is divided into sub-basins that are further divided into hydrological response units (HRUs). The HRUs represent homogeneous areas of land use and soil properties within the sub-basin.

The model simulates the hydrological processes in each HRU separately, and then aggregates the results to obtain the overall response of the catchment. The model includes different algorithms for simulating the different hydrological processes, such as infiltration, surface runoff, and groundwater recharge.

Hydrological Simulation Program-FORTRAN (HSPF)

The Hydrological Simulation Program-FORTRAN (HSPF) is a comprehensive and complex physically-based model for simulating the hydrological processes in a catchment. The model uses a continuous simulation approach, where the hydrological processes are simulated at a fine temporal resolution, typically on an hourly or sub-hourly basis.

The HSPF model includes different components that represent the physical processes of the catchment, such as land use, soil properties, vegetation, and channel routing. The model also includes a weather generator that generates synthetic rainfall data for the catchment.

The HSPF model uses a hierarchical approach, where the catchment is divided into sub-basins that are further divided into reaches. The reaches represent the channels within the catchment. The model simulates the hydrological processes in each reach separately, and then aggregates the results to obtain the overall response of the catchment.

The model includes different algorithms for simulating the different hydrological processes, such as infiltration, surface runoff, groundwater recharge, and channel routing. The model also includes a water quality component that simulates the transport and fate of different pollutants in the catchment.

Applications of Rainfall-Runoff Models

Rainfall-runoff models are used in a wide range of applications, including water resources management, flood forecasting, and water quality management.

Water Resources Management

Rainfall-runoff models are essential for water resources management, as they provide information on the amount and timing of water availability in a catchment. The models can be used to estimate the water balance of a catchment, which is the difference between the water input and output over a given period of time.

The water balance can be used to determine the amount of water that is available for different uses, such as irrigation, domestic use, and industrial use. The models can also be used to simulate the effects of different land use scenarios on the water balance of a catchment, and to assess the impacts of climate change on water availability.

Flood Forecasting

Rainfall-runoff models are essential for flood forecasting, as they provide information on the timing and magnitude of floods in a catchment. The models can be used to simulate the hydrological response of a catchment to a rainfall event, and to predict the resulting flood peak and volume.

The models can also be used to simulate the effects of different flood mitigation measures, such as the construction of detention basins and the implementation of land use regulations. The models can also be used to assess the flood risk in a catchment and to develop flood warning systems.

Water Quality Management

Rainfall-runoff models are essential for water quality management, as they provide information on the transport and fate of pollutants in a catchment. The models can be used to simulate the effects of different land use scenarios on water quality, and to assess the impacts of different pollution sources on the water quality of a catchment.

The models can also be used to develop management strategies to reduce the impacts of pollution on water quality. For example, the models can be used to assess the effectiveness of different agricultural practices in reducing nutrient and sediment runoff, and to evaluate the impacts of wastewater treatment plants on receiving waters.

Limitations of Rainfall-Runoff Models

Despite their usefulness, rainfall-runoff models have some limitations that should be considered when using them for hydrological simulations.

Data Requirements

Rainfall-runoff models require a large amount of data to calibrate and validate the model parameters. The data requirements include meteorological data, streamflow data, topographic data, land use data, soil data, and vegetation data. In many cases, the required data are not available or are of poor quality, which can affect the accuracy of the model simulations.

Model Complexity

Rainfall-runoff models can be very complex and require a high level of expertise to set up and run the models. The models require a detailed understanding of the physical processes of the catchment, as well as the mathematical and computational skills to implement the models.

Calibration and Validation

Rainfall-runoff models require calibration and validation to ensure that the model simulations are accurate and reliable. Calibration involves adjusting the model parameters to fit the observed data, while validation involves testing the model with independent data to assess its performance. Calibration and validation can be time-consuming and difficult, and require a good understanding of the physical processes of the catchment.

Uncertainty

Rainfall-runoff models are subject to uncertainties, which can affect the accuracy and reliability of the model simulations. The uncertainties can arise from different sources, such as errors in the input data, errors in the model structure, and errors in the model parameters. The uncertainties can also arise from the limitations of the models themselves, such as their inability to simulate all of the physical processes of the catchment.

Conclusion

Rainfall-runoff models are essential tools for simulating the hydrological processes in a catchment. The models can be used for a wide range of applications, including water resources management, flood forecasting, and water quality management.

The different types of rainfall-runoff models have different advantages and disadvantages, and the choice of model depends on the specific requirements of the application.

Despite their usefulness, rainfall-runoff models have some limitations that should be considered when using them for hydrological simulations. The limitations include the data requirements, model complexity, calibration and validation, and uncertainty.

Overall, rainfall-runoff models are valuable tools for understanding the hydrological processes in a catchment and for developing management strategies to ensure sustainable use of water resources.

64 - Modelling Snow Melt: Techniques and Applications

Abstract

Snow melt is an important process that affects various aspects of the environment, including hydrology, ecology, and climate. Modelling snow melt is crucial for predicting water availability, flood risk, and ecosystem health. This article provides an overview of snow melt modelling techniques and their applications. We begin by discussing the physical processes involved in snow melt and the factors that affect it. We then review the different types of snow melt models, including degree-day models, energy balance models, and physically-based models. We also discuss the input data required for snow melt modelling, such as snow depth, air temperature, and solar radiation. Next, we examine the applications of snow melt models, including water resource management, flood forecasting, and climate change studies. Finally, we discuss some of the challenges and limitations of snow melt modelling, such as data uncertainty, parameterization, and model validation. This article aims to provide a comprehensive overview of snow melt modelling that will be useful for researchers, practitioners, and policymakers.

Introduction

Snow is a vital component of the global water cycle, and it is estimated that about one-sixth of the world's population depends on snowmelt for their water supply (Barnett et al., 2005). Snow accumulation and melt also have important impacts on the environment, including hydrology, ecology, and climate. The amount and timing of snowmelt can affect the water balance of watersheds, streamflow, and groundwater recharge, and can also impact the health of aquatic ecosystems. Moreover, snow cover and albedo play important roles in regulating the Earth's energy balance and influencing regional and global climate.

Modelling snowmelt is essential for understanding and predicting these impacts. Snowmelt models use input data such as snow depth, air temperature, and solar radiation to estimate the rate and timing of snowmelt. These models can be used for a range of applications, including water resource management, flood forecasting, and climate change studies. In this article, we will review the physical processes involved in snowmelt, the different types of snowmelt models, the input data required for snowmelt modelling, and the applications and limitations of these models.

Physical Processes of Snow Melt

Snowmelt is a complex process that depends on various physical and environmental factors. The two main processes involved in snowmelt are energy transfer and water transport. Energy transfer refers to the transfer of heat from the atmosphere to the snow surface, which can cause snow to melt. Water transport refers to the movement of water within the snowpack, which can also contribute to snowmelt.

Energy Transfer

The energy transfer process involves three main components: solar radiation, sensible heat, and latent heat. Solar radiation is the primary source of energy that drives snowmelt. The amount of solar radiation that reaches the snow surface depends on several factors, including the sun's angle of incidence, cloud cover, and snow albedo. Snow albedo is the reflectivity of the snow surface and is influenced by snow age, grain size, and impurities such as dust and soot. Darker snow surfaces absorb more solar radiation and melt faster than lighter surfaces.

Sensible heat refers to the transfer of heat from the atmosphere to the snow surface due to differences in temperature. When the air temperature is warmer than the snow surface, heat is transferred from the air to the snow, causing it to melt. Conversely, when the air temperature is colder than the snow surface, heat is transferred from the snow to the air, causing it to cool.

Latent heat refers to the transfer of heat due to the phase change of water. When snow melts, the heat energy required to change the solid snow into liquid water is called the latent heat of fusion. The opposite process occurs when the temperature falls below freezing point, and the snow refreezes, forming a snowpack that can accumulate over time.

Water Transport

Water transport within the snowpack can also contribute to snowmelt. Snow is a porous material that can hold water, and the amount of water that snow can hold depends on its density and temperature. When the snowpack is heated by solar radiation or air temperature, the water within the snowpack can become liquid and flow through the snow layers, contributing to snowmelt. This process is called infiltration-excess overland flow.

Another important process related to water transport is the refreezing of water within the snowpack. When the temperature drops below freezing, the water within the snowpack can freeze, creating ice layers that can slow down or stop snowmelt. This process is important in regions where snowmelt occurs over a long period, as the refreezing of water can delay the onset of spring runoff.

Snow Melt Models

There are several types of snow melt models that use different approaches to estimate the rate and timing of snowmelt. These models can be divided into three main categories: degree-day models, energy balance models, and physically-based models.

Degree-Day Models

Degree-day models are the simplest type of snow melt models and are based on the assumption that the rate of snowmelt is proportional to the degree-days above a certain threshold temperature. The threshold temperature is usually set at 0°C, as this is the temperature at which snow starts to melt. Degree-day models use daily average air temperature data and a snowmelt coefficient to estimate the amount of snowmelt per day. The snowmelt coefficient is a parameter that represents the sensitivity of snowmelt to air temperature and is usually calibrated using observed snowmelt data.

Degree-day models are easy to implement and require only a few input variables, making them suitable for applications where data is limited. However, degree-day models are less accurate than other types of models, especially in regions with complex terrain, where air temperature can vary significantly over short distances.

Energy Balance Models

Energy balance models are more complex than degree-day models and take into account the energy transfer processes involved in snowmelt. Energy balance models use input variables such as air temperature, solar radiation, wind speed, and humidity to calculate the net energy available for snowmelt. The net energy is then used to estimate the amount of snowmelt per day.

Energy balance models can be divided into two types: single-layer models and multi-layer models. Single-layer models assume that the snowpack is homogeneous and that all the snow layers receive the same amount of energy. Multi-layer models, on the other hand, take into account the vertical distribution of snow layers and the differences in energy transfer between them.

Energy balance models are more accurate than degree-day models and can be used in regions with complex terrain and variable snow conditions. However, energy balance models require more input variables and are more computationally intensive, making them more difficult to implement and calibrate.

Physically-based Models

Physically-based models are the most complex type of snow melt models and simulate the physical processes involved in snowmelt using mathematical equations. Physically-based models take into account the energy transfer processes, water transport processes, and the properties of the snowpack, such as density and temperature.

Physically-based models can be divided into two types: lumped models and distributed models. Lumped models simulate the snowpack as a single entity and assume that all the snow layers have the same properties. Distributed models, on the other hand, simulate the snowpack as a collection of interconnected cells, each with its own properties and energy balance.

Physically-based models are the most accurate type of snow melt models and can be used in regions with complex terrain, variable snow conditions, and different snowpack properties. However, physically-based models require a large amount of input data and are computationally intensive, making them more difficult to implement and calibrate.

In addition to the types of models, there are also different approaches to modelling snowmelt. These include deterministic models and stochastic models. Deterministic models assume that the input variables are known with certainty and that the model output is deterministic. Stochastic models, on the other hand, take into account the uncertainty in the input variables and model output using probability distributions.

Software Programs

There are several software programs available for snow melt modelling, ranging from simple degree-day models to complex physically-based models. Some of the commonly used snow melt modelling software programs are:

Snowmelt: Snowmelt is a software program developed by the USDA Agricultural Research Service for simulating snowmelt runoff from mountainous watersheds. It includes several snowmelt models, such as the degree-day model, temperature-index model, and energy-balance model. Snowmelt also includes several hydrological models, such as the Soil and Water Assessment Tool (SWAT) and the Hydrological Simulation Program-Fortran (HSPF).

SnowModel: SnowModel is a software program developed by the USDA Forest Service for simulating snow accumulation and melt in mountainous watersheds. It includes several snowmelt models, such as the degree-day model, temperature-index model, and energy-balance model. SnowModel also includes several modules for simulating snowpack evolution, snow density, and snow albedo.

iSnobal: iSnobal is a physically-based snowmelt model developed by the USDA Agricultural Research Service. It uses a detailed energy balance approach to simulate snow accumulation and melt. iSnobal also includes several modules for simulating snow density, snow albedo, and snow cover fraction.

Snowy Range Hydrologic Simulation Program (SRHSP): SRHSP is a software program developed by the University of Wyoming for simulating snowmelt runoff from mountainous watersheds. It includes several snowmelt models, such as the degree-day model, temperature-index model, and energy-balance model. SRHSP also includes several hydrological models, such as the Snowmelt Runoff Model (SRM) and the Kinematic Wave Model (KWM).

TOPKAPI-ETH: TOPKAPI-ETH is a distributed hydrological model developed by the Swiss Federal Institute of Technology (ETH) for simulating snowmelt runoff from mountainous watersheds. It uses a physically-based approach to simulate snow accumulation and melt, and includes several modules for simulating snow cover fraction, snow density, and snow albedo.

WATFLOOD: WATFLOOD is a software program developed by the University of Waterloo for simulating snowmelt runoff from mountainous watersheds. It includes several snowmelt models, such as the degree-day model, temperature-index model, and energy-balance model. WATFLOOD also includes several hydrological models, such as the Soil and Water Assessment Tool (SWAT) and the Hydrological Simulation Program-Fortran (HSPF).

These software programs are designed for different types of users, from researchers to water resource managers, and offer different levels of complexity and sophistication. The choice of software program depends on the user's needs and the available data and resources.

Input Data Required for Snow Melt Modelling

Snow melt modelling requires several input data to estimate the rate and timing of snowmelt. The most important input data include snow depth, air temperature, and solar radiation. Other relevant data may include wind speed, humidity, and precipitation.

Snow Depth

Snow depth is a critical input variable for snow melt models as it determines the amount of snow that is available to melt. Snow depth can be measured using snow gauges or remote sensing techniques, such as satellite or airborne sensors. Snow depth measurements should be representative of the entire snowpack and should be taken regularly throughout the winter season to capture changes in the snowpack.

Air Temperature

Air temperature is another critical input variable for snow melt models as it determines the rate of energy transfer from the atmosphere to the snowpack. Air temperature data can be obtained from weather stations, which may be located within or near the study area. Alternatively, gridded meteorological data can be used to interpolate air temperature data across the study area.

Solar Radiation

Solar radiation is the primary source of energy that drives snowmelt, making it a crucial input variable for snow melt models. Solar radiation data can be obtained from ground-based measurements, such as pyranometers, or from satellite-based sensors, such as the Moderate Resolution Imaging Spectroradiometer (MODIS). Solar radiation data should be representative of the entire study area and should be available at a temporal resolution that matches the time step of the snow melt model.

Wind Speed and Humidity

Wind speed and humidity can also affect the rate of snow melt by influencing the transfer of energy from the atmosphere to the snowpack. Wind speed can increase the rate of energy transfer by enhancing the turbulent exchange of heat between the atmosphere and the snowpack. Humidity can affect snow albedo by altering the surface energy balance. These variables can be obtained from weather stations or interpolated from gridded meteorological data.

Precipitation

Precipitation can affect the snowpack by adding to the snow depth or by altering the properties of the snowpack, such as its density or grain size. Precipitation data can be obtained from weather stations or interpolated from gridded meteorological data. Precipitation data should be representative of the entire study area and should be available at a temporal resolution that matches the time step of the snow melt model.

Snow Melt Model Applications

Snow melt models are used in a variety of applications, including hydrological forecasting, water resource management, climate change impact assessment, and avalanche forecasting. Hydrological forecasting involves predicting the timing and magnitude of snowmelt runoff, which is important for flood control and water supply management. Snow melt models are also used to estimate the water balance of mountainous watersheds, which is important for water resource management.

Climate change impact assessment involves predicting the effects of climate change on snowpack dynamics and snowmelt runoff. Snow melt models can be used to simulate different climate scenarios and predict the effects of temperature and precipitation changes on snowmelt. Avalanche forecasting involves predicting the stability of snow layers and the likelihood of avalanches. Snow melt models can be used to simulate the temperature and water content of snow layers and predict the potential for avalanche formation.

Snow Melt Model Evaluation

The accuracy of snow melt models depends on the quality and availability of input data, the model structure and parameters, and the validation method used. Model validation is an important step in model development and involves comparing model outputs with observed data.

Validation methods include statistical measures such as the coefficient of determination (R2), root mean square error (RMSE), and Nash-Sutcliffe efficiency (NSE). These measures are used to evaluate the accuracy of model predictions and compare different models. The choice of validation method depends on the type of model and the available data.

Challenges and Limitations of Snow Melt Modelling

Despite the usefulness of snow melt models, there are several challenges and limitations associated with them.

Data Uncertainty

Snow melt models require input data, such as air temperature, solar radiation, and precipitation, which are subject to uncertainty. Errors in these input data can propagate through the model, leading to errors in the output. Moreover, input data may not be available for certain areas or may be limited in their temporal or spatial resolution, further increasing the uncertainty of the model output.

Parameterization

Snow melt models also require the specification of several parameters, such as the snowmelt coefficient in degree-day models or the snow albedo in energy balance models. These parameters may vary spatially and temporally and may be difficult to estimate accurately. Moreover, different parameter values may lead to different model output, making it challenging to select appropriate parameter values for a given application.

Model Validation

Finally, snow melt models need to be validated using observed snow melt data to ensure that they are accurate and reliable. However, obtaining accurate snow melt data can be challenging, particularly in remote or inaccessible areas. Moreover, snow melt models may not capture all the physical processes involved in snowmelt, leading to discrepancies between model output and observed data.

Conclusion

Snow melt modelling is an important tool for predicting the timing and magnitude of snowmelt runoff, which is important for water resource management, flood control, and climate change impact assessment. There are different types of snow melt models, including degree-day models, energy balance models, and physically-based models, which use different approaches to estimate the rate and timing of snowmelt. The accuracy of snow melt models depends on the quality and availability of input data, the model structure and parameters, and the validation method used. Snow melt models are used in a variety of applications and can help improve our understanding of snowpack dynamics and the water balance of mountainous watersheds.

Reference

Barnett, T. P., Adam, J. C., & Lettenmaier, D. P. (2005). Potential impacts of a warming climate on water availability in snow-dominated regions. Nature, 438(7066), 303-309.

65 - An Overview of Unsaturated Flow Modelling for Water Resources Management

Abstract

Unsaturated flow modelling is a valuable tool for understanding and managing water resources in the unsaturated zone of soil and rock. This article provides an overview of unsaturated flow modelling, including the principles of unsaturated flow, mathematical models used to simulate unsaturated flow, data requirements for unsaturated flow modelling, and applications of unsaturated flow modelling in water resources management. The article highlights the importance of unsaturated flow modelling for predicting the movement of water through the soil which can help in evaluating the effectiveness of water management strategies, and mitigating the impacts of natural disasters such as floods and droughts.

Introduction

Unsaturated flow modelling is the process of simulating the movement of water through the unsaturated zone, or vadose zone, of the soil. The unsaturated zone is the area between the ground surface and the water table, where the pores in the soil are partially filled with water and partially filled with air. Understanding unsaturated flow is important for managing water resources and mitigating the impacts of natural disasters such as floods and droughts. Unsaturated flow modelling can help predict the movement of water through the soil, including the infiltration, percolation, and evapotranspiration processes. It can also help evaluate the effectiveness of various water management strategies, such as irrigation, drainage, and infiltration practices.

Unsaturated flow modelling requires detailed information about the soil properties, such as soil texture, porosity, and hydraulic conductivity, as well as the climate conditions, such as rainfall, temperature, and humidity. Field measurements, such as soil moisture content, soil water potential, and soil hydraulic conductivity, are also important for calibrating and validating the model.

The outputs of unsaturated flow modelling include the spatial and temporal distribution of water content, water potential, and flow velocity in the soil. These outputs can be used to evaluate the impacts of different scenarios, such as changes in land use, climate conditions, or water management practices. Unsaturated flow modelling can also be used in combination with other models, such as groundwater models and surface water models, to create a comprehensive water resources management system.

Principles of Unsaturated Flow

Unsaturated flow occurs when the moisture content in the soil or rock is less than the saturation level, and the water in the soil is held by capillary forces. The movement of water in the unsaturated zone is governed by the hydraulic conductivity and the water retention properties of the soil or rock. Hydraulic conductivity is the measure of the ability of the soil to transmit water, while water retention properties describe the relationship between the moisture content and the pressure of the water in the soil.

The hydraulic conductivity and water retention properties of the soil or rock can vary widely depending on the soil type, texture, structure, and other factors. The unsaturated zone is also influenced by other factors such as rainfall, evapotranspiration, temperature, and land use.

Mathematical Models of Unsaturated Flow

There are several mathematical models and software packages available for simulating unsaturated flow, including the Richards equation, the Green-Ampt model, and HYDRUS. These models differ in their complexity and the type of data required for input. The Richards equation is a partial differential equation that describes the flow of water through porous media, and is the most commonly used model for unsaturated flow modelling. The Green-Ampt model is a simpler model that is based on the concept of infiltration, and is often used for estimating infiltration rates in agricultural and urban areas. HYDRUS is a software package that uses the Richards equation to simulate unsaturated flow, and is known for its user-friendly interface and visualization tools.

Unsaturated flow modelling involves the development and application of mathematical models to simulate the movement of water in the unsaturated zone. The most commonly used models are based on the Richards equation, which is a partial differential equation that describes the conservation of mass and momentum in the unsaturated zone.

The Richards equation can be solved using numerical methods such as finite difference, finite element, or boundary element methods. These methods divide the unsaturated zone into discrete elements or nodes and calculate the water content and pressure at each location based on the hydraulic conductivity and water retention properties of the soil or rock.

The Richards equation is highly complex and requires significant computational resources and expertise. To simplify the model, several assumptions are often made, such as assuming that the soil is homogeneous or neglecting the effects of root water uptake.

Data Requirements for Unsaturated Flow Modelling

Unsaturated flow modelling requires a significant amount of data, including soil properties, soil moisture retention characteristics, meteorological data, and hydrological data. Soil properties such as texture, structure, hydraulic conductivity and retention curve can be obtained from soil surveys and laboratory tests. Meteorological data such as rainfall, temperature, and evapotranspiration can be obtained from weather stations or satellite data. Hydrological data such as groundwater levels and streamflow can be obtained from monitoring wells or gauges.

The accuracy and reliability of unsaturated flow modelling depend on the quality and quantity of the data used for model calibration and validation. Model calibration involves adjusting the model parameters to match the observed data, while model validation involves testing the model using independent data sets to ensure that the model performs well under different conditions.

Applications of Unsaturated Flow Modelling

Unsaturated flow modelling has numerous applications in water resources management, including:

Contaminant Transport: Unsaturated flow modelling can be used to simulate the transport and fate of contaminants in the unsaturated zone, such as pesticides or pollutants from landfills. Contaminant transport models can help authorities to develop strategies to manage and control water pollution.

Irrigation Management: Unsaturated flow modelling can be used to optimize irrigation management practices by simulating the movement of water in the soil and the uptake of water by plants. Irrigation management models can help farmers to minimize water use and maximize crop yields.

Soil Conservation: Unsaturated flow modelling can be used to evaluate the effectiveness of soil conservation measures, such as terracing or vegetative cover, in reducing soil erosion and improving soil health.

Groundwater Recharge: Unsaturated flow modelling can be used to evaluate the potential for groundwater recharge by simulating the infiltration of rainfall or surface water into the soil. Groundwater recharge models can help water resource managers to identify suitable locations for artificial recharge facilities and estimate the potential recharge rates.

Climate Change: Unsaturated flow modelling can be used to assess the impacts of climate change on water resources by simulating changes in rainfall patterns, temperature, and evapotranspiration. Climate change models can help water resource managers to develop adaptation strategies to cope with changes in water availability and quality.

Unsaturated Flow Modelling Codes

Unsaturated flow modelling codes are widely used to simulate water movement in unsaturated soils. These codes are critical tools in various fields, including agriculture, environmental science, and geotechnical engineering, where the accurate modelling of water movement is essential. Unsaturated flow modelling codes help scientists and engineers to predict the impact of various factors such as precipitation, irrigation, and groundwater recharge on soil moisture content, plant growth, and water resource management. An overview of few widely used unsaturated flow modelling codes is gen below.

SWAP

Soil, Water, and Plant (SWAP) is widely used unsaturated flow modelling code developed by the Alterra Research Centre in the Netherlands. SWAP simulates water movement, solute transport, and root water uptake in unsaturated soils. SWAP includes several soil hydraulic models, including van Genuchten, Brooks-Corey, and Campbell. The code has been used for various applications, including irrigation management, land-use planning, and soil remediation.

VS2DHI

The Vadose Zone 2D Hydrological Model with Inverse Estimation of Hydraulic Properties (VS2DHI) is a two-dimensional unsaturated flow modelling code developed by the Waterloo Hydrogeologic Inc. VS2DHI simulates water flow, heat transfer, and solute transport in unsaturated soils. The code uses finite difference methods to solve the governing equations of flow and transport. VS2DHI includes several soil hydraulic models, including van Genuchten, Mualem, and Brooks-Corey. The code has been used for various applications, including groundwater recharge estimation, land-use planning, and irrigation management.

HYDRUS-1D/2D/3D

HYDRUS-1D/2D/3D is a suite of codes developed by the Czech Technical University in Prague. The codes simulate water movement, heat transfer, and solute transport in unsaturated porous media. The codes use finite element methods to solve the governing equations of flow and transport. HYDRUS-1D simulates one-dimensional flow, while HYDRUS-2D and HYDRUS-3D simulate two-dimensional and three-dimensional flow, respectively. The codes have been extensively used for various applications, including crop water management, vadose zone hydrology, and contaminant transport.

Unsaturated flow modelling codes are essential tools for predicting the movement of water, heat, and solutes in unsaturated soils. These codes offer a variety of soil hydraulic models and functions for various environmental conditions, which allow scientists and engineers to simulate various applications, including crop water management, vadose zone hydrology, and contaminant transport.

Challenges and Future Directions

Despite its many applications, unsaturated flow modelling faces several challenges and limitations. These include the complexity and uncertainty of the models, the availability and quality of data, and the high computational requirements.

To overcome these challenges, there is a need for further research and development of unsaturated flow modelling techniques. This includes improving the accuracy and reliability of the models, developing new data sources and collection methods, and exploring new approaches such as machine learning and artificial intelligence.

Conclusion

Unsaturated flow modelling is a powerful tool for understanding and managing water resources in the unsaturated zone. The use of mathematical models to simulate unsaturated flow can help to predict the movement of water and solutes in the soil and rock, evaluate the effectiveness of management practices, and assess the impacts of climate change on water resources.

However, unsaturated flow modelling faces several challenges and limitations, including the complexity and uncertainty of the models, the availability and quality of data, and the high computational requirements. Further research and development of unsaturated flow modelling techniques are needed to overcome these challenges and enhance the effectiveness and efficiency of water resources management.

66 - Understanding Groundwater Modelling: Approaches and Methodology

Abstract

This article provides an overview of groundwater modelling, a tool used to understand and manage groundwater resources. The article explains the three main types of groundwater models: analytical, numerical, and hybrid models, and outlines the stepwise methodology of groundwater modelling. The components of groundwater models are also discussed, including the geological and hydrogeological framework, boundary conditions, flow equations, contaminant transport, and model calibration and validation. Groundwater modelling is essential for sustainable groundwater use and protection and can provide insights into complex groundwater systems that cannot be easily observed through direct measurements.

Introduction

Groundwater modelling is a powerful tool for understanding and managing groundwater resources. Groundwater modelling involves the use of mathematical and numerical models to simulate the behaviour of groundwater systems under different conditions. Groundwater modelling can be used for various purposes, including groundwater resource assessment, prediction of groundwater levels and quality, and evaluation of the impacts of human activities and climate change on groundwater resources.

Groundwater modelling can provide insights into the complex behaviour of groundwater systems that cannot be easily observed through direct measurements. Groundwater modelling can help to identify the most effective management strategies for sustainable groundwater use and protection.

Groundwater Modelling Approaches

Groundwater modelling can be classified into three broad categories: analytical, numerical, and hybrid models.

Analytical models are based on mathematical equations that describe the flow of groundwater through porous media. Analytical models can provide fast and accurate solutions for simple groundwater problems, such as steady-state flow and flow in confined aquifers. However, analytical models are limited in their ability to represent complex groundwater systems and processes.

Numerical models are based on the discretization of groundwater systems into smaller units, such as cells or nodes. Numerical models solve the flow equations for each unit using iterative methods. Numerical models can represent complex groundwater systems and processes, including variable aquifer properties, transient flow, and contaminant transport. However, numerical models require significant computational resources and expertise.

Hybrid models combine the advantages of analytical and numerical models. Hybrid models use simplified analytical solutions for some parts of the groundwater system and numerical methods for others. Hybrid models can reduce the computational requirements of numerical models while still representing complex groundwater systems and processes.

Stepwise Methodology of Groundwater Modelling

Develop a conceptual model: Before beginning the actual modelling process, it is important to develop a clear and comprehensive understanding of the hydrogeological system being modelled. This involves gathering and analyzing data on the geology, hydrology, and other relevant factors, and developing a conceptual model that accurately represents the system.

Choose an appropriate modelling software: There are many different groundwater modelling software packages available, each with their own strengths and weaknesses. It is important to choose a software that is appropriate for the specific objectives of the study, and that can handle the complexity and scale of the hydrogeological system being modelled.

Gather high-quality data: Accurate data is crucial for developing an accurate model. This includes data on groundwater levels, aquifer properties, hydraulic conductivity, and other relevant parameters. It is important to ensure that the data is collected using appropriate methods and is of high quality.

Validate the model: Once the model is developed, it is important to validate it using independent data sets, such as data from monitoring wells. This helps to ensure that the model accurately represents the hydrogeological system being modelled.

Conduct sensitivity analyses: Sensitivity analyses can help to identify the most important parameters in the model, and can help to assess the robustness of the model. This involves varying input parameters and observing the effect on the model outputs.

Communicate the results: The results of the modelling study should be communicated clearly and effectively to stakeholders, such as regulatory agencies, water resource managers, and the general public. This can involve presenting the results in a visual format, such as maps or graphs, and providing a clear explanation of the implications of the results.

Update the model as necessary: As new data becomes available, or as the hydrogeological system changes over time, the model should be updated to ensure that it remains accurate and useful.

Groundwater Modelling Components

Groundwater models consist of several components that represent the different aspects of groundwater systems.

Geological and Hydrogeological Framework

The geological and hydrogeological framework represents the physical properties of the groundwater system, such as the distribution and properties of aquifers and aquitards. The geological and hydrogeological framework is based on geological maps, borehole data, and geophysical surveys.

The geological and hydrogeological framework is represented in groundwater models using a grid or mesh of cells or nodes. Each cell or node represents a volume of the groundwater system with specific hydraulic properties, such as hydraulic conductivity and porosity.

Boundary Conditions

Boundary conditions represent the inputs and outputs of groundwater systems, such as recharge, discharge, and water use. Boundary conditions can be either constant or time-varying.

Boundary conditions are essential for predicting the behaviour of groundwater systems under different conditions. Boundary conditions can be based on measurements, such as groundwater levels and streamflow, or estimated using hydrological models.

Flow Equations

Flow equations represent the flow of groundwater through porous media. The most common flow equations used in groundwater modelling are the Dupuit-Forchheimer equation and the groundwater flow equation.

The Dupuit-Forchheimer equation is a simplified equation that assumes steady-state, horizontal flow in unconfined aquifers. The groundwater flow equation is a more general equation that can represent transient, multidirectional flow in confined and unconfined aquifers.

Flow equations can be solved using analytical or numerical methods. Numerical methods, such as the finite difference and finite element methods, are the most commonly used methods for solving flow equations.

Contaminant Transport

Contaminant transport represents the movement and fate of contaminants in groundwater systems. Contaminant transport is represented by advection, dispersion, and diffusion processes.

Contaminant transport can be affected by various factors, such as the properties of the contaminants, the hydraulic properties of the aquifer, and the chemical and biological reactions that occur in the groundwater system.

Contaminant transport can be modelled using analytical or numerical methods. Numerical methods, such as the finite difference and finite element methods, are the most commonly used methods for solving contaminant transport equations.

Model Calibration and Validation

Model calibration and validation are essential components of groundwater modelling. Model calibration involves adjusting the model parameters to match observed data, such as groundwater levels or pumping rates. Model validation involves testing the model's ability to predict the behaviour of groundwater systems under different conditions.

Model calibration and validation are critical for ensuring the accuracy and reliability of groundwater models. Model calibration and validation can be challenging due to the complexity and uncertainty of groundwater systems and the limitations of data.

Uncertainty and Sensitivity Analysis

Groundwater models are subject to various sources of uncertainty, such as errors in data, model parameters, and conceptual models. Uncertainty and sensitivity analysis are essential components of groundwater modelling that can help to quantify and reduce uncertainty.

Uncertainty analysis involves evaluating the range and probability of outcomes of groundwater models under different scenarios. Sensitivity analysis involves evaluating the sensitivity of groundwater models to different model parameters.

Uncertainty and sensitivity analysis can help to identify the most significant sources of uncertainty and the most critical parameters in groundwater models. Uncertainty and sensitivity analysis can also help to improve the robustness and reliability of groundwater models.

Applications of Groundwater Modelling

Groundwater modelling has numerous applications in groundwater resource management and environmental protection.

Groundwater Resource Assessment

Groundwater modelling can be used to assess the quantity and quality of groundwater resources in different regions. Groundwater modelling can provide insights into the potential impacts of human activities and climate change on groundwater resources.

Groundwater modelling can help to identify the most effective management strategies for sustainable groundwater use and protection. Groundwater modelling can also help to evaluate the impacts of groundwater pumping on surface water resources and ecosystems.

Groundwater Remediation

Groundwater modelling can be used to design and evaluate the effectiveness of groundwater remediation strategies. Groundwater modelling can help to predict the behaviour of contaminants in groundwater systems and the effectiveness of different remediation techniques.

Groundwater modelling can also help to optimize the design and operation of remediation systems, such as pumping and treatment systems.

Land Use Planning

Groundwater modelling can be used to evaluate the impacts of land use changes on groundwater resources. Groundwater modelling can help to predict the impacts of land use changes, such as urbanization and agricultural intensification, on groundwater recharge, discharge, and quality.

Groundwater modelling can also help to identify the most effective land use management strategies for sustainable groundwater use and protection.

Groundwater Modelling Codes

There are several modelling codes available for groundwater modelling. Here are some of the most commonly used codes:

MODFLOW (MODular Finite-Difference Flow model): MODFLOW is a software package developed by the U.S. Geological Survey (USGS) that is used to simulate groundwater flow in three dimensions. It includes a range of features, such as the ability to simulate variable-density flow and solute transport.

FEFLOW (Finite Element subsurface FLOW system): FEFLOW is a software package developed by the German Research Centre for Geosciences that is used to simulate groundwater flow and transport in three dimensions. It includes a range of features, such as the ability to simulate heat transport and the interaction between groundwater and surface water.

SEAWAT (SEAwater-intrusion and Tidal effects model): SEAWAT is a software package developed by the USGS that is used to simulate groundwater flow and transport in coastal aquifers. It includes a range of features, such as the ability to simulate saltwater intrusion and the effects of tides on groundwater levels.

MODPATH (MODular PATHline analysis): MODPATH is a software package developed by the USGS that is used to simulate the movement of particles in groundwater flow. It is often used in conjunction with MODFLOW to study contaminant transport and other groundwater-related issues.

PEST (Parameter ESTimation): PEST is a software package developed by the Watermark Numerical Computing company that is used to calibrate groundwater models. It includes a range of optimization algorithms that can be used to adjust model parameters to better match observed data.

These are just a few examples of the many modelling codes available for groundwater modelling. The choice of code depends on the specific needs and requirements of the modelling project, as well as the expertise and resources available to the modelling team.

Aquaveo: Aquaveo is a company that provides software and services related to water resources modelling. They offer online training courses on MODFLOW and related software, such as GMS (Groundwater Modelling System).

Universities: Many universities offer courses on groundwater modelling using MODFLOW as part of their environmental engineering or hydrology programs. Check with your local universities to see if they offer any courses on this topic.

It is important to note that some of these training courses may require prior knowledge or experience in hydrogeology and groundwater modelling.

Conclusion

MODFLOW is a critical tool for hydrogeologists and water resource managers to manage groundwater resources effectively. Understanding MODFLOW's data requirements and modelling process is essential to ensure accurate model results. The modularity and flexibility of MODFLOW make it a valuable tool for simulating a wide range of hydrogeological settings and scenarios, and its capability to be extended with additional modules enhances its functionality. Ultimately, the article emphasizes the importance of using MODFLOW to develop robust groundwater management strategies that consider the complexities of the hydrogeological system being modeled.

70 - Overcoming Challenges and Errors in Using MODFLOW Groundwater Modelling Software

Abstract

MODFLOW is a widely used software tool for simulating and analyzing groundwater flow and transport. However, like any software, it has its challenges and errors that users face. This article discusses the challenges and errors commonly faced by MODFLOW users and their solutions. The challenges and errors discussed in this article include complex model setup, input data quality, model calibration, numerical instability, data processing and management, model output analysis, and lack of documentation and support. Solutions to these challenges and errors are provided, including using graphical user interfaces, collecting reliable data, combining automated and manual calibration techniques, carefully selecting time-step size and grid resolution, efficiently managing and processing data, using visualization and statistical analysis tools, and taking advantage of available documentation and support resources. By understanding these challenges and errors and their solutions, users can effectively use MODFLOW and obtain accurate results.

Introduction

MODFLOW (MODular FLOW) is a popular software tool for simulating and analyzing groundwater flow and transport. The software is widely used by hydrologists, water resource managers, and environmental engineers to model groundwater systems and assess the impacts of various activities on the environment. However, like any software, MODFLOW has its challenges and errors that users face. This article discusses the challenges and errors commonly faced by MODFLOW users and their solutions.

Challenges and Errors Faced in Using MODFLOW

Complex Model Setup

One of the most significant challenges that users face when using MODFLOW is the complex model setup. Setting up a groundwater model can be a daunting task, especially for users who are not familiar with the software. There are many inputs required, and it can be difficult to determine the appropriate values for the various parameters. Moreover, the user must define the model domain, specify the boundary conditions, and assign properties to each cell.

Solution: The solution to this challenge is to have a good understanding of groundwater systems and how to represent them in the software. It is also essential to have a good knowledge of the software's input requirements and how to use the software's tools and features effectively. To simplify the model setup process, users can use graphical user interfaces (GUIs) that come with MODFLOW packages such as GMS (Groundwater Modelling System), which provides a visual representation of the model domain and enables users to define the model inputs with ease.

Conclusion

Groundwater modelling is a powerful tool for understanding and managing groundwater resources. Groundwater modelling can provide insights into the complex behaviour of groundwater systems and the impacts of human activities and climate change on groundwater resources.

Groundwater modelling requires a comprehensive understanding of the physical and chemical processes that occur in groundwater systems, as well as the uncertainties and limitations of data and models. Groundwater modelling also requires collaboration between water managers, policymakers, scientists, and the public to ensure the effective use and protection of groundwater resources.

By adopting a comprehensive and integrated approach to groundwater modelling, we can ensure the sustainable use and protection of groundwater resources for generations to come.

67 - An Overview of Widely Used Groundwater Modelling Codes

Abstract

Groundwater modelling is an essential tool for understanding and managing groundwater resources. There are various groundwater modelling codes available, each with its own strengths and weaknesses. In this article, we provide an overview of some of the most widely used groundwater modelling codes, including MODFLOW, FEFLOW, HYDRUS, SUTRA, SEAWAT, MODFLOW-USG, PEST, PFLOTRAN, MODPATH, MT3DMS, FEMWATER, PHREEQC, and RT3D. We also discuss the features, strengths, and applications of the codes to help readers choose the most appropriate one for their specific modelling needs.

Introduction

Groundwater is a vital resource that provides drinking water to billions of people worldwide and supports various economic activities such as agriculture, industry, and mining. However, groundwater resources are under increasing pressure due to population growth, urbanization, and climate change. The complex nature of groundwater systems makes it challenging to understand and manage them effectively. Groundwater modelling is a critical tool for studying the behavior of groundwater systems and developing sustainable management strategies.

Numerical groundwater modelling codes are computer programs that simulate groundwater flow and transport in the subsurface. These codes use mathematical equations to represent the physical processes that govern groundwater movement and solute transport. There are various groundwater modelling codes available, each with its own strengths and weaknesses. Choosing the most appropriate code for a specific modelling application can be challenging, especially for those who are new to groundwater modelling.

In this article, we provide an overview of some of the most widely used groundwater modelling codes, including MODFLOW, FEFLOW, HYDRUS, SUTRA, SEAWAT, MODFLOW-USG, PEST, PFLOTRAN, MODPATH, MT3DMS, FEMWATER, PHREEQC, and RT3D. We also discuss the features, strengths, and applications of the codes to help readers choose the most appropriate one for their specific modelling needs. By providing an overview of these modelling codes, we hope to facilitate the selection of the most suitable code for groundwater modelling applications.

MODFLOW

MODFLOW (Modular Groundwater Flow) is a finite-difference numerical modelling code that simulates groundwater flow using a system of partial differential equations. It is developed and maintained by the United States Geological Survey (USGS) and is one of the most widely used groundwater modelling codes in the world. MODFLOW is modular, meaning that it consists of several independent modules that can be used separately or together to model different aspects of groundwater flow and transport.

One of the key strengths of MODFLOW is its versatility. It can be used to model a wide range of hydrogeological settings, including confined and unconfined aquifers, heterogeneous aquifers, and karst systems. It also has a wide range of boundary conditions that can be used to model various scenarios. MODFLOW can be used to model both steady-state and transient flow, as well as solute transport.

Another strength of MODFLOW is its user-friendly interface, which allows users to easily define model inputs, visualize results, and analyze output. It also has a large and active user community, which provides support and resources for users.

FEFLOW

FEFLOW is a finite element numerical modelling code that simulates groundwater flow and solute transport. It is developed and maintained by the German software company DHI-WASY and is widely used in Europe and other parts of the world.

FEFLOW has several advantages over other modelling codes. It uses a flexible meshing approach that allows for detailed representation of complex hydrogeological features, such as fault zones, fractures, and aquifer heterogeneity. It also has advanced capabilities for modelling coupled processes, such as groundwater-surface water interactions and heat transport.

One of the main strengths of FEFLOW is its ability to handle large and complex models. It can be used to model systems with millions of cells and is optimized for parallel computing, which allows for faster simulations.

HYDRUS

HYDRUS is a finite element numerical modelling code that simulates water flow, heat transport, and solute transport in variably saturated porous media. It is developed and maintained by the Czech Academy of Sciences and is widely used in academia and industry.

HYDRUS has several unique features that make it well-suited for certain applications. One of these features is its ability to model root water uptake and plant growth in the soil-plant-atmosphere continuum. This makes it a valuable tool for studying plant water use efficiency and irrigation management.

Another strength of HYDRUS is its ability to model biogeochemical reactions, such as nutrient cycling and microbial activity, in the subsurface. This makes it a valuable tool for studying soil remediation and contaminant attenuation.

SUTRA

SUTRA (Saturated-Unsaturated Transport) is a finite-element groundwater modelling code that is specifically designed to handle unsaturated flow conditions. It is often used for simulating flow in unsaturated zones, such as vadose zones or wetlands. SUTRA has the ability to simulate flow under both steady-state and transient conditions, as well as solute transport.

SEAWAT

SEAWAT is a finite-difference groundwater modelling code that is specifically designed to simulate saltwater intrusion in coastal aquifers. It is based on the MODFLOW code and adds the ability to simulate density-dependent flow and transport in coastal zones. SEAWAT is often used for managing freshwater resources in coastal areas, where saltwater intrusion can be a significant problem.

MODFLOW-USG

MODFLOW-USG (Unstructured Grid) is a newer version of MODFLOW that uses an unstructured grid to model groundwater flow. It can handle more complex geometries and has the ability to model surface water-groundwater interactions, as well as unsaturated flow and solute transport.

PEST

PEST (Parameter ESTimation) is not a groundwater modelling code, but rather a software package that is used to calibrate groundwater models. It is often used in conjunction with other codes, such as MODFLOW, to help improve the accuracy of model predictions. PEST uses an optimization algorithm to find the best combination of model parameters that fit observed data.

PFLOTRAN (Petascale Flow and Transport in Porous Media)

PFLOTRAN is an open-source scientific software package for modelling subsurface flow and transport processes in porous media. It is primarily used for simulating groundwater flow, solute transport, and heat transfer in geological formations. It includes a wide range of physical and chemical processes, including flow in unsaturated and saturated media, heat transfer, multiphase flow, reactive transport, and geochemistry.

MODPATH

MODPATH is designed to work with MODFLOW and uses MODFLOW's output files as input. It is used to simulate the movement of groundwater particles, known as pathlines or streamlines, in a three-dimensional aquifer system. The code is particularly useful for analyzing groundwater flow and contaminant transport, as it can be used to track the movement of specific particles or contaminants through the aquifer over time.

MT3DMS (Modflow Transport in 3D Multi-Species)

MT3DMS is a software package developed by the U.S. Geological Survey (USGS) that is used to simulate groundwater solute transport in three dimensions. It can simulate multiple solutes with different properties, and includes a range of features for studying reactions, such as degradation and adsorption.

FEMWATER (Finite Element Model for Surface-Water and Ground-Water Flow)

FEMWATER is a software package developed by the USGS that is used to simulate groundwater flow and solute transport in three dimensions. It includes a range of features, such as the ability to simulate surface water and groundwater interactions and contaminant transport.

PHREEQC (PH Redox Equilibria in Aqueous Systems)

PHREEQC is a software package developed by the USGS that is used to simulate chemical reactions in groundwater and surface water systems. It includes a range of geochemical models that can be used to simulate a variety of processes, such as mineral dissolution and precipitation, and sorption reactions.

RT3D (Reactive Transport in 3 Dimensions)

RT3D is a software package developed by the USGS that is used to simulate reactive transport in three dimensions. It can simulate the transport of multiple solutes with different properties, and includes a range of features for studying reactions, such as degradation and adsorption.

PHT3D (PHreeqc Transport in 3 Dimensions)

PHT3D is a software package developed by the USGS that is used to simulate groundwater flow and solute transport, with the added ability to simulate complex geochemical reactions using PHREEQC. It includes a range of features, such as the ability to simulate advection, dispersion, and sorption.

These are just a few examples of the many modelling codes available for groundwater flow and quality modelling. The choice of groundwater modelling code depends on the specific needs of the project. Factors such as the complexity of the aquifer, the type of hydrogeological conditions, the available data, and the required level of detail all need to be considered when selecting a modelling code. It is important to note that these codes are often used in combination with other software packages, such as GIS tools and data analysis software, to create a comprehensive groundwater management system.

Conclusion

Groundwater modelling is a critical tool for understanding and managing groundwater resources, and there are many different groundwater modelling codes available to meet different modelling needs. MODFLOW, FEFLOW, and HYDRUS are three of the most widely used groundwater modelling codes, each with its own unique strengths and applications.

MODFLOW is versatile and user-friendly, making it a good choice for a wide range of hydrogeological settings and scenarios. FEFLOW excels at handling large and complex models and can model coupled processes such as groundwater-surface water interactions and heat transport. HYDRUS is well-suited for modelling variably saturated porous media and can model root water uptake and biogeochemical reactions.

Other groundwater modelling codes, such as SEAWAT, SUTRA, PFLOTRAN, PHREEQC, and MODPATH, also have their own unique strengths and applications.

It is important to choose the appropriate groundwater modelling code for the specific modelling needs, taking into consideration the hydrogeological setting, desired level of detail, computational resources, and availability of user support and resources. In addition, it is important to have a thorough understanding of the underlying hydrogeology and hydrological processes to ensure accurate and reliable modelling results.

Overall, groundwater modelling codes are essential tools for managing and protecting groundwater resources, and ongoing research and development in this field will continue to improve our understanding of groundwater systems and inform better management practices.

68 - MODFLOW: An Overview of the Widely Used Groundwater Modelling Code

Abstract

Groundwater modelling is a process of simulating the behavior of groundwater systems using mathematical models to manage and predict the response of groundwater systems. These models are essential for studying groundwater-related processes and can simulate a wide range of groundwater activities such as aquifer recharge and discharge, groundwater flow and transport, and interaction between surface water and groundwater. Groundwater models typically consist of a set of mathematical equations solved using computer software to inform groundwater management decisions. MODFLOW is a popular groundwater Modelling software widely used to simulate groundwater flow in porous media. This article provides a step-by-step guide to groundwater modelling with MODFLOW, including understanding the conceptual model, defining the model domain, specifying hydrogeologic parameters, setting up boundary conditions, calibrating the model, and analyzing model results. The article also provides an overview of MODFLOW's features, capabilities, and applications in groundwater Modelling.

Introduction

Groundwater Modelling is an essential tool for groundwater management, which involves simulating groundwater systems using mathematical models. The models describe the flow of water through the subsurface and the transport of solutes in groundwater. The models are solved numerically using computer software, and the results are used to inform groundwater management decisions. MODFLOW (Modular Groundwater Flow Model) is a widely-used groundwater Modelling software that simulates groundwater flow in porous media. The groundwater Modelling with MODFLOW involves several steps, including developing a conceptual model of the groundwater system, defining the model domain, specifying hydrogeologic parameters, setting up boundary conditions, calibrating the model to field data, and analyzing the results. MODFLOW is a finite-difference numerical Modelling code that simulates groundwater flow using a system of partial differential equations. The code is modular, meaning that it consists of several independent modules that can be used separately or together to model different aspects of groundwater flow and transport.

Groundwater Modelling

Groundwater modelling is the process of simulating the behaviour of groundwater systems using mathematical models. Groundwater modelling is an essential tool for groundwater management and is used to predict the response of groundwater systems to changes in hydrogeological conditions, such as changes in recharge rates, pumping rates, and land use.

Groundwater models typically consist of a set of mathematical equations that describe the flow of water through the subsurface and the transport of solutes in groundwater. These equations are solved numerically using computer software, and the results of the simulations are used to inform groundwater management decisions.

Groundwater models are used to simulate a wide range of groundwater-related processes, including aquifer recharge and discharge, groundwater flow and transport, and the interaction between surface water and groundwater. They can also be used to predict the impacts of groundwater abstraction and contamination, and to identify suitable locations for new wells and other groundwater infrastructure.

There are several software packages available for groundwater modelling, including MODFLOW (MODular Finite-Difference Flow model), FEFLOW (Finite Element subsurface FLOW system), and SEAWAT (SEAwater-intrusion and Tidal effects model). The choice of software package depends on the specific needs and requirements of the modelling project, as well as the expertise and resources available to the modelling team.

Overall, groundwater modelling is an important tool for groundwater management and is used to inform decision-making related to groundwater resources. It allows for better understanding and prediction of groundwater behavior, which can help to ensure sustainable use of this valuable resource.

Groundwater Modelling Protocol

Groundwater Modelling involves several steps, including understanding the conceptual model, defining the model domain, specifying hydrogeologic parameters, setting up boundary conditions, and calibrating the model to field data.

Understand the Conceptual Model: The first step in developing a groundwater model is to develop a conceptual model of the groundwater system. This involves understanding the geologic and hydrologic features of the area being studied, such as the aquifer properties, recharge and discharge zones, and the boundary conditions. The conceptual model provides the basis for creating the numerical model.

Define the Model Domain: Once the conceptual model has been developed, the next step is to define the model domain. This involves defining the extent and boundaries of the area being studied, as well as the horizontal and vertical discretization of the model grid.

Specify Hydrogeologic Parameters: The hydrogeologic properties of the aquifer, such as hydraulic conductivity, porosity, and storage coefficients, need to be specified. This involves collecting data from various sources, including geological maps, well logs, and field measurements, and then assigning these values to the corresponding grid cells in the model.

Set up Boundary Conditions: The boundary conditions define the interaction between the groundwater system and its surroundings. This involves specifying the locations and types of boundary conditions, such as recharge, discharge, and hydraulic head, and assigning the appropriate values.

Calibrate the Model: The next step is to calibrate the model to field data. This involves adjusting the model parameters to match observed data, such as groundwater levels or discharge rates. Calibration is an iterative process that involves adjusting the model parameters and rerunning the model until an acceptable fit between the model and field data is achieved.

Analyze Model Results: Once the model has been calibrated, the results can be analyzed to evaluate the behavior of the groundwater system and to test various scenarios. This involves generating contour maps, time series plots, and other visualizations to help interpret the results.

Overview of MODFLOW

Groundwater models are useful for studying the behavior of groundwater systems, predicting the impact of human activities on groundwater resources, and evaluating the effectiveness of water management strategies. There are several resources available for learning groundwater Modelling with MODFLOW. These include online tutorials, textbooks, and courses offered by universities and professional organizations. The U.S. Geological Survey provides a wealth of information on MODFLOW, including the MODFLOW and more software packages, user guides, and technical reports. The MODFLOW forum, hosted by the National Ground Water Association, is also a valuable resource for asking questions and sharing experiences with other users.

MODFLOW (Modular Groundwater Flow) is a widely used numerical modelling code for simulating groundwater flow and solute transport in three dimensions. It is developed and maintained by the United States Geological Survey (USGS) and is used by researchers, engineers, and water resource managers around the world. An overview of the MODFLOW code, its features and capabilities, and its applications in groundwater modelling is presented below.

MODFLOW is a finite-difference numerical modelling code that simulates groundwater flow using a system of partial differential equations. The code is modular, meaning that it consists of several independent modules that can be used separately or together to model different aspects of groundwater flow and transport. The basic modules of MODFLOW include the following.

Flow Package

The Flow Package is the core of the MODFLOW code and simulates groundwater flow using the groundwater flow equation. This equation describes the flow of water in porous media based on the principles of conservation of mass and Darcy's Law. The Flow Package can be used to model a variety of boundary conditions, including constant-head boundaries, specified flux boundaries, and general-head boundaries.

Transport Package

The Transport Package simulates the transport of solutes in groundwater using the advection-dispersion equation. This equation describes the transport of solutes based on the principles of conservation of mass and Fick's Law. The Transport Package can be used to model a variety of boundary conditions, including constant-concentration boundaries, specified flux boundaries, and general-head boundaries.

Recharge Package

The Recharge Package simulates the infiltration of water into the groundwater system from precipitation and other sources. It can be used to model both spatially and temporally variable recharge rates.

Drain Package

The Drain Package simulates the removal of water from the groundwater system through drains or other drainage features. It can be used to model both vertical and horizontal drains.

Well Package

The Well Package simulates the pumping of water from the groundwater system through wells. It can be used to model both pumping rates and well depths.

Applications of MODFLOW

MODFLOW is widely used in groundwater modelling for a variety of applications, including:

Groundwater Resource Management: MODFLOW can be used to evaluate the sustainable yield of groundwater resources, assess the impacts of groundwater pumping on surface water resources, and develop strategies for managing groundwater resources.

Contaminant Transport: MODFLOW can be used to model the transport of contaminants in groundwater, including the migration of pollutants from contaminated sites and the transport of contaminants in aquifers.

Land Subsidence: MODFLOW can be used to model land subsidence caused by groundwater pumping, which can have significant impacts on infrastructure and the environment.

Saltwater Intrusion: MODFLOW can be used to model saltwater intrusion into coastal aquifers, which can have significant impacts on freshwater resources and ecosystems.

Climate Change: MODFLOW can be used to assess the impacts of climate change on groundwater resources, including changes in recharge rates, groundwater availability, and water quality.

Challenges and Future Directions

MODFLOW is a powerful tool for groundwater modelling, but it faces several challenges and limitations. These include the complexity and uncertainty of the models, the availability and quality of data, and the high computational requirements.

To overcome these challenges, there is a need for further research and development of MODFLOW and other groundwater modelling tools. This includes improving the accuracy and reliability of the models, developing new data sources and collection methods, and exploring new approaches such as machine learning and artificial intelligence.

Conclusion

MODFLOW is a widely used numerical modelling code for simulating groundwater flow and solute transport in three dimensions. It is modular, meaning that it consists of several independent modules that can be used separately or together to model different aspects of groundwater flow and transport. The code has been applied to a wide range of applications, including groundwater resource management, contaminant transport, land subsidence, saltwater intrusion, and climate change.

Despite its capabilities, MODFLOW faces several challenges and limitations, including the complexity and uncertainty of the models, the availability and quality of data, and the high computational requirements. Overcoming these challenges requires continued research and development of groundwater modelling tools, including improving the accuracy and reliability of the models, developing new data sources and collection methods, and exploring new approaches such as machine learning and artificial intelligence.

Overall, MODFLOW has played a critical role in advancing our understanding of groundwater flow and transport and has provided valuable insights for groundwater resource management and environmental protection. As we continue to face growing challenges related to groundwater availability and quality, the use of advanced modelling tools like MODFLOW will be essential for developing effective solutions to these challenges.

69 - Understanding MODFLOW: A Comprehensive Guide to Groundwater Modelling

Abstract

MODFLOW is a widely-used groundwater modelling software developed by the USGS that helps hydrogeologists and water resource managers simulate and analyze groundwater flow and transport processes. The software is based on the finite-difference method and is highly modular, allowing it to be extended with additional modules to simulate other processes such as solute transport and heat transport. However, accurate results depend on appropriate model-specific and site-specific data inputs, including grid specification, boundary conditions, aquifer properties, and sources and sinks of water or solutes. This article provides a comprehensive guide to understanding MODFLOW, including its data requirements and modelling process, to help hydrogeologists and water resource managers effectively manage groundwater resources.

Introduction

The article provides an overview of MODFLOW, its application in hydrogeology and water resource management, and its capability to simulate complex hydrogeological systems, steady-state and transient flow conditions, and interactions between surface water and groundwater. The modularity and flexibility of MODFLOW are highlighted as significant advantages for hydrogeologists and water resource managers in simulating a wide range of hydrogeological settings and scenarios.

MODFLOW (MODular Finite-difference FLOW model) is a widely-used groundwater modelling software developed by the United States Geological Survey (USGS). It is a comprehensive tool that helps hydrogeologists and water resource managers to simulate and analyze groundwater flow and transport processes in a wide range of hydrogeological settings.

MODFLOW is based on the finite-difference method, which discretizes the groundwater flow equations into a grid of cells. Each cell represents a small volume of the subsurface and is interconnected with adjacent cells through a network of nodal points. MODFLOW provides various options for modelling complex hydrogeological systems, including steady-state and transient flow conditions, confined and unconfined aquifers, and interactions between surface water and groundwater.

MODFLOW is highly modular, which means that it can be extended with additional modules to simulate other processes such as solute transport, heat transport, and groundwater-surface water interactions. The software also includes a variety of tools for visualizing and analyzing model results, including contour maps, cross-section plots, and time-series graphs.

Overall, MODFLOW is an essential tool for hydrogeologists and water resource managers to understand and manage groundwater resources effectively. Its flexibility and modularity make it a valuable tool for simulating a wide range of hydrogeological settings and scenarios.

MODFLOW Versions

MODFLOW has undergone several updates and revisions since its initial release in 1984. Here are some of the major versions of MODFLOW that have been developed over the years:

MODFLOW-88: This was the original version of MODFLOW, which was released in 1988. It included support for steady-state and transient groundwater flow simulation, but did not include support for solute transport.

MODFLOW-96: This version of MODFLOW was released in 1996 and included several new features, such as support for solute transport, heat transport, and well-bore storage.

MODFLOW-2000: This version of MODFLOW was released in 2000 and included several new features, such as support for unstructured grids, generalized boundary conditions, and advanced solvers.

MODFLOW-2005: This version of MODFLOW was released in 2005 and included several new features, such as support for the new UZF (unsaturated zone flow) package, improved solvers, and support for MODFLOW-LGR (local grid refinement).

MODFLOW-NWT: This version of MODFLOW was released in 2009 and included several new features, such as improved solvers, support for Newton-Raphson formulation, and improved handling of dry cells.

MODFLOW-USG: This version of MODFLOW was released in 2016 and included several new features, such as support for unstructured grids, support for multiple models in a single simulation, and improved support for simulating flow in karst systems.

Each version of MODFLOW includes new features and improvements over previous versions, and users may choose to use the version that is best suited for their specific modelling needs. Additionally, many MODFLOW GUIs and pre- and post-processing tools have been developed to work with specific versions of MODFLOW.

GUI for MODFLOW

There are several graphical user interfaces (GUIs) available for MODFLOW, which provide a user-friendly interface for setting up, running, and visualizing MODFLOW models. Here are some of the most popular GUIs for MODFLOW:

Groundwater Vistas: Groundwater Vistas is a commercial software package that provides a complete suite of tools for groundwater modelling, including support for MODFLOW and related software. It includes a user-friendly interface for setting up and running MODFLOW models, as well as tools for visualizing and analyzing model output.

MODFLOW-GUI: MODFLOW-GUI is a free GUI developed by the U.S. Geological Survey (USGS) that provides a graphical interface for setting up and running MODFLOW models. It includes tools for visualizing and analyzing model output, and supports a range of MODFLOW packages and solvers.

GMS (Groundwater Modelling System): GMS is a commercial software package developed by Aquaveo that provides a complete suite of tools for groundwater modelling, including support for MODFLOW and related software. It includes a user-friendly interface for setting up and running MODFLOW models, as well as tools for visualizing and analyzing model output.

Visual MODFLOW: Visual MODFLOW is a commercial software package developed by Waterloo Hydrogeologic that provides a user-friendly interface for setting up and running MODFLOW models. It includes a range of tools for visualizing and analyzing model output, as well as support for a range of MODFLOW packages and solvers.

FloPy: FloPy is a free Python package that provides a programming interface for MODFLOW. While not strictly a GUI, it allows users to create and modify MODFLOW models using Python code, which can be more flexible and customizable than a traditional GUI.

These GUIs can greatly simplify the process of setting up, running, and analyzing MODFLOW models, and may be particularly useful for users who are less familiar with command-line interfaces or programming.

Data Requirement

MODFLOW requires a variety of data inputs to simulate groundwater flow and transport processes. The data requirements can be divided into two categories: model-specific data and site-specific data.

Model-Specific Data

The model-specific data includes the parameters and properties that are required to set up and run the MODFLOW model. These include:

Grid Specification: The model requires a 3D grid that represents the subsurface domain. The grid must be specified with the appropriate size and resolution to accurately represent the hydrogeological system.

Boundary Conditions: The model requires boundary conditions that define the hydraulic head, water flux, or solute concentration at the model boundaries. These can be specified as constant head, specified flux, or mixed boundary conditions.

Aquifer Properties: The model requires information on the hydraulic properties of the aquifer, such as hydraulic conductivity, specific yield, and specific storage. These properties can be estimated from field measurements, laboratory tests, or literature values.

Pumping and Recharge: The model requires information on the pumping and recharge rates within the model domain. These rates can be estimated from historical data, field measurements, or hydrological analyses.

Site-Specific Data

The site-specific data includes the information that is specific to the study area and the hydrogeological system being modeled. These include:

Geology and Hydrostratigraphy: The model requires information on the geology and hydrostratigraphy of the study area to accurately represent the subsurface geology and hydrogeology.

Hydraulic Head: The model requires data on the hydraulic head distribution within the model domain. This data can be obtained from field measurements, borehole data, or remote sensing data.

Water Quality: If the model includes the transport of solutes, then water quality data is required. This data can be obtained from field measurements, laboratory analyses, or literature values.

Climate Data: The model requires information on the climate of the study area, including precipitation, temperature, and evapotranspiration rates. This data can be obtained from meteorological stations or from remote sensing data.

Overall, the data requirements for MODFLOW are complex and varied. It is important to ensure that the model inputs are representative of the hydrogeological system being modeled to obtain accurate results. A thorough understanding of the study area and the hydrogeological processes involved is critical for successful modelling with MODFLOW.

Modelling Process

Setting up a MODFLOW model involves defining the model domain, specifying the boundary conditions, assigning aquifer properties, and defining the sources and sinks of water or solutes. The following steps provide a general overview of the model setup process:

Conceptual Model Development

The first step in the modelling process is to develop a conceptual model of the study area. This involves defining the boundaries of the groundwater system, identifying the various hydrogeological units, and characterizing the properties of these units, such as hydraulic conductivity, porosity, and storativity. The conceptual model can be developed using various sources of information, including geologic maps, well logs, geophysical surveys, and field observations.

Grid Design

The next step is to design a numerical grid that will be used to represent the study area in the model. The grid should be fine enough to capture the spatial variability of the hydraulic properties of the hydrogeological units, but not so fine that the model becomes computationally expensive. The grid can be designed using software tools such as MODFLOW's Gridgen, or other commercial or open-source software.

Boundary Conditions

The next step is to define the boundary conditions of the model, which include specifying the locations and characteristics of sources and sinks of water, such as rivers, lakes, recharge areas, and pumping wells. The boundary conditions also include defining the hydraulic head values at the model boundaries, which can be estimated from nearby observation wells or based on regional hydrologic data.

Assigning Aquifer Properties

The next step is to assign values for hydraulic conductivity, porosity, specific yield, and other aquifer properties to the cells within the grid. These properties can be estimated from field measurements, laboratory tests, or literature values. The aquifer properties should be representative of the hydrogeological system being modeled.

Defining Sources and Sinks

Once the aquifer properties have been assigned, the next step is to define the sources and sinks of water or solutes within the model domain. These may include wells, rivers, recharge areas, or contaminant sources. The rates of pumping or recharge can be estimated from historical data, field measurements, or hydrological analyses.

Overall, setting up a MODFLOW model involves a series of interrelated steps that require careful planning and attention to detail. The model setup should be representative of the hydrogeological system being modeled and should be calibrated to match field observations. With proper model setup, MODFLOW can be a powerful tool for simulating and analyzing groundwater flow and transport processes.

Solution Techniques

MODFLOW offers several solution techniques for solving the groundwater flow equation. The choice of solution technique depends on the size and complexity of the groundwater model and the accuracy required for the simulation. The most commonly used solution techniques in MODFLOW include:

Direct Solver: This technique involves solving the groundwater flow equation using a direct matrix solver. Direct solvers are efficient for small to medium-sized models with fewer unknowns.

Iterative Solver: This technique involves solving the groundwater flow equation iteratively until a solution is achieved. Iterative solvers are more efficient for large and complex models with many unknowns. MODFLOW uses several iterative solvers, including the Conjugate Gradient (CG) method, the Preconditioned Conjugate Gradient (PCG) method, and the Generalized Minimum Residual (GMRES) method.

Preconditioning: This technique involves applying a preconditioner to the system of equations to improve the efficiency of the iterative solver. MODFLOW offers various preconditioning options, including the Jacobi, Incomplete Cholesky, and Incomplete LU preconditioners.

Multigrid: This technique involves solving the groundwater flow equation using a multigrid solver that applies a series of coarse and fine grids to solve the equation. Multigrid solvers are particularly efficient for large and complex models with many unknowns.

Hybrid Solver: This technique combines the benefits of both direct and iterative solvers. It involves solving the groundwater flow equation using a direct solver for the largest part of the system and an iterative solver for the remaining part.

MODFLOW also provides options for parallel processing to further enhance the solution efficiency of large and complex models. Parallel processing involves dividing the model into smaller parts and solving each part simultaneously on multiple processors.

Overall, MODFLOW offers a range of solution techniques to choose from, depending on the size and complexity of the groundwater model and the accuracy required for the simulation. The choice of solution technique can significantly impact the computational efficiency and accuracy of the simulation.

Model Calibration and Validation

Calibration, validation, and sensitivity analysis are important steps in the modelling process for MODFLOW. These steps ensure that the model accurately represents the actual groundwater system and that the model outputs are reliable and useful for decision-making.

Calibration: Calibration is the process of adjusting the model parameters to match the observed groundwater behavior in the real world. The model parameters that are typically calibrated include hydraulic conductivity, specific yield, and recharge rates. The calibration process involves comparing the simulated groundwater levels or flows to the observed data from monitoring wells, and adjusting the parameters until the simulated and observed values match. This process is typically iterative, with the modeler adjusting the parameters and running the model multiple times until an acceptable match is achieved.

Validation: Validation is the process of testing the model's ability to predict future groundwater behavior based on data that was not used during the calibration process. The model is run with input data for a future time period, and the resulting groundwater levels or flows are compared to observed data from that same time period. If the model accurately predicts the observed behavior, then it is considered validated.

Sensitivity Analysis: Sensitivity analysis is the process of evaluating how changes in the model inputs affect the model outputs. This step is important for understanding the uncertainty associated with the model outputs and identifying which inputs have the greatest impact on the results. For example, a sensitivity analysis may involve varying the hydraulic conductivity parameter over a range of values and observing how the groundwater levels or flows change in response.

Scenario Analysis: Once the model has been calibrated and validated, it can be used to simulate different scenarios, such as changes in groundwater recharge rates, pumping rates, or land use. Scenario analysis can help to predict the effects of different management or policy options on the groundwater system and guide decision-making.

These are essential steps in the modelling process for MODFLOW. Calibration involves adjusting the model parameters to match the observed groundwater behavior, validation involves testing the model's ability to predict future behavior, and sensitivity analysis involves evaluating the uncertainty associated with the model inputs and outputs. By performing these steps, modelers can ensure that the model outputs are reliable and useful for decision-making.

Modelling Errors

There are several types of errors that one may encounter while using MODFLOW. Here are some of the most common errors and their potential causes:

Input errors: Input errors occur when the input data provided to MODFLOW is incorrect or incomplete. For example, if the dimensions of the model grid are not specified correctly, MODFLOW will produce an error message.

Numerical errors: Numerical errors occur when the model equations cannot be solved accurately due to numerical instability. This can happen if the model grid is too coarse or if the time step is too large. In these cases, MODFLOW may produce an error message or the simulation may fail to converge.

Convergence refers to the point at which the numerical solution has reached a steady state and the changes in the solution between iterations are negligible. In MODFLOW simulations, convergence is typically assessed by monitoring the changes in water levels or fluxes between iterations. If the changes fall below a specified threshold, the solution is considered converged. Convergence is important because it indicates that the solution has reached a stable state and that further iterations will not significantly improve the accuracy of the results.

Stability refers to the numerical stability of the solution, or the ability of the simulation to produce accurate results over a range of time steps. Unstable solutions can produce unrealistic or erratic behavior, such as sudden changes in water levels or mass balance errors. Stability in MODFLOW simulations can be affected by factors such as the size of the time step, the spatial discretization of the model, and the presence of boundary conditions or wells. In order to ensure numerical stability, it is important to use appropriate time steps and model discretization, and to carefully calibrate the simulation based on field observations.

Oscillations refer to the numerical oscillations or fluctuations that can occur in the solution during the simulation. Oscillations can result from a variety of factors, including numerical instability, nonlinearities in the system, or abrupt changes in boundary conditions or recharge. Oscillations can be particularly problematic in transient simulations, where they can produce unrealistic results and make it difficult to accurately model the system behavior. To address oscillations, it is often necessary to adjust the numerical parameters of the simulation, such as the time step or solver tolerance, or to modify the model inputs to better reflect the actual system behavior.

Memory errors: Memory errors occur when the amount of memory allocated for the model is insufficient. This can happen if the model is too large or if the computer does not have enough memory to run the model. In these cases, MODFLOW may produce an error message or the simulation may crash.

Output errors: Output errors occur when the output files produced by MODFLOW are incorrect or incomplete. This can happen if the output files are not written correctly or if the output file format is not compatible with the software used to view the output.

Software errors: Software errors occur when there is a problem with the MODFLOW software itself. This can happen if there is a bug in the software or if the software is not compatible with the computer hardware or operating system.

To avoid these errors, it is important to carefully review and validate the input data, select appropriate numerical settings, and use compatible software to view the output. It is also important to consult the MODFLOW user manual and seek assistance from experienced users or technical support if needed.

Applications of MODFLOW

MODFLOW is a widely used groundwater modelling software that has a range of applications in various fields. Some of the key applications and uses of MODFLOW are:

Groundwater management: MODFLOW is commonly used in groundwater management studies to evaluate the impacts of water use and to develop strategies for sustainable groundwater management. It can be used to simulate the behavior of groundwater systems and to analyze the impacts of various management scenarios.

Water supply planning: MODFLOW can be used to model groundwater resources and to assess the potential impacts of new water supply projects, such as wells or recharge facilities. It can also be used to simulate the impacts of climate change on groundwater availability and quality.

Contaminant transport: MODFLOW can be used to model the transport of contaminants in groundwater systems. It can be used to simulate the transport of both organic and inorganic contaminants, and can be used to evaluate the impacts of different remediation strategies.

Mine dewatering: MODFLOW can be used to model the behavior of groundwater systems in mining areas. It can be used to simulate the impacts of mine dewatering on groundwater resources and to develop strategies for managing mine water.

Environmental impact assessment: MODFLOW can be used in environmental impact assessments to evaluate the potential impacts of new developments, such as landfills, oil and gas operations, or industrial facilities, on groundwater resources.

Geothermal energy development: MODFLOW can be used to model the behavior of geothermal systems and to assess the potential for geothermal energy development.

Climate change studies: MODFLOW can be used to simulate the impacts of climate change on groundwater resources. It can be used to evaluate the impacts of changes in precipitation, temperature, and other climate variables on groundwater availability and quality.

Overall, MODFLOW is a powerful tool for modelling and understanding groundwater systems, and its flexibility and wide range of applications make it a valuable tool for groundwater management and research.

Training Courses

There are several organizations and institutions that offer training courses specifically focused on MODFLOW, which is a popular groundwater modelling software. Here are some options to consider:

International Groundwater Modelling Center (IGWMC): IGWMC offers short courses and workshops on MODFLOW, as well as other groundwater modelling software. Their courses cover topics such as MODFLOW-USG, MODFLOW-2005, and more.

The National Ground Water Association (NGWA): NGWA offers online courses on MODFLOW, as well as other groundwater modelling software. Their courses cover topics such as groundwater modelling basics, model calibration, and model validation using MODFLOW.

Environmental Modelling & Software (EMS): EMS hosts an annual conference that includes training workshops on MODFLOW, as well as other groundwater modelling software. Their workshops cover a wide range of topics related to groundwater modelling, such as uncertainty analysis, model calibration, and more.

Input Data Quality

The quality of the input data is crucial for the accuracy of the groundwater model. The model inputs must be accurate, consistent, and representative of the groundwater system being modelled. Inaccurate or inconsistent input data can lead to erroneous results.

Solution: To overcome this challenge, users should collect as much data as possible about the groundwater system being modelled. This data should be reliable and should include information on the hydrogeological properties of the aquifer, the hydraulic boundary conditions, and the sources of recharge and discharge. Users should also verify the accuracy of the data by comparing it with field observations or other available data sources.

Model Calibration

Calibrating the model is essential to ensure that the model accurately represents the groundwater system being modelled. Calibration involves adjusting the model parameters to match the observed data. However, this can be a challenging task, as there may be many model parameters that need to be adjusted, and the process can be time-consuming.

Solution: To calibrate the model effectively, users should use a combination of automated and manual calibration techniques. Automated calibration techniques such as PEST (Parameter ESTimation) can help to reduce the number of model runs required for calibration. However, manual calibration is still necessary to ensure that the calibrated model accurately represents the groundwater system being modelled.

Numerical Instability

Numerical instability can occur when using MODFLOW, resulting in inaccurate or unrealistic results. This can be caused by a variety of factors, such as inappropriate time-step size, poor grid resolution, or high model complexity.

Solution: To prevent numerical instability, users should carefully select the time-step size and grid resolution to ensure that the model runs smoothly. Users should also avoid using overly complex models, which can lead to numerical instability. To identify and fix numerical instability issues, users can use the software's diagnostic tools, such as the mass balance check, to identify inconsistencies in the model results.

Data Processing and Management

Managing and processing data can be a significant challenge when using MODFLOW. Large data sets may need to be managed, and complex data processing tasks may need to be performed to create the required input data for the model.

Solution: To overcome this challenge, users should use software tools that can efficiently manage and process large data sets. For example, GIS (Geographic Information System) software can be used to store and manage spatial data, and tools such as Excel can be used to process and manipulate data. Users should also take advantage of MODFLOW's data pre-processing tools, which can automate some of the data processing tasks required for model setup.

Model Output Analysis

Interpreting and analyzing the model output can be a significant challenge for users. MODFLOW generates a large amount of output data, which can be difficult to interpret and analyze.

Solution: To overcome this challenge, users should use visualization tools to help interpret and analyze the model output. Graphical representations of the output data can help to identify trends and patterns in the data. Users should also use statistical analysis tools to analyze the output data and identify any outliers or anomalies.

Lack of Documentation and Support

The lack of documentation and support can be a significant challenge for users, especially for new users who are not familiar with the software.

Solution: To overcome this challenge, users should take advantage of the available documentation and support resources. MODFLOW comes with a comprehensive user manual that provides detailed information on how to use the software. There are also many online resources, such as forums and discussion groups, where users can ask questions and get help from other users. Additionally, some software vendors offer training courses and consulting services to help users learn how to use the software effectively.

Conclusion

In conclusion, MODFLOW is a powerful software tool for simulating and analyzing groundwater flow and transport. However, like any software, it has its challenges and errors that users face. Understanding these challenges and errors and their solutions is essential for users to effectively use the software and obtain accurate results. By using the appropriate tools and techniques, and taking advantage of available resources, users can overcome these challenges and effectively model groundwater systems using MODFLOW.

71 - Modelling Seawater Intrusion in Coastal Aquifers: An Overview of Methods and Approaches

Abstract

Seawater intrusion is a common phenomenon that occurs in coastal aquifers, leading to degradation of water quality, reduction in water availability, and increased cost of water treatment. Accurate modelling of seawater intrusion is essential for effective management and conservation of coastal aquifers. This article provides an overview of the various methods and approaches used for modelling seawater intrusion, including analytical, semi-analytical, numerical, integrated, and data-driven models. The advantages and disadvantages of each modelling approach are discussed, and recommendations for future research are provided. The article aims to provide a comprehensive understanding of the different methods available for modelling seawater intrusion and help researchers and practitioners in selecting the appropriate approach based on the complexity of the problem, available data, and computational resources.

Introduction

Seawater intrusion is a common phenomenon that occurs in coastal aquifers, where saline water infiltrates the freshwater aquifer due to natural or anthropogenic reasons. This process can result in the degradation of water quality, reduction in water availability, and increase in the cost of water treatment. Therefore, the development of accurate models to predict the behavior of seawater intrusion is essential for effective management and conservation of coastal aquifers. This article provides an overview of the various methods used for modelling seawater intrusion and their advantages and disadvantages.

Analytical models

Analytical models are based on the solution of mathematical equations that describe the physical processes that govern seawater intrusion. These models are simple, fast, and require minimal data input. The most commonly used analytical model for seawater intrusion is the Ghyben-Herzberg model, which assumes a two-layer system, with a freshwater layer overlaying a saltwater layer of constant density. The Ghyben-Herzberg model provides a simple and conservative estimate of the seawater-freshwater interface location, but it has limitations as it does not account for vertical and horizontal heterogeneity, and the effect of groundwater pumping.

- ❖ Ghyben-Herzberg model can be easily implemented using Excel or any other spreadsheet software that can perform simple calculations.

Semi-analytical models

Semi-analytical models are based on the solution of partial differential equations that describe the physical processes that govern seawater intrusion. These models are more complex than analytical models and require more data input, but they provide more accurate predictions. The most commonly used semi-analytical model for seawater intrusion is the Sharp Interface Model (SIM), which assumes a sharp interface between the freshwater and saltwater zones. The SIM model takes into account vertical and horizontal heterogeneity, and the effect of groundwater pumping. However, it has limitations as it does not account for dispersion and mixing.

- Sharp Interface Model (SIM) can be implemented using the SEAWAT code, which is part of the USGS MODFLOW groundwater modelling software package.
- Another option is to use the SEAWAT-MT3DMS package, which includes the SEAWAT code for simulating variable-density flow and transport in porous media.

Numerical models

Numerical models are based on the discretization of the domain into a grid, and the solution of partial differential equations that describe the physical processes that govern seawater intrusion. These models are the most accurate and flexible, but they require a large amount of data input and computational resources. The most commonly used numerical models for seawater intrusion are the Finite Difference Method (FDM), the Finite Element Method (FEM), and the Boundary Element Method (BEM). These models take into account vertical and horizontal heterogeneity, the effect of groundwater pumping, dispersion, and mixing. The choice of the numerical method depends on the complexity of the problem, the available data, and the computational resources.

- Finite Difference Method (FDM) can be implemented using MODFLOW, which is a popular groundwater flow modelling software package.
- Finite Element Method (FEM) can be implemented using FEFLOW, which is a powerful groundwater modelling software that uses the finite element method for simulating groundwater flow and transport.
- Boundary Element Method (BEM) can be implemented using the BEMMOD software, which is a boundary element-based groundwater modelling software package.

Integrated models

Integrated models are combinations of two or more modelling approaches, and they aim to provide a more accurate representation of the physical processes that govern seawater intrusion. Integrated models can be classified into two categories: deterministic and stochastic. Deterministic integrated models use a single set of input data to provide a single output, while stochastic integrated models use multiple sets of input data to provide a range of possible outputs. The most commonly used integrated models for seawater intrusion are the Analytical-Numerical Model (ANM), the Analytical- Stochastic Model (ASM), and the Numerical-Stochastic Model (NSM).

- Analytical-Numerical Model (ANM) can be implemented using a combination of Excel and MODFLOW.
- Analytical-Stochastic Model (ASM) can be implemented using the analytical model in Excel and a stochastic simulation tool such as Monte Carlo simulation in MATLAB.

- ❖ Numerical-Stochastic Model (NSM) can be implemented using FEFLOW and Monte Carlo simulation in MATLAB.

Data-driven models

Data-driven models are based on statistical and machine learning techniques, and they aim to predict the behavior of seawater intrusion based on historical data. Data-driven models require a large amount of data input, but they can provide accurate predictions without the need for detailed knowledge of the physical processes that govern seawater intrusion. The most commonly used data-driven models for seawater intrusion are the Artificial Neural Network (ANN) model and the Support Vector Regression (SVR) model.

- ❖ Artificial Neural Network (ANN) model can be implemented using software packages such as MATLAB, Python (using libraries like Keras or Tensorflow), or commercial software packages like Neural Designer or NeuroSolutions.
- ❖ Support Vector Regression (SVR) model can be implemented using MATLAB or Python (using libraries like scikit-learn).

Conclusion

Seawater intrusion is a complex process that requires accurate modelling for effective management and conservation of coastal aquifers. Various modelling approaches have been developed, ranging from simple analytical models to complex integrated and data-driven models. The choice of modelling approach depends on the complexity of the problem, the available data, and the computational resources. Analytical models provide a simple estimate of the seawater-freshwater interface location, but they have limitations in accounting for vertical and horizontal heterogeneity and the effect of groundwater pumping. Semi-analytical models provide a more accurate estimate of seawater intrusion, but they also have limitations in accounting for dispersion and mixing. Numerical models are the most accurate and flexible, but they require a large amount of data input and computational resources. Integrated models combine two or more modelling approaches and provide a more accurate representation of the physical processes that govern seawater intrusion. Data-driven models are based on statistical and machine learning techniques and require a large amount of data input, but they can provide accurate predictions without the need for detailed knowledge of the physical processes.

In conclusion, modelling of seawater intrusion is a complex process that requires the integration of various modelling approaches. The development of accurate models is essential for effective management and conservation of coastal aquifers. Therefore, it is important to select the appropriate modelling approach based on the complexity of the problem, the available data, and the computational resources. Future research should focus on the development of more accurate and efficient models that can provide reliable predictions of seawater intrusion under different scenarios.

72 - Challenges and Advances in Groundwater Modelling of Hard Rock Aquifers: Case Studies and Applications

Abstract

This article discusses groundwater modelling in hard rock aquifers, which are characterized by low permeability and complex geological structures. It covers the types of models used, data requirements, challenges involved, steps involved in modelling, and applications of groundwater modelling. The article also provides details of groundwater modelling software that can be used for modelling hard rock aquifers, such as MODFLOW and FEFLOW. The article includes case studies demonstrating the use of groundwater modelling in a fractured rock aquifer in Sweden, assessing the impact of climate change on groundwater resources in a hard rock aquifer in India, and modelling mine dewatering in a hard rock aquifer in Canada.

Introduction

Groundwater modelling is a technique used to study the behavior of groundwater systems, including the flow and transport of water and contaminants in the subsurface. It is an essential tool for managing water resources in hard rock aquifers, which are characterized by low permeability and complex geological structures. In this article, we will discuss the groundwater modelling of hard rock aquifers in detail, including the types of models used, the data required, and the challenges involved.

Groundwater Models

Groundwater models can be classified into two types: analytical models and numerical models. Analytical models are based on mathematical equations that describe the flow of groundwater in the subsurface. They are simple and easy to use but are limited in their ability to represent complex geological structures. Numerical models, on the other hand, use computers to solve complex equations that describe the flow of groundwater in the subsurface. They can represent complex geological structures and are widely used in groundwater modelling of hard rock aquifers.

There are several groundwater modelling software available for modelling of hard rock aquifers. Here are some popular options:

MODFLOW: Developed by the United States Geological Survey (USGS), MODFLOW is one of the most widely used groundwater modelling software. It is a finite-difference numerical model that can simulate groundwater flow in three dimensions. It is capable of handling complex geological structures and is widely used in groundwater modelling of hard rock aquifers.

FEFLOW: Developed by DHI-WASY GmbH, FEFLOW is a finite element numerical model that can simulate groundwater flow and transport in three dimensions. It can handle complex geological structures and is widely used in groundwater modelling of hard rock aquifers. FEFLOW also has a user-friendly interface and powerful visualization tools.

Visual MODFLOW: Developed by Waterloo Hydrogeologic, Visual MODFLOW is a graphical user interface for MODFLOW. It allows users to build numerical models using a simple drag-and-drop interface and provides powerful visualization and data management tools. Visual MODFLOW is widely used in groundwater modelling of hard rock aquifers.

Groundwater Vistas: Developed by Environmental Simulations Inc., Groundwater Vistas is a graphical user interface for MODFLOW and other groundwater modelling software. It allows users to build numerical models using a simple drag-and-drop interface and provides powerful visualization and data management tools. Groundwater Vistas is widely used in groundwater modelling of hard rock aquifers.

GMS: Developed by Aquaveo LLC, GMS is a graphical user interface for MODFLOW and other groundwater modelling software. It allows users to build numerical models using a simple drag-and-drop interface and provides powerful visualization and data management tools. GMS is widely used in groundwater modelling of hard rock aquifers.

These software packages are widely used in the groundwater modelling community and have proven to be effective for modelling of hard rock aquifers. However, the selection of software depends on several factors such as user experience, budget, project requirements, and the complexity of the aquifer system.

Data Requirements for Groundwater Modelling of Hard Rock Aquifers

Groundwater modelling of hard rock aquifers requires a significant amount of data, including geological, hydrological, and hydrogeological data. Geological data includes information on the geological structure of the aquifer, such as the type and orientation of the rock formations. Hydrological data includes information on the surface water and groundwater recharge rates, as well as the evapotranspiration rates. Hydrogeological data includes information on the hydraulic properties of the aquifer, such as the permeability and porosity of the rock formations.

Challenges Involved in Groundwater Modelling of Hard Rock Aquifers

Groundwater modelling of hard rock aquifers is challenging due to the low permeability and complex geological structures of these aquifers. In addition, the lack of data on the hydraulic properties of the aquifer and the uncertainties in the data can make modelling difficult. The presence of fractures and faults in the aquifer can also make modelling challenging, as these features can significantly influence the flow of groundwater.

Steps Involved in Groundwater Modelling of Hard Rock Aquifers

Groundwater modelling of hard rock aquifers involves several steps, including data collection, model development, calibration, and validation. The following is a brief overview of each step:

Data collection: This step involves collecting geological, hydrological, and hydrogeological data on the aquifer. This includes information on the geological structure of the aquifer, surface water and groundwater recharge rates, and the hydraulic properties of the aquifer.

Model development: This step involves developing a conceptual model of the aquifer, which includes the geological structure, hydraulic properties, and boundary conditions. The conceptual model is then used to develop a numerical model of the aquifer.

Calibration: This step involves adjusting the parameters of the numerical model to match the observed data. This includes adjusting the hydraulic conductivity and recharge rates to match the observed water levels and flow rates.

Validation: This step involves testing the numerical model against independent data sets to verify its accuracy. This includes comparing the predicted water levels and flow rates to the observed data from monitoring wells.

Applications of Groundwater Modelling of Hard Rock Aquifers

Groundwater modelling of hard rock aquifers has several applications, including water resources management, contamination assessment, and mine dewatering. The following is a brief overview of each application:

Water resources management: Groundwater modelling can be used to assess the sustainable yield of the aquifer and to optimize the placement of wells for water supply.

Contamination assessment: Groundwater modelling can be used to assess the transport of contaminants in the subsurface and to develop remediation strategies.

Mine dewatering: Groundwater modelling can be used to design dewatering systems for mines to prevent flooding and to manage water resources in the mine.

Case Studies

The following are some case studies that demonstrate the use of groundwater modelling in hard rock aquifers.

1. Modelling groundwater flow in a fractured rock aquifer

In this study, a numerical model was developed to simulate groundwater flow in a fractured rock aquifer in Sweden. The aquifer was characterized by a complex geological structure with multiple fracture sets. The model was calibrated using water level data from monitoring wells and was validated using tracer test data. The model results showed that the fracture sets had a significant impact on groundwater flow and that the hydraulic conductivity of the fractures was the most important parameter controlling groundwater flow.

2. Assessing the impact of climate change on groundwater resources

In this study, a numerical model was developed to assess the impact of climate change on groundwater resources in a hard rock aquifer in India. The model was calibrated using water level data from monitoring wells and was validated using independent data sets. The model was used to simulate groundwater flow under different climate scenarios and to predict the future changes in groundwater resources. The results showed that the groundwater recharge rates would decrease under future climate scenarios, which would lead to a reduction in groundwater resources.

3. Modelling mine dewatering in a hard rock aquifer

In this study, a numerical model was developed to simulate mine dewatering in a hard rock aquifer in Canada. The aquifer was characterized by a complex geological structure with multiple faults and fractures. The model was calibrated using water level data from monitoring wells and was validated using independent data sets. The model was used to design a dewatering system for the mine, which involved pumping groundwater from the mine and discharging it to the surface. The model results showed that the dewatering system would be effective in preventing flooding in the mine.

These studies are just a few examples of the many research papers that have been published on groundwater modelling of hard rock aquifers.

Conclusion

Groundwater modelling is an essential tool for managing water resources in hard rock aquifers. It can be used to assess the sustainable yield of the aquifer, to optimize the placement of wells for water supply, to assess the transport of contaminants in the subsurface, and to design dewatering systems for mines. Groundwater modelling of hard rock aquifers is challenging due to the low permeability and complex geological structures of these aquifers. However, with the availability of advanced modelling software and improved data collection techniques, it is now possible to develop accurate numerical models of hard rock aquifers.

References

Bhattacharya, P., Singh, P., & Mukherjee, A. (2012). Assessing the impact of climate change on groundwater resources in a hard rock aquifer in India. Hydrogeology Journal, 20(5), 977-990.

Fernandez, G., Mackay, R., & Hendry, M. J. (2011). Modelling of mine dewatering and inflow in a hard rock underground mine using numerical and empirical methods. Mine Water and the Environment, 30(4), 266-276.

Sjöberg, J., Neretnieks, I., & Lindgren, S. (2005). Groundwater flow modelling in a fractured rock aquifer using discrete fracture networks. Journal of Hydrology, 309(1-4), 305-318.

73 - Groundwater Modelling and Management Strategies for Sustainable Mining Operations in Mining Areas

Abstract

This article discusses the importance of groundwater modelling and management in mining areas to protect groundwater resources and the environment. The article provides an overview of the hydrological behavior of mining areas, the potential impacts of mining activities on groundwater, and the groundwater modelling techniques used to simulate groundwater flow and transport. The article also describes the groundwater management strategies that can be used to prevent, mitigate, and remediate the potential impacts of mining activities on groundwater resources. A case study of the Mount Polley mine in British Columbia, Canada, is presented to illustrate the application of groundwater modelling and management in a mining context. The article concludes that groundwater modelling and management should be an integral part of mining operations to ensure sustainable mining practices that protect groundwater resources and the environment.

Introduction

Groundwater is a valuable natural resource that plays a vital role in supporting the ecosystems and providing clean drinking water to communities. However, the mining activities can impact groundwater quality and quantity due to the excavation and extraction of minerals, which can alter the hydrological balance of the area. Therefore, it is crucial to understand the hydrogeological processes of mining areas and develop effective groundwater management strategies to mitigate the potential impacts.

Groundwater modelling is a powerful tool that can provide insights into the hydrological behavior of mining areas and help to develop sustainable groundwater management plans. This article aims to provide an overview of groundwater modelling and management in mining areas, including the principles of groundwater flow and transport, groundwater modelling techniques, and management strategies for sustainable mining operations.

Principles of Groundwater Flow and Transport

Groundwater is the water that occurs below the ground surface in the spaces between the soil particles and rocks. It is replenished by rainfall and snowmelt, and it moves through the soil and rocks to replenish rivers, lakes, and oceans. Groundwater flow is influenced by the properties of the aquifer, such as porosity, permeability, and hydraulic conductivity, and the hydrological conditions of the area, such as recharge and discharge rates.

Groundwater transport refers to the movement of water and solutes through the porous media of the aquifer. The transport of solutes can be affected by various physical, chemical, and biological processes, including diffusion, dispersion, advection, and reactions with the mineral matrix.

The principles of groundwater flow and transport are essential to understanding the hydrogeological behavior of mining areas and developing groundwater management strategies.

Groundwater Modelling Techniques

Groundwater modelling is a process of developing mathematical models that simulate the hydrological behavior of the groundwater system. The groundwater models can be used to predict the changes in groundwater levels, the movement of groundwater, and the transport of solutes in response to different hydrological scenarios.

There are two types of groundwater models, including analytical and numerical models. Analytical models use mathematical equations to solve for the hydraulic head and groundwater flow rate. Analytical models are relatively simple and can be used for small-scale problems that have simple geometries and boundary conditions.

Numerical models, on the other hand, use a discretized grid system to solve for the hydraulic head and groundwater flow rate. Numerical models are more complex and can be used for large-scale problems that have complicated geometries and boundary conditions. The numerical models can also simulate the transport of solutes, making them more useful for groundwater management in mining areas.

Numerical models require input data, including geologic data, hydraulic properties, boundary conditions, and hydrological data. The input data can be obtained from various sources, including field observations, laboratory experiments, and remote sensing data. The accuracy of the input data is crucial for the reliability of the numerical model predictions.

Groundwater Management Strategies

Groundwater management in mining areas is crucial to ensure the sustainability of groundwater resources and the protection of the environment. Groundwater management strategies can be divided into three categories, including prevention, mitigation, and remediation.

Prevention strategies aim to prevent groundwater contamination or depletion before the mining operations begin. Prevention strategies can include the selection of mining sites with low potential for groundwater contamination or depletion, the development of best management practices to minimize the impact of mining activities on groundwater, and the use of alternative mining methods that have less impact on groundwater.

Mitigation strategies aim to reduce the impact of mining activities on groundwater quality and quantity. Mitigation strategies can include the installation of monitoring wells to track changes in groundwater levels and quality, the development of groundwater management plans to ensure sustainable groundwater use, and the implementation of remediation technologies to treat contaminated groundwater.

Remediation strategies aim to clean up contaminated groundwater after mining activities have ceased. Remediation strategies can include the use of various technologies, such as pump-and-treat systems, in-situ remediation techniques, and natural attenuation. The selection of remediation technologies depends on the characteristics of the contaminant, the hydrogeological properties of the site, and the remediation goals.

Groundwater management strategies in mining areas should also consider the potential impacts of climate change on groundwater resources. Climate change can alter the hydrological cycle and affect the availability and quality of groundwater. Therefore, it is crucial to develop adaptive management strategies that can respond to the changing hydrological conditions.

Case Study: Groundwater Modelling and Management in the Mount Polley Mine, British Columbia, Canada

The Mount Polley mine is a copper and gold mine located in British Columbia, Canada. The mine has been operating since 1997, and in 2014, a tailings dam breach released millions of cubic meters of wastewater and tailings into nearby lakes and rivers, resulting in one of the largest environmental disasters in Canadian history.

Following the disaster, the mine owners developed a groundwater management plan to prevent future incidents and protect groundwater resources. The groundwater management plan included a groundwater model that simulated the hydrological behavior of the site and predicted the potential impacts of mining activities on groundwater.

The groundwater model was a three-dimensional numerical model that used MODFLOW software to simulate the groundwater flow and transport. The model included the geological, hydrogeological, and hydrological data of the site, including the bedrock geology, hydraulic conductivity, recharge, and discharge rates. The model was calibrated and validated using field observations of groundwater levels and quality.

The groundwater model predicted that the mining activities would cause a decline in groundwater levels and an increase in the concentration of some contaminants, such as sulfate and nitrate. To mitigate the potential impacts, the groundwater management plan included the installation of monitoring wells, the implementation of best management practices to minimize the impact of mining activities on groundwater, and the development of a contingency plan in case of an emergency.

Conclusion

Groundwater modelling and management are essential for sustainable mining operations that protect groundwater resources and the environment. Groundwater modelling provides insights into the hydrological behavior of mining areas and can be used to develop effective groundwater management strategies. Groundwater management strategies can include prevention, mitigation, and remediation strategies, depending on the stage of the mining operations and the potential impacts on groundwater resources.

The case study of the Mount Polley mine highlights the importance of groundwater modelling and management in preventing environmental disasters and protecting groundwater resources.

The groundwater model developed for the mine provided valuable information that informed the groundwater management plan and ensured the sustainability of the mining operations.

In conclusion, groundwater modelling and management should be an integral part of mining operations, and effective groundwater management strategies should be developed to protect groundwater resources and the environment. The use of advanced technologies, such as remote sensing and artificial intelligence, can further enhance the accuracy and reliability of groundwater modelling and management in mining areas.

Reference

Norris, R., & Rich, J. (2016). Groundwater management at the Mount Polley mine: a case study. Hydrogeology Journal, 24(6), 1461-1471.

74 - Groundwater Models as Interactive Management Tools: Applications, Benefits, and Limitations

Abstract

Groundwater management is a complex issue in regions where groundwater is the primary source of water for human consumption and irrigation. Groundwater models have been developed as interactive management tools that allow for the simulation of groundwater flow, the movement of contaminants, and the response of the aquifer to changes in pumping rates and recharge rates. This article explores groundwater models as interactive management tools, discussing their applications, benefits, and limitations. The article also presents two case studies to illustrate the application of groundwater models in different contexts.

Introduction

Groundwater management is a complex issue, particularly in regions where groundwater is the primary source of water for human consumption and irrigation. Groundwater models have been developed as tools for interactive management of groundwater resources. These models allow for the simulation of groundwater flow, the movement of contaminants, and the response of the aquifer to changes in pumping rates and recharge rates. In this article, we will explore groundwater models as interactive management tools, discussing their applications, benefits, and limitations.

Groundwater Models

Groundwater models are mathematical representations of the hydrological processes that occur in aquifers. These models use mathematical equations to simulate the movement of groundwater and the associated transport of solutes. The equations used in groundwater models are based on fundamental principles of fluid mechanics and mass transport. The most commonly used groundwater models are based on numerical methods that discretize the groundwater system into small elements, allowing for the simulation of the flow of water and the transport of contaminants between these elements.

Groundwater models can be used for a range of purposes, including aquifer characterization, evaluation of management scenarios, and prediction of future conditions. These models are particularly useful in situations where direct measurement of groundwater flow and transport is difficult or expensive. Groundwater models can also be used to predict the response of an aquifer to changes in pumping rates, recharge rates, or land use.

Interactive Management

Groundwater models can be used as interactive management tools to aid in the management of groundwater resources. Interactive management involves the use of models to evaluate alternative management scenarios and to facilitate stakeholder engagement in the decision-making process. Interactive management allows stakeholders to explore the consequences of different management strategies and to make informed decisions based on the best available information.

One of the main advantages of using groundwater models as interactive management tools is that they allow stakeholders to evaluate the long-term impacts of management decisions. For example, a groundwater model can be used to evaluate the impacts of different pumping rates on the groundwater system over several decades. This allows stakeholders to evaluate the sustainability of different management strategies and to identify the most effective strategies for achieving their goals.

Benefits of Groundwater Models as Interactive Management Tools

There are several benefits to using groundwater models as interactive management tools. Some of the most significant benefits include:

Improved Decision-Making: Groundwater models provide stakeholders with the information needed to make informed decisions about the management of groundwater resources. By simulating the impacts of different management scenarios, stakeholders can evaluate the likely outcomes of each scenario and make decisions based on the best available information.

Stakeholder Engagement: Interactive management involves stakeholders in the decision-making process, allowing them to provide input and feedback on the management strategies being considered. This engagement improves the likelihood of successful implementation of management strategies, as stakeholders are more likely to support strategies that they have been involved in developing.

Long-Term Planning: Groundwater models allow for long-term planning, enabling stakeholders to evaluate the sustainability of different management strategies over several decades. This long-term planning is particularly important in regions where groundwater is the primary source of water for human consumption and irrigation.

Cost-Effective: Groundwater models are often more cost-effective than traditional monitoring approaches. By simulating the behavior of the groundwater system, groundwater models can provide information on the response of the system to different management strategies, reducing the need for expensive field measurements.

Limitations of Groundwater Models as Interactive Management Tools

While there are many benefits to using groundwater models as interactive management tools, there are also some limitations that must be considered. Some of the most significant limitations include:

Data Requirements: Groundwater models require a significant amount of data, including hydrogeological parameters, pumping rates, and recharge rates. Obtaining this data can be time-consuming and expensive.

Uncertainty: Groundwater models are subject to uncertainty, particularly in situations where data is limited or uncertain. This uncertainty can make it difficult to predict the response of the groundwater system to management strategies.

Complexity: Groundwater models can be complex and difficult to understand, particularly for stakeholders who are not familiar with hydrological modelling. This complexity can make it difficult to effectively engage stakeholders in the decision-making process.

Model Validity: Groundwater models are only as valid as the data used to develop them. If the data is inaccurate or incomplete, the model results may be unreliable.

Applications of Groundwater Models as Interactive Management Tools

Groundwater models have a wide range of applications as interactive management tools. Some of the most significant applications include:

Water Resources Management: Groundwater models can be used to evaluate different management scenarios for water resources. For example, a groundwater model can be used to evaluate the impacts of different pumping rates on the availability of water for irrigation and human consumption.

Contaminant Transport: Groundwater models can be used to evaluate the transport of contaminants in groundwater systems. For example, a groundwater model can be used to evaluate the potential impacts of a chemical spill on the groundwater system.

Land Use Planning: Groundwater models can be used to evaluate the impacts of land use changes on the groundwater system. For example, a groundwater model can be used to evaluate the impacts of urbanization on groundwater recharge rates.

Climate Change Adaptation: Groundwater models can be used to evaluate the impacts of climate change on groundwater resources. For example, a groundwater model can be used to evaluate the impacts of changes in precipitation patterns on groundwater recharge rates.

Case Studies

There are many examples of the successful use of groundwater models as interactive management tools. Two case studies are presented below to illustrate the application of groundwater models in different contexts.

Case Study 1: Central Valley Groundwater Model, California, USA

The Central Valley Groundwater Model is a regional groundwater model developed to aid in the management of groundwater resources in the Central Valley of California, USA. The Central Valley is one of the most productive agricultural regions in the world, and groundwater is the primary source of water for irrigation.

The Central Valley Groundwater Model was developed to simulate groundwater flow and transport in the Central Valley aquifer system. The model was used to evaluate the impacts of different management scenarios on groundwater resources over a 100-year planning horizon. The management scenarios evaluated included changes in pumping rates, changes in land use, and implementation of water conservation measures.

The use of the Central Valley Groundwater Model as an interactive management tool allowed stakeholders to evaluate the long-term impacts of different management strategies. The model results were used to inform the development of a sustainable groundwater management plan for the Central Valley.

Case Study 2: North China Plain Groundwater Model, China

The North China Plain is one of the most productive agricultural regions in China, and groundwater is the primary source of water for irrigation. The North China Plain Groundwater Model was developed to aid in the management of groundwater resources in the region.

The North China Plain Groundwater Model was used to evaluate the impacts of different management scenarios on groundwater resources over a 50-year planning horizon. The management scenarios evaluated included changes in pumping rates, changes in land use, and implementation of water conservation measures.

The use of the North China Plain Groundwater Model as an interactive management tool allowed stakeholders to evaluate the impacts of different management strategies on the sustainability of groundwater resources. The model results were used to inform the development of a sustainable groundwater management plan for the North China Plain.

Conclusion

Groundwater models offer several benefits as interactive management tools, including improved decision-making, stakeholder engagement, long-term planning, and cost-effectiveness. However, groundwater models have some limitations, such as data requirements, uncertainty, complexity, and model validity. Despite these limitations, groundwater models have a wide range of applications in water resources management, contaminant transport, land use planning, and climate change adaptation. The case studies presented in this article demonstrate the successful use of groundwater models as interactive management tools in real-world situations. Groundwater models are valuable tools for ensuring the sustainable management of groundwater resources.

References

Faunt, C. C., & Sneed, M. (2016). *Groundwater availability of the Central Valley Aquifer, California*. US Geological Survey.

Harter, T. (2016). *The California groundwater revolution: a model for sustainable groundwater management*. Groundwater, 54(3), 297-308.

Wang, J., Li, Y., & Li, W. (2019). *A coupled water and nitrogen transport model for the North China Plain*. Journal of Environmental Management, 231, 42-51.

Wang, X., Fan, H., & Zheng, C. (2016). *Assessment of groundwater resources and groundwater management in the North China Plain*. Environmental Earth Sciences, 75(4), 1-15.

75 - Soil and Water Assessment Tool (SWAT): Simulating Water Quality and Quantity in Agricultural Watersheds

Abstract

The Soil and Water Assessment Tool (SWAT) model is a powerful tool for simulating the movement of water, sediment, nutrients, and pesticides in agricultural watersheds. This article provides an overview of the components of the SWAT model, including weather, land use, soil, hydrology, sediment, nutrients, and pesticides, and their applications in managing water resources in agricultural watersheds. The article also discusses the limitations of the SWAT model and its importance in understanding and managing water resources in agricultural watersheds.

Introduction

Water is a critical resource for agriculture, and its availability and quality can significantly impact crop production and the environment. With the increasing demand for food and water resources, managing water resources in agricultural watersheds has become a priority for policymakers and researchers. To effectively manage water resources in agricultural watersheds, it is essential to understand the movement of water, sediment, nutrients, and pesticides in the watershed.

The Soil and Water Assessment Tool (SWAT) model is a widely used tool for simulating water quality and quantity in agricultural watersheds. The SWAT model simulates the hydrologic cycle in the watershed, including the movement of water, sediment, nutrients, and pesticides. The model consists of several components, including weather, land use, soil, hydrology, sediment, nutrients, and pesticides. Each component simulates a different aspect of the water cycle and allows for a comprehensive understanding of water quality and quantity in the watershed. It is widely used by researchers, watershed managers, and policy-makers to assess the impact of land use, climate change, and management practices on water resources.

History

The development of SWAT began in the late 1980s as a collaboration between the USDA Agricultural Research Service (ARS) and Texas A&M University. The initial goal was to develop a tool that could evaluate the impact of land use change on water quality in agricultural watersheds. The first version of the model was released in 1996, and since then, several updates and modifications have been made to improve the model's accuracy and usability.

Model Overview

The SWAT model is a semi-distributed, continuous-time, and physically based model that uses daily time steps to simulate the water quality and quantity in a watershed. The model is divided into sub-basins, and each sub-basin is further divided into hydrological response units (HRUs) based on land use, soil type, and slope. The HRUs are the smallest unit of the model, and each HRU represents a unique combination of land use, soil, and management practices.

The model simulates the movement of water, sediment, nutrients, and pesticides through a watershed using a combination of hydrologic and hydraulic processes. The hydrologic processes include surface runoff, subsurface flow, lateral flow, and groundwater flow. The hydraulic processes include channel routing and reservoir routing.

SWAT uses a variety of input data, including climate data, soil data, land use data, and management practices data. The model requires a minimum of ten years of climate data to run, but longer datasets can be used to improve the accuracy of the model. The model also requires information about the land use and management practices in the watershed, including crop rotations, tillage practices, and nutrient application rates.

The model output includes daily streamflow, sediment yield, nutrient loadings, and pesticide loadings at the watershed outlet. The model also provides information about the hydrologic balance of the watershed, including the water balance, soil water balance, and groundwater balance. The model output can be used to evaluate the impact of land use change, climate change, and management practices on water resources.

Model Components

Weather generator

The weather generator component of the SWAT model simulates the daily weather data needed for the model. The component uses statistical methods to generate daily weather data, including precipitation, temperature, solar radiation, and wind speed. The weather generator can generate weather data for multiple years, and the generated data can be used to simulate long-term hydrological conditions.

Soil and water assessment tool watershed editor

The SWAT watershed editor is a graphical user interface that allows users to create and edit the SWAT model input files. The editor is used to define the watershed boundaries, sub-basins, and HRUs. The editor also allows users to input land use data, soil data, management practices data, and climate data.

Land use

The land use component of the SWAT model represents the different types of land use in the watershed, including agricultural land, forest land, urban land, and water bodies. The model uses land use data to determine the amount of water and nutrients that are available for plant growth, as well as the amount of runoff and erosion that occurs in each HRU.

Soil

The soil component of the SWAT model represents the different types of soil in the watershed. The model uses soil data to determine the physical and chemical properties of the soil, including its water holding capacity, infiltration rate, and nutrient availability. The soil data is used to calculate the water balance in each HRU, including the amount of water that is stored in the soil and the amount of water that is lost to evapotranspiration or runoff.

Hydrology

The hydrology component of the SWAT model simulates the movement of water through the watershed. The model uses a combination of surface runoff, subsurface flow, lateral flow, and groundwater flow to simulate the hydrologic processes in each HRU. The surface runoff component of the model calculates the amount of water that runs off the surface of the land and enters the stream channel. The subsurface flow component of the model calculates the amount of water that infiltrates into the soil and then moves through the soil profile to the groundwater. The lateral flow component of the model calculates the amount of water that moves laterally across the land surface before entering the stream channel. The groundwater flow component of the model calculates the movement of water through the groundwater system.

Sediment

The sediment component of the SWAT model simulates the transport of sediment through the watershed. The model uses a combination of erosion, deposition, and transport processes to simulate the movement of sediment in the watershed. The model calculates the amount of sediment that is eroded from each HRU and then transported downstream to the stream channel. The model also calculates the amount of sediment that is deposited in the stream channel and on the floodplain.

Nutrients

The nutrient component of the SWAT model simulates the transport of nutrients through the watershed. The model uses a combination of nutrient uptake, nutrient transport, and nutrient deposition processes to simulate the movement of nutrients in the watershed. The model calculates the amount of nutrients that are taken up by plants, the amount of nutrients that are transported through the soil profile and into the stream channel, and the amount of nutrients that are deposited in the stream channel and on the floodplain.

Pesticides

The pesticide component of the SWAT model simulates the transport of pesticides through the watershed. The model uses a combination of adsorption, desorption, and degradation processes to simulate the movement of pesticides in the watershed. The model calculates the amount of pesticides that are applied to the land surface, the amount of pesticides that are adsorbed to the soil particles, and the amount of pesticides that are transported through the soil profile and into the stream channel.

Applications

The SWAT model has been widely used for a variety of applications, including:

- Evaluating the impact of land use change on water resources.
- Assessing the impact of climate change on water resources.
- Developing and evaluating management practices to reduce water pollution.
- Assessing the impact of agricultural practices on water quality and quantity.
- Developing watershed management plans.
- Evaluating the effectiveness of conservation programs.
- Developing water allocation plans.
- Developing water quality standards.

Limitations

While the SWAT model is a powerful tool for simulating water quality and quantity in agricultural watersheds, there are some limitations to its use. Some of these limitations include:

- ➢ The model requires a significant amount of input data, including climate data, land use data, soil data, and management practices data.
- ➢ The model can be computationally intensive and requires a significant amount of computing power to run.
- ➢ The model requires calibration and validation using field data, which can be time-consuming and costly.
- ➢ The model assumes that the watershed is in a steady-state condition, which may not always be the case.
- ➢ The model does not account for the impact of groundwater pumping on the water balance of the watershed.

Conclusion

The Soil and Water Assessment Tool (SWAT) model is a powerful tool for simulating the movement of water, sediment, nutrients, and pesticides in agricultural watersheds. The model has been widely used for a variety of applications, including evaluating the impact of land use change, assessing the impact of climate change, and developing and evaluating management practices to reduce water pollution.

The SWAT model consists of several components, including weather, land use, soil, hydrology, sediment, nutrients, and pesticides. Each component simulates a different aspect of the water cycle and allows for a comprehensive understanding of water quality and quantity in the watershed.

Despite its many strengths, the SWAT model has some limitations that should be considered when using it for watershed management. These limitations include the significant amount of input data required, the computational intensity of the model, and the need for calibration and validation using field data.

In conclusion, the SWAT model is an important tool for managing water resources in agricultural watersheds. By simulating the movement of water, sediment, nutrients, and pesticides, the model can help identify areas of the watershed that are at risk of water pollution and develop management practices to reduce that risk. While there are limitations to its use, the SWAT model remains a valuable tool for understanding and managing water resources in agricultural watersheds.

76 - Understanding SWAT-MODFLOW: An Integrated Modelling System for Surface Water and Groundwater Processes

Abstract

SWAT-MODFLOW is a powerful modelling system that integrates the SWAT and MODFLOW models to simulate surface water and groundwater processes. This article provides an overview of the model, including its key features, benefits, and limitations. We discuss the data requirements, model complexity, and computational resources needed to use the model effectively, as well as the potential applications for water resource management and environmental assessment. By understanding the capabilities and limitations of SWAT-MODFLOW, researchers and water resource managers can make more informed decisions about water allocation, land use planning, and water quality management. In this article, we will explore the key features of SWAT-MODFLOW, its applications, and the benefits it offers for water resource management.

Introduction

Water resources management is a critical issue in many regions of the world, particularly in areas where water is scarce. Effective management of water resources requires an understanding of the complex interactions between surface water and groundwater. Historically, these two systems have been modeled separately, which can lead to inaccurate predictions and suboptimal management decisions. To address this challenge, researchers have developed coupled modelling systems that integrate surface water and groundwater models. SWAT-MODFLOW is one such system that has gained widespread use in recent years.

Overview of SWAT-MODFLOW

SWAT-MODFLOW is a coupled modelling system that integrates the SWAT and MODFLOW models. SWAT is a widely used model for simulating surface water and watershed processes, while MODFLOW is a groundwater flow model. The integration of these two models allows for the simultaneous simulation of surface water and groundwater processes, making it possible to capture the complex interactions between these two systems.

SWAT-MODFLOW operates on a daily time step and simulates water balance components, such as precipitation, evapotranspiration, runoff, infiltration, and groundwater recharge. The model also accounts for land use, soil characteristics, topography, and climate data. The spatial resolution of SWAT-MODFLOW can range from a small watershed to a large river basin, depending on the available data and the research objectives.

One of the key features of SWAT-MODFLOW is its ability to represent groundwater-surface water interactions. The model simulates groundwater recharge, which is the process by which water infiltrates the soil and replenishes the groundwater table. This recharge can then discharge into surface water bodies, such as rivers and lakes. The model also accounts for the impact of surface water on groundwater, such as the effects of streamflow on the water table elevation and the exchange of water between the two systems.

SWAT-MODFLOW also includes the ability to simulate water management scenarios. For example, the model can simulate the effects of irrigation on groundwater levels, the impact of groundwater pumping on surface water, and the effects of land use changes on water availability. These scenarios can help water resource managers make informed decisions about water allocation and land use planning.

Applications of SWAT-MODFLOW

SWAT-MODFLOW has a wide range of applications in water resources management and environmental assessment. Some of the key applications of SWAT-MODFLOW include:

Groundwater management: SWAT-MODFLOW can be used to simulate the effects of groundwater pumping on surface water and vice versa. This can help water resource managers make informed decisions about groundwater use and allocation.

Water quality assessment: SWAT-MODFLOW can be used to simulate the fate and transport of pollutants in surface water and groundwater. This can help identify areas of water contamination and inform remediation efforts.

Climate change impacts: SWAT-MODFLOW can be used to simulate the impacts of climate change on water resources. For example, the model can simulate changes in precipitation patterns, evapotranspiration rates, and groundwater recharge rates.

Land use planning: SWAT-MODFLOW can be used to simulate the effects of land use changes on water resources. This can help inform land use planning decisions and identify areas where water availability may be limited.

Water allocation: SWAT-MODFLOW can be used to simulate different water allocation scenarios, such as the effects of different water management strategies on water availability and quality.

Benefits of SWAT-MODFLOW

SWAT-MODFLOW offers several benefits for water resource management and environmental assessment. Some of the key benefits of the model include:

Comprehensive analysis: SWAT-MODFLOW allows for the comprehensive analysis of surface water and groundwater processes. This allows for a more accurate representation of the complex interactions between these two systems.

Integrated approach: The integration of the SWAT and MODFLOW models allows for the simultaneous simulation of surface water and groundwater processes. This makes it possible to capture the feedbacks and interactions between these two systems.

Enhanced accuracy: The integrated modelling approach of SWAT-MODFLOW leads to more accurate predictions of water availability, quality, and management. This can help inform water resource management decisions and improve environmental assessment.

Improved decision-making: The ability to simulate water management scenarios with SWAT-MODFLOW can help inform decision-making related to water allocation, land use planning, and water quality management. This can lead to more effective and sustainable water resource management.

Flexible spatial resolution: SWAT-MODFLOW can be used to simulate water resources at different spatial resolutions, depending on the available data and research objectives. This allows for a wide range of applications, from small watersheds to large river basins.

Limitations of SWAT-MODFLOW

While SWAT-MODFLOW offers many benefits for water resource management and environmental assessment, there are also some limitations to the model. Some of the key limitations include:

Data requirements: SWAT-MODFLOW requires a significant amount of input data, including land use, soil characteristics, topography, climate data, and groundwater characteristics. Obtaining this data can be challenging, particularly in data-poor regions.

Model complexity: SWAT-MODFLOW is a complex model that requires significant expertise to set up and calibrate. This can be a barrier to entry for some users, particularly those without experience in hydrological modelling.

Computational resources: The simulation of surface water and groundwater processes at high spatial resolutions can require significant computational resources. This can be a limiting factor for some users, particularly those with limited access to high-performance computing resources.

Model uncertainty: As with any modelling approach, there is uncertainty associated with the predictions generated by SWAT-MODFLOW. This uncertainty can arise from errors in input data, model structure, and calibration.

Conclusion

SWAT-MODFLOW is a powerful modelling system that integrates the SWAT and MODFLOW models to simulate surface water and groundwater processes. The model allows for the comprehensive analysis of water resources, including the complex interactions between surface water and groundwater. SWAT-MODFLOW has a wide range of applications in water resource management and environmental assessment, including groundwater management, water quality assessment, climate change impacts, land use planning, and water allocation. While there are some limitations to the model, including data requirements, model complexity, and computational resources, the benefits of SWAT-MODFLOW make it a valuable tool for water resource managers and researchers.

77 - Exploring the Capabilities and Limitations of the MIKE SHE Model in Hydrological Modelling

Abstract

The MIKE SHE model is a widely used hydrological model that simulates the movement of water through the hydrological system, from precipitation to surface runoff, infiltration, evapotranspiration, and groundwater flow. It is a spatially-distributed model that takes into account variations in soil properties, vegetation cover, topography, and climate across the landscape. The model has a wide range of applications in flood forecasting, water resources management, land-use planning, and environmental modelling. However, it also has limitations and challenges, including data requirements, model complexity, parameter uncertainty, and model validation. This article provides an overview of the MIKE SHE, its components, and its applications, as well as a discussion of its limitations and challenges.

Introduction

The MIKE SHE model is a well-known and widely used integrated hydrological model developed by the Danish Hydraulic Institute (DHI). It is a three-dimensional, spatially-distributed model that simulates water movement through the soil, vegetation, and the atmosphere. MIKE SHE was first developed in the 1990s and has since undergone several revisions and updates, becoming one of the most sophisticated and widely used hydrological models in the world.

The MIKE SHE model is used in a variety of applications, including flood forecasting, land-use planning, and water resources management. It has been used to model water flow and quality in a variety of settings, including urban and rural environments, as well as in wetlands and other natural systems.

This article provides an overview of the MIKE SHE, its main features, and its applications. We will explore the basic principles of the model, how it works, and how it is used to simulate hydrological processes.

Basic Principles of MIKE SHE

The MIKE SHE model is based on the principle of mass balance, which means that the total amount of water in the system must be conserved. The model simulates the movement of water through different compartments of the hydrological system, including the atmosphere, the soil, and the vegetation.

The model is designed to simulate a wide range of hydrological processes, including surface runoff, infiltration, evapotranspiration, groundwater flow, and subsurface flow. It uses a complex set of equations to simulate the interactions between these processes and the various compartments of the hydrological system.

The MIKE SHE model is a spatially-distributed model, meaning that it simulates water movement and hydrological processes across a three-dimensional grid. This allows the model to take into account variations in soil properties, vegetation cover, topography, and climate across the landscape.

How the MIKE SHE works

The MIKE SHE model is made up of several components, each of which simulates a different aspect of the hydrological system. These components include the soil water balance component, the surface water balance component, the groundwater flow component, and the surface-subsurface interaction component.

Soil water balance component

The soil water balance component of the MIKE SHE simulates the movement of water through the soil. It takes into account the soil properties, including soil texture, porosity, and hydraulic conductivity, as well as the soil moisture content.

The model calculates the soil water balance by considering the inflows and outflows of water in the soil. Inflows include precipitation, irrigation, and groundwater recharge, while outflows include evapotranspiration, surface runoff, and subsurface flow.

The soil water balance component also takes into account the effect of vegetation on soil moisture content. Vegetation can significantly affect the amount of water in the soil, as plants take up water through their roots and release it through transpiration.

Surface water balance component

The surface water balance component of the MIKE SHE simulates the movement of water across the land surface. It takes into account the topography, soil properties, and vegetation cover, as well as the amount and intensity of rainfall.

The surface water balance component calculates the surface runoff by considering the amount of rainfall that exceeds the soil infiltration capacity. The model also takes into account the effect of vegetation cover on surface runoff, as well as the effect of land use and land cover on surface water flow.

Groundwater flow component

The groundwater flow component of the MIKE SHE simulates the movement of water through the subsurface. It takes into account the geology, soil properties, and groundwater recharge and discharge.

The model calculates the groundwater flow by considering the inflows and outflows of water in the subsurface. Inflows include groundwater recharge, while outflows include groundwater discharge.

Surface-subsurface interaction component

The surface-subsurface interaction component of the MIKE SHE simulates the interaction between the surface water and groundwater systems. It takes into account the hydraulic properties of the soil and the subsurface, as well as the surface water and groundwater flow rates.

The model calculates the surface-subsurface interactions by considering the hydraulic gradient between the surface and subsurface systems. It also takes into account the effect of groundwater on the soil moisture content and the effect of surface water on groundwater recharge.

Overall, the MIKE SHE integrates these different components to simulate the movement of water through the hydrological system, from precipitation to surface runoff, infiltration, evapotranspiration, and groundwater flow.

Applications of the MIKE SHE

The MIKE SHE has a wide range of applications in hydrology, water resources management, and environmental modelling. Some of its main applications include:

Flood forecasting: The MIKE SHE model is used to simulate the flow of water in rivers and other waterways, which can be used to predict and mitigate flooding events.

Water resources management: The MIKE SHE model is used to simulate water availability and usage in a variety of settings, from urban water supplies to agricultural irrigation systems.

Land-use planning: The MIKE SHE model is used to simulate the effects of land-use changes, such as deforestation or urbanization, on the hydrological system.

Environmental modelling: The MIKE SHE model is used to simulate the movement of pollutants through the hydrological system, including nutrients, sediment, and other contaminants.

In addition to these applications, the MIKE SHE model is also used in research to better understand the processes that govern the hydrological system, including the interactions between surface water and groundwater, the effects of climate change on hydrology, and the role of vegetation in regulating the water balance.

Limitations and Challenges of MIKE SHE

Like all models, the MIKE SHE has some limitations and challenges that must be taken into account when interpreting its results. Some of the main limitations of the model include:

Data requirements: The MIKE SHE model requires a significant amount of data to be inputted, including soil properties, vegetation cover, topography, and climate data. The quality and availability of this data can affect the accuracy of the model's results.

Model complexity: The MIKE SHE model is a complex and computationally intensive model, which can make it difficult to run and interpret. It also requires significant expertise to use and configure properly.

Parameter uncertainty: The MIKE SHE includes many parameters that must be calibrated to fit the specific characteristics of the study area. The uncertainty in these parameters can affect the accuracy of the model's results.

Model validation: The MIKE SHE model must be validated against observed data to ensure that its results are accurate. This can be challenging, as there may be discrepancies between the model's results and observed data due to measurement errors or other factors.

Conclusion

The MIKE SHE model is a sophisticated and widely used hydrological model that simulates the movement of water through the hydrological system. It is a spatially-distributed model that takes into account variations in soil properties, vegetation cover, topography, and climate across the landscape.

The MIKE SHE has a wide range of applications, including flood forecasting, water resources management, land-use planning, and environmental modelling. However, the model also has some limitations and challenges, including data requirements, model complexity, parameter uncertainty, and model validation.

Overall, the MIKE SHE model is an important tool for understanding and managing the hydrological system, and it is likely to continue to be used in a wide range of applications in the future.

78 - The Causes, Effects, and Preparedness for Natural Disasters: Understanding the Threats to Human Life and the Environment

Abstract

This article provides an overview of natural disasters, including their types, causes, effects, and how to prepare for them. The types of natural disasters covered include earthquakes, floods, hurricanes, tornadoes, volcanic eruptions, and wildfires. The causes of natural disasters include climate change, human activities, tectonic activity, and weather patterns. The effects of natural disasters include loss of life, property damage, environmental damage, and economic effects. To prepare for natural disasters, individuals and communities can develop emergency preparedness kits and plans, adhere to building codes and regulations, and obtain insurance coverage.

Introduction

Natural disasters are catastrophic events that occur naturally and have the potential to cause severe harm to human life, property, and the environment. These disasters can happen anywhere in the world and can be triggered by several natural phenomena such as earthquakes, floods, hurricanes, tornadoes, volcanic eruptions, and wildfires, among others. Natural disasters have been a part of human history since time immemorial, and their occurrence and intensity have only increased in recent years due to climate change, population growth, and urbanization. In this article, we will explore the different types of natural disasters, their causes, effects, and how to prepare for them.

Types of Natural Disasters

Earthquakes

An earthquake is a sudden shaking or trembling of the earth's surface caused by the movement of tectonic plates. Earthquakes can occur anywhere, but they are most common in areas with high tectonic activity, such as the Pacific Rim. The intensity of an earthquake is measured on the Richter scale, which ranges from 1 to 10. A 10 on the Richter scale is the most severe and can cause massive destruction and loss of life.

Floods

Floods are the most common natural disasters globally and occur when water levels rise beyond their normal boundaries. They can be caused by heavy rains, melting snow, or tsunamis. Floods can cause widespread damage to infrastructure, homes, and businesses, and can also lead to loss of life.

Hurricanes

A hurricane is a severe tropical cyclone that occurs in the Atlantic Ocean and Northeast Pacific Ocean. They are known by different names in different regions, such as typhoons in the Western Pacific and cyclones in the Indian Ocean. Hurricanes are characterized by high winds, heavy rains, and storm surges, which can cause severe flooding and damage to homes, buildings, and infrastructure.

Tornadoes

Tornadoes are violent rotating columns of air that touch the ground and can cause severe damage to homes, buildings, and infrastructure. They are most common in the United States but can occur anywhere in the world. Tornadoes are characterized by their funnel shape, high winds, and the destruction they cause in their path.

Volcanic Eruptions

Volcanic eruptions occur when magma, ash, and gas are released from a volcano. They can cause severe damage to homes, buildings, and infrastructure and can also lead to loss of life. Volcanic eruptions are most common in areas with high volcanic activity, such as the Pacific Ring of Fire.

Wildfires

Wildfires are uncontrolled fires that occur in areas with dry vegetation and can be caused by lightning strikes or human activities. They can cause massive destruction to homes, buildings, and infrastructure, and can also lead to loss of life. Wildfires are most common in areas with dry climates, such as the western United States and Australia.

Causes of Natural Disasters

Climate Change

Climate change has been identified as a significant contributor to the increase in natural disasters globally. The rise in global temperatures has led to more frequent and intense weather events such as floods, hurricanes, and wildfires. Climate change is also causing sea levels to rise, which can lead to severe flooding and damage to coastal areas.

Human Activities

Human activities such as deforestation, urbanization, and mining can lead to natural disasters. Deforestation, for example, can increase the risk of landslides and floods, while urbanization can lead to more significant flooding and water pollution. Mining can cause landslides and earthquakes, which can be devastating to communities.

Tectonic Activity

Tectonic activity, such as earthquakes and volcanic eruptions, is a natural cause of natural disasters. These events occur when tectonic plates move, causing seismic activity and volcanic eruptions. These events can be unpredictable, and their occurrence and intensity can cause severe damage to infrastructure and loss of life.

Weather Patterns

Weather patterns, such as El Niño and La Niña, can also contribute to natural disasters. El Niño, for example, can cause droughts in some regions and heavy rains in others, leading to floods and landslides. La Niña, on the other hand, can lead to increased hurricane activity in the Atlantic.

Effects of Natural Disasters

Loss of Life

Natural disasters can cause significant loss of life, either directly or indirectly. For example, earthquakes can cause buildings to collapse, leading to casualties, while floods and hurricanes can cause drowning and injuries.

Property Damage

Natural disasters can also cause significant damage to infrastructure, homes, and businesses. This damage can be costly to repair, and it can take years for communities to recover fully.

Environmental Damage

Natural disasters can cause significant environmental damage, such as deforestation, erosion, and water pollution. This damage can have long-lasting effects on ecosystems and the environment.

Economic Effects

Natural disasters can have a significant impact on the economy, such as loss of productivity, job losses, and increased government spending on disaster relief efforts.

Preparing for Natural Disasters

Emergency Preparedness Kits

It is crucial to have an emergency preparedness kit that includes essential items such as water, food, a first aid kit, flashlights, and batteries. This kit should be easily accessible in case of an emergency.

Emergency Plans

Having an emergency plan in place can help minimize the damage caused by natural disasters. This plan should include evacuation routes, communication plans, and emergency contact information.

Building Codes and Regulations

Building codes and regulations can help ensure that buildings are designed and constructed to withstand natural disasters. These codes should be strictly enforced to reduce the risk of damage and loss of life.

Insurance

Having insurance coverage can help individuals and communities recover from the financial impacts of natural disasters. It is essential to ensure that insurance policies cover natural disasters and that coverage is adequate.

Conclusion

Natural disasters are a significant threat to human life, property, and the environment. Understanding the causes and effects of natural disasters is crucial in preparing for and mitigating their impact. It is essential to have emergency plans and preparedness kits in place, enforce building codes and regulations, and ensure adequate insurance coverage. By taking these steps, communities can reduce the impact of natural disasters and recover more quickly. However, it is crucial to note that natural disasters can be unpredictable and that preparation is key in minimizing their impact.

79 - Mitigating the Impact of Cloud Bursts: Prevention, Preparedness, and Response Measures

Abstract

This article explores the phenomenon of cloud bursts, a natural event that can cause significant damage to infrastructure, property, and human life. The article examines the complex causes of cloud bursts and the severe impacts they can have, particularly in developing countries where infrastructure and emergency response systems are often inadequate. The article discusses measures to mitigate the effects of cloud bursts, which can be broadly classified into prevention, preparedness, and response measures. These measures include early warning systems, emergency response plans, emergency shelter and supplies, and emergency services. The article emphasizes the need for a coordinated and sustained effort from governments, non-governmental organizations, and communities to build more resilient communities that are better prepared to withstand the impacts of cloud bursts and other extreme weather events.

Introduction

Cloud bursts are one of the most destructive natural phenomena that can cause significant damage to infrastructure, human life, and property. Cloud bursts are characterized by heavy rainfall over a small area for a short duration, typically less than six hours. The sudden and intense rainfall can cause flash floods, landslides, and damage to infrastructure like roads, bridges, and buildings. Cloud bursts are a significant concern for regions with hilly terrain and steep slopes, where the runoff can accumulate quickly and lead to widespread damage. This article will provide an in-depth analysis of cloud bursts, their causes, impacts, and measures to mitigate their effects.

What are Cloud Bursts?

Cloud bursts are intense and localized rainfall events that occur when warm and humid air rises and cools, leading to the condensation of water vapor into clouds. When the clouds become saturated with water, they release the excess moisture in the form of rain. Cloud bursts typically occur during the monsoon season or in regions with a high frequency of thunderstorms. Cloud bursts are characterized by their short duration and high intensity, with rainfall rates often exceeding 100 mm per hour.

Causes of Cloud Bursts

Cloud bursts are caused by a combination of atmospheric and geographical factors. The most common cause of cloud bursts is the interaction of warm and humid air with cool air masses. This interaction leads to the formation of cumulonimbus clouds, which are capable of producing heavy rainfall. The warm and humid air rises due to convection, which is the transfer of heat energy from the Earth's surface to the atmosphere. As the warm air rises, it cools and condenses into clouds. If the cooling is rapid enough, it can lead to the formation of ice particles, which can further enhance the rainfall intensity.

Another factor that contributes to cloud bursts is the topography of the region. Regions with steep slopes and hilly terrain are more prone to cloud bursts as the runoff from the rainfall accumulates quickly and leads to flash floods and landslides. The soil in these regions is often unable to absorb the excess water, which can cause erosion and damage to infrastructure.

Impacts of Cloud Bursts

Cloud bursts can cause significant damage to infrastructure, property, and human life. The sudden and intense rainfall can lead to flash floods, landslides, and damage to roads, bridges, and buildings. The runoff from the rainfall can cause erosion, which can lead to the loss of fertile soil and damage to crops. The floodwater can also carry pollutants and debris, which can contaminate water sources and pose a risk to human health.

The impacts of cloud bursts can be particularly severe in developing countries, where infrastructure and emergency response systems are often inadequate. In these regions, cloud bursts can lead to widespread displacement, loss of livelihoods, and economic losses.

Measures to Mitigate the Effects of Cloud Bursts

There are several measures that can be taken to mitigate the effects of cloud bursts. These measures can be broadly classified into three categories: prevention, preparedness, and response.

Prevention

Prevention measures are aimed at reducing the likelihood of cloud bursts occurring in the first place. One effective prevention measure is the implementation of sustainable land use practices. These practices can help reduce soil erosion, increase the water-holding capacity of the soil, and promote the infiltration of rainwater. Examples of sustainable land use practices include agroforestry, terracing, and the use of cover crops.

Another prevention measure is the implementation of effective stormwater management systems. These systems can help capture and store rainwater, which can then be used for irrigation, groundwater recharge, or other purposes. Effective stormwater management systems can also help reduce the risk of flash floods and landslides by providing a safe and controlled path for the runoff.

Preparedness

Preparedness measures are aimed at reducing the impact of cloud bursts and improving the ability of communities to respond to the event. One key preparedness measure is the development of early warning systems. Early warning systems can provide advance notice of impending cloud bursts, allowing communities to prepare and evacuate if necessary. Early warning systems can be based on a range of technologies, including satellite imagery, weather radar, and ground-based sensors.

Another preparedness measure is the development of emergency response plans. Emergency response plans should outline the roles and responsibilities of different stakeholders, including government agencies, non-governmental organizations, and community members. The plans should also identify the resources needed to respond to the event, such as equipment, personnel, and emergency supplies.

Response

Response measures are aimed at providing immediate assistance to affected communities and minimizing the impact of the event. One key response measure is the provision of emergency shelter and supplies. Emergency shelters can provide a safe place for displaced individuals and families to stay, while emergency supplies like food, water, and medical supplies can help meet basic needs.

Another response measure is the provision of emergency services like search and rescue, medical assistance, and debris removal. These services can help address immediate needs and prevent further damage to infrastructure and property.

Conclusion

Cloud bursts are a significant natural phenomenon that can cause significant damage to infrastructure, property, and human life. The causes of cloud bursts are complex, involving a combination of atmospheric and geographical factors. The impacts of cloud bursts can be severe, particularly in developing countries where infrastructure and emergency response systems are often inadequate.

Measures to mitigate the effects of cloud bursts can be broadly classified into prevention, preparedness, and response. Prevention measures aim to reduce the likelihood of cloud bursts occurring in the first place, while preparedness measures aim to reduce the impact of the event and improve the ability of communities to respond. Response measures aim to provide immediate assistance to affected communities and minimize the impact of the event.

Given the increasing frequency and severity of extreme weather events, it is essential to continue developing and implementing measures to mitigate the effects of cloud bursts. This will require a coordinated and sustained effort from governments, non-governmental organizations, and communities. By working together, we can build more resilient communities that are better prepared to withstand the impacts of cloud bursts and other extreme weather events.

80 - Melting Glaciers: Causes, Effects, and Potential Solutions for a Sustainable Future

Abstract

Glaciers are melting at an unprecedented rate due to human-induced climate change, causing a variety of environmental and socio-economic impacts. This article explores the causes, effects, and potential solutions to melting glaciers. The primary cause of melting glaciers is global warming, which is caused by human activities such as burning fossil fuels and deforestation. Changes in precipitation patterns and ocean warming also contribute to melting glaciers. The effects of melting glaciers include sea-level rise, changes in water availability, changes in ecosystems, and an increase in natural disasters. To address this issue, solutions such as reducing greenhouse gas emissions, adapting to changes, geoengineering, and international cooperation are discussed. The article concludes that urgent action is needed to prevent the worst impacts of melting glaciers and ensure a sustainable future for all.

Introduction

Glaciers are large masses of ice that accumulate over time from snowfall and are found in high-altitude areas such as mountains, polar regions, and even in some low-altitude areas like Patagonia. They play a significant role in the world's ecosystem by providing freshwater for irrigation, hydroelectric power, and domestic consumption. However, with the advent of climate change, glaciers are melting at an unprecedented rate, leading to a variety of environmental and socio-economic impacts. This article will explore the causes, effects, and potential solutions to the melting of glaciers.

Causes of Melting Glaciers

Global Warming

Global warming is the primary cause of melting glaciers. Due to the burning of fossil fuels, deforestation, and other human activities, the concentration of greenhouse gases in the atmosphere has increased. These gases trap heat from the sun, causing the planet's average temperature to rise. As a result, glaciers in many parts of the world are rapidly melting.

Changes in precipitation patterns

Climate change has also caused changes in precipitation patterns, which can affect the accumulation and melting of glaciers. In some areas, there has been a decrease in precipitation, leading to reduced snowfall and glacial retreat. In other areas, there has been an increase in precipitation, leading to increased snowfall and glacial advance. However, increased precipitation can also lead to an increase in the number and intensity of glacial lake outburst floods.

Ocean warming

Ocean warming is another significant cause of melting glaciers, particularly in polar regions. Warmer ocean temperatures cause the melting of ice shelves, which can lead to the accelerated melting of glaciers. As the ice shelves melt, they release more ice into the ocean, causing sea levels to rise.

Effects of Melting Glaciers

Sea Level Rise

Melting glaciers contribute to sea-level rise, which is a significant concern for coastal communities worldwide. As glaciers melt, they release water into the ocean, causing sea levels to rise. This rise in sea level can cause flooding, erosion, and coastal land loss. The Intergovernmental Panel on Climate Change (IPCC) predicts that global sea levels could rise by up to 1.1 meters by the end of the century if greenhouse gas emissions are not significantly reduced.

Changes in Water Availability

Melting glaciers also affect the availability of freshwater, which is a vital resource for human societies and ecosystems. Many communities around the world depend on glaciers for their water supply, especially during the dry season. As glaciers melt, this source of water becomes less reliable, leading to water shortages and conflicts over water resources.

Changes in Ecosystems

Melting glaciers also affect ecosystems, particularly those in high-altitude regions. As glaciers melt, the habitats of many species, such as mountain goats and snow leopards, are threatened. Glacial meltwater also provides nutrients for downstream ecosystems, and changes in the timing and quantity of this water can affect the growth of plants and animals.

Increase in Natural Disasters

Melting glaciers can also increase the frequency and severity of natural disasters. Glacial meltwater can cause floods, landslides, and avalanches, which can be dangerous to people living in nearby areas. The release of glacial lakes can also lead to catastrophic floods, such as the one that occurred in Bhutan in 1994, which killed 21 people and caused significant damage to infrastructure.

Potential Solutions to Melting Glaciers

Reduction of Greenhouse Gas Emissions

The most effective solution to melting glaciers is the reduction of greenhouse gas emissions. This can be achieved by using renewable energy sources, reducing fossil fuel use, promoting energy efficiency, and improving transportation systems. Governments and individuals can also take steps to reduce their carbon footprint, such as by reducing meat consumption, using public transportation or electric vehicles, and switching to energy-efficient appliances.

Adapting to Changes

Adapting to the changes brought about by melting glaciers is also essential. Communities that rely on glacial meltwater for their water supply can develop alternative water sources or improve water management practices. Infrastructure can also be designed to withstand the increased risk of floods and landslides caused by melting glaciers. In addition, ecosystems can be protected through the creation of protected areas and the restoration of degraded ecosystems.

Geoengineering

Geoengineering, which involves manipulating the Earth's climate to counteract the effects of global warming, is another potential solution to melting glaciers. However, this approach is still in the experimental stage, and there are concerns about unintended consequences and ethical implications.

International Cooperation

Finally, international cooperation is essential in addressing the issue of melting glaciers. The effects of melting glaciers are felt globally, and no single country can solve the problem alone. Cooperation is needed to reduce greenhouse gas emissions, share water resources, and develop adaptation strategies that benefit everyone.

Conclusion

The melting of glaciers is a significant issue that requires urgent action. It affects the availability of freshwater, ecosystems, and increases the risk of natural disasters. Addressing this issue will require a combination of mitigation strategies to reduce greenhouse gas emissions, adaptation strategies to prepare for the impacts of melting glaciers, and international cooperation to ensure a sustainable future for all. By taking action now, we can prevent the worst impacts of melting glaciers and ensure a healthy and resilient planet for future generations.

81 - Glacial Lake Outburst Floods (GLOFs): Understanding the Threat, Preventive Measures, and Long-Term Strategies for a Sustainable Future

Abstract

Glacial Lake Outburst Floods (GLOFs) pose a growing threat to communities living downstream from glacial lakes. These floods can have severe impacts on the environment, economy, and human life. Preventing GLOFs requires a combination of monitoring, early warning systems, infrastructure development, and long-term strategies. Mitigating climate change, building infrastructure such as dams, and capacity building for local communities are some of the measures that can be implemented to prevent GLOFs. This article provides an overview of the causes and impacts of GLOFs and examines the preventive measures that can be taken to protect communities downstream. By taking preventive measures, we can safeguard communities from the devastating effects of GLOFs and ensure a sustainable future for our planet.

Introduction

Glacial Lake Outburst Floods (GLOFs) are a natural phenomenon that occurs when water in a glacial lake is released suddenly, causing a catastrophic flood downstream. GLOFs are a result of climate change and are becoming increasingly common in many parts of the world, particularly in high-altitude regions.

GLOFs can have devastating effects on communities downstream, including loss of life, destruction of infrastructure, and damage to agricultural land. In this article, we will explore the causes, effects, and prevention measures of GLOFs.

What are GLOFs?

GLOFs are floods that occur when water is released suddenly from a glacial lake. Glacial lakes are formed when meltwater from glaciers collects in depressions on the surface of the ice. These lakes can be very large, holding billions of liters of water. When the water in the lake reaches a critical level, it can breach the natural or artificial dam holding it in and be released downstream.

The water released in a GLOF can reach enormous speeds and cause massive destruction. It can destroy buildings, bridges, roads, and other infrastructure, as well as agricultural land and forests. GLOFs are a result of climate change, which is causing glaciers to melt faster than usual, resulting in more water being collected in glacial lakes.

Causes of GLOFs

There are several causes of GLOFs, including natural causes and human activities. The most significant cause of GLOFs is climate change, which is causing glaciers to melt at an alarming rate. As glaciers melt, more water is collected in glacial lakes, increasing the risk of GLOFs.

Another cause of GLOFs is earthquakes. Earthquakes can cause the natural or artificial dams holding the water in the glacial lakes to rupture, releasing the water downstream. The earthquake can also cause landslides, which can block the flow of the river, causing the lake to overflow.

Human activities can also cause GLOFs. The construction of dams, roads, and other infrastructure can block the flow of the river, causing the water to back up and form a glacial lake. If the dam or other infrastructure fails, the water in the lake can be released suddenly, causing a GLOF.

Effects of GLOFs

The effects of GLOFs can be devastating. The water released in a GLOF can cause massive destruction downstream, destroying buildings, bridges, roads, and other infrastructure. It can also cause landslides and mudflows, which can bury entire villages and towns.

The economic impact of GLOFs can also be significant. The destruction of infrastructure and agricultural land can disrupt the local economy, causing a loss of income for farmers and businesses. It can also lead to food shortages and increased food prices.

GLOFs can also have a significant impact on the environment. The sudden release of water can erode the soil, causing sedimentation and pollution downstream. It can also destroy habitats and ecosystems, affecting the biodiversity of the area.

Prevention of GLOFs

Preventing GLOFs is essential to protect communities downstream and minimize the damage caused by these floods. There are several prevention measures that can be taken, including monitoring, early warning systems, and infrastructure development.

Monitoring

Monitoring glacial lakes is critical in preventing GLOFs. Monitoring systems can provide information on the water level in the lake, the rate of glacier melt, and the stability of the natural or artificial dam holding the water in the lake. This information can be used to predict when a GLOF is likely to occur, allowing authorities to take preventive measures.

Early warning systems

Early warning systems can alert communities downstream when a GLOF is imminent. These systems can use a variety of sensors, including seismic sensors, water level sensors, and weather sensors, to provide real-time data on the conditions that may lead to a GLOF. The warning systems can trigger an alarm or notification to the community downstream, allowing them to evacuate to safety before the flood arrives.

Infrastructure development

The development of infrastructure, such as dams, can help to prevent GLOFs. Dams can be built to regulate the flow of water downstream, preventing the formation of glacial lakes. They can also be used to store water for agriculture, hydropower, and other purposes.

However, the construction of dams can also have negative environmental impacts, such as altering river ecosystems and disrupting fish migration patterns. Therefore, it is essential to carefully assess the environmental and social impacts of any infrastructure development before proceeding.

In addition to these measures, there are also several long-term strategies that can be implemented to prevent GLOFs. These include:

Glacier monitoring and research

Long-term glacier monitoring and research can provide information on the rate of glacier melt and the formation of glacial lakes. This information can be used to predict when a GLOF is likely to occur and take preventive measures.

Climate change mitigation

Addressing the root cause of GLOFs, which is climate change, is essential in preventing these floods. Mitigating climate change through reducing greenhouse gas emissions and promoting sustainable development can slow down the rate of glacier melt and reduce the risk of GLOFs.

Capacity building

Building the capacity of local communities to prepare for and respond to GLOFs is critical in preventing these floods. Capacity building can include training in early warning systems, emergency response, and evacuation procedures.

Conclusion

GLOFs are a growing threat to communities downstream from glacial lakes. These floods can have devastating effects on the environment, the economy, and human life. Preventing GLOFs requires a combination of monitoring, early warning systems, infrastructure development, and long-term strategies.

Addressing the root cause of GLOFs, which is climate change, is essential in preventing these floods. Mitigating climate change through reducing greenhouse gas emissions and promoting sustainable development can slow down the rate of glacier melt and reduce the risk of GLOFs.

It is also essential to build the capacity of local communities to prepare for and respond to GLOFs. This can include training in early warning systems, emergency response, and evacuation procedures. By taking these preventive measures, we can protect communities downstream from the devastating effects of GLOFs and ensure a sustainable future for our planet.

82 - The Importance and Key Elements of a Feasibility Study for Water Resources

Abstract

This article discusses the importance of conducting a feasibility study for water resources before embarking on any water project. A feasibility study is a critical step in identifying potential sources of water, analyzing the quality and quantity of available water, assessing the feasibility of treatment and delivery options, and estimating the cost of the project. The article outlines the key elements of a feasibility study, including water demand analysis, water supply analysis, water treatment and delivery options analysis, regulatory and environmental analysis, and financial analysis. The article also highlights the potential benefits of conducting a feasibility study, including ensuring financial feasibility, identifying the best water sources and treatment options, identifying and mitigating potential risks and challenges, and ensuring that the project is sustainable and environmentally friendly. A proper feasibility study can save time and resources in the long run and can help ensure the success and sustainability of water projects.

Introduction

Water resources are an essential part of human life, and it is important to ensure that they are used effectively and efficiently. A feasibility study for water resources is a critical step in determining the viability of a water project. This study is aimed at identifying potential sources of water, analyzing the quality and quantity of available water, assessing the feasibility of treatment and delivery options, and estimating the cost of the project. In this article, we will discuss the key elements of a feasibility study for water resources, the importance of conducting such a study, and the potential benefits that can be derived from it.

The Importance of a Feasibility Study for Water Resources

A feasibility study for water resources is important for several reasons. Firstly, it helps to identify the best water sources for a particular project. This is critical because the quality and quantity of water vary depending on the source. The study helps to determine the most appropriate source of water for the project, taking into account the quality and quantity of water required, the proximity of the source to the project site, and the cost of extraction.

Secondly, a feasibility study helps to assess the viability of the project. This is important because water projects can be expensive, and it is essential to ensure that they are financially feasible before embarking on them. The study helps to estimate the cost of the project, including the cost of construction, operation, maintenance, and financing. It also helps to identify potential sources of funding, such as grants, loans, or partnerships.

Thirdly, a feasibility study helps to identify potential risks and challenges associated with the project. Water projects can be complex, and there are several factors that can impact their success, including environmental factors, regulatory requirements, and community support. A feasibility study helps to identify these risks and challenges and provides recommendations on how to mitigate them.

Key Elements of a Feasibility Study for Water Resources

A feasibility study for water resources typically consists of the following key elements:

Water demand analysis

The first step in a feasibility study is to analyze the water demand for the project. This involves determining the quantity and quality of water required for the project, as well as the anticipated water usage patterns. The water demand analysis should take into account the projected growth of the project and the surrounding community, as well as any potential changes in water usage patterns.

Water supply analysis

The next step is to analyze the potential sources of water for the project. This involves identifying all available sources of water, such as groundwater, surface water, and rainwater. The analysis should also include an assessment of the quality and quantity of the available water, as well as an evaluation of the reliability and sustainability of each source.

Water treatment and delivery options analysis

Once potential sources of water have been identified, the next step is to analyze the treatment and delivery options. This involves evaluating the feasibility of different treatment methods, such as filtration, disinfection, and desalination. The analysis should also include an assessment of the most appropriate delivery method, such as pipelines, tankers, or trucks.

Regulatory and environmental analysis

The feasibility study should also include an analysis of regulatory and environmental factors that could impact the project. This includes an assessment of the legal requirements for water extraction and treatment, as well as an evaluation of the potential environmental impact of the project.

Financial analysis

The final step in the feasibility study is to conduct a financial analysis. This involves estimating the cost of the project, including the cost of construction, operation, and maintenance. The analysis should also include an assessment of potential sources of funding, such as grants, loans, or partnerships.

Potential Benefits of a Feasibility Study for Water Resources

There are several potential benefits to conducting a feasibility study for water resources:

It can help to ensure that the water project is financially feasible, which can save time and resources in the long run. By identifying potential sources of funding and estimating the cost of the project, the feasibility study can help project managers to make informed decisions about the viability of the project and to secure financing.

It can help to identify the best water sources and treatment options for the project. By analyzing the quantity and quality of available water sources and evaluating different treatment options, the feasibility study can help to ensure that the project is using the most effective and efficient methods for providing clean water.

It can help to identify and mitigate potential risks and challenges associated with the project. By analyzing regulatory and environmental factors, the feasibility study can help to identify potential roadblocks that could impact the success of the project. This allows project managers to plan accordingly and to take steps to mitigate potential risks and challenges.

It can help to ensure that the project is sustainable and environmentally friendly. By evaluating the potential environmental impact of the project and identifying ways to mitigate any negative effects, the feasibility study can help to ensure that the project is sustainable and beneficial to the environment.

Conclusion

In conclusion, a feasibility study for water resources is an essential step in determining the viability of a water project. It helps to identify potential sources of water, analyze the quality and quantity of available water, assess the feasibility of treatment and delivery options, and estimate the cost of the project. Conducting a feasibility study can save time and resources in the long run, ensure that the project is financially feasible, identify the best water sources and treatment options, identify and mitigate potential risks and challenges, and ensure that the project is sustainable and environmentally friendly. Therefore, it is essential to conduct a feasibility study before embarking on any water project to ensure its success and sustainability.

83 - Hydrological Area Drainage Study: Understanding and Managing Water Resources

Abstract

Hydrological area drainage study is a critical process of water resource management that involves collecting and analyzing data related to the hydrological characteristics of a particular area, including the size and shape of the drainage basin, the location of water bodies, and the distribution of rainfall. This information is used to develop strategies for the management and conservation of water resources within the area, such as the construction of new dams, the implementation of water conservation programs, and the development of flood management plans. By understanding and managing our water resources through hydrological area drainage study, we can ensure the sustainable use of this vital resource for future generations.

Introduction

Hydrological Area Drainage Study is an important aspect of water resource management. It is the process of studying the hydrological characteristics of a particular area, such as the size and shape of the drainage basin, the location of water bodies, and the distribution of rainfall. The information gathered in the study is used to develop strategies for the management and conservation of water resources within the area.

The Hydrological Cycle

The hydrological cycle is the process of how water circulates through the environment. It involves the movement of water through various stages, including evaporation, precipitation, infiltration, and runoff. The hydrological cycle is driven by the sun's energy, which causes water to evaporate from the surface of the earth and enter the atmosphere. This water then condenses and falls back to the earth's surface as precipitation, where it can be stored in lakes, rivers, and groundwater.

Drainage Basins

A drainage basin is an area of land where all water flows into a particular water body, such as a lake or river. It is also known as a watershed. Drainage basins are defined by natural features such as mountains, hills, and ridges that separate one basin from another. The size and shape of a drainage basin determine the amount of water that will flow into a particular water body.

Hydrological Area Drainage Study

A hydrological area drainage study involves the collection and analysis of data related to the hydrological characteristics of a particular area. The data collected includes information on the size and shape of the drainage basin, the location of water bodies, and the distribution of rainfall. The study also considers the impact of human activities on the water resources within the area.

The data collected in the study is used to develop strategies for the management and conservation of water resources within the area. The strategies developed may include the construction of new dams, the implementation of water conservation programs, and the development of flood management plans.

Data Collection

The first step in a hydrological area drainage study is to collect data related to the hydrological characteristics of the area. The data collected includes information on the size and shape of the drainage basin, the location of water bodies, and the distribution of rainfall.

The size and shape of the drainage basin can be determined using topographic maps and satellite imagery. These tools provide information on the elevation of the land and the location of rivers and streams within the basin.

The location of water bodies can also be determined using topographic maps and satellite imagery. In addition, on-site surveys may be conducted to collect more detailed information on the water bodies within the basin.

Rainfall data can be obtained from meteorological stations located within or near the drainage basin. This data provides information on the amount and distribution of rainfall within the area.

Data Analysis

Once the data has been collected, it is analyzed to determine the hydrological characteristics of the area. The analysis may include the calculation of runoff coefficients, peak flow rates, and flood volumes.

Runoff coefficients are used to estimate the amount of rainfall that will become runoff within the drainage basin. Peak flow rates are used to estimate the maximum amount of water that will flow into a particular water body during a storm event. Flood volumes are used to estimate the amount of water that will be stored in a reservoir during a flood event.

The analysis of the data also considers the impact of human activities on the water resources within the area. Human activities such as urbanization, agriculture, and industrial development can have a significant impact on the hydrological characteristics of the area. For example, the construction of impervious surfaces such as roads and buildings can increase the amount of runoff within the drainage basin, leading to an increased risk of flooding.

Management Strategies

The data collected and analyzed in the hydrological area drainage study is used to develop strategies for the management and conservation of water resources within the area. The strategies developed may include the construction of new dams, the implementation of water conservation programs, and the development of flood management plans.

New Dams

The construction of new dams is one strategy that may be developed based on the data collected in the hydrological area drainage study. Dams can be used to store water during periods of high rainfall, which can then be released during periods of low rainfall to maintain a consistent water supply. Dams can also be used to regulate the flow of water into rivers and streams, which can help to reduce the risk of flooding downstream.

Water Conservation Programs

Water conservation programs may also be developed based on the data collected in the hydrological area drainage study. These programs may include public education campaigns to encourage water conservation practices, such as using low-flow showerheads and toilets, fixing leaky faucets, and watering lawns and gardens during off-peak hours.

Flood Management Plans

The development of flood management plans is another strategy that may be developed based on the data collected in the hydrological area drainage study. These plans may include the construction of levees and other flood control structures, the development of warning systems to alert residents of impending floods, and the implementation of evacuation plans for areas at high risk of flooding.

Conclusion

Hydrological area drainage study is an important aspect of water resource management. It involves the collection and analysis of data related to the hydrological characteristics of a particular area, including the size and shape of the drainage basin, the location of water bodies, and the distribution of rainfall. The data collected in the study is used to develop strategies for the management and conservation of water resources within the area. These strategies may include the construction of new dams, the implementation of water conservation programs, and the development of flood management plans. By implementing these strategies, we can ensure the sustainable use of our water resources for future generations.

84 - Environmental Impact Assessment Study for Water Resources Projects: Importance and Process

Abstract

Water resources projects, such as dams and hydropower plants, can have significant environmental and social impacts. Environmental Impact Assessment (EIA) is a critical tool for assessing and mitigating these impacts, ensuring that such projects are implemented in a sustainable manner. This article provides an overview of the EIA process for water resources projects, including the scoping process, baseline studies, impact assessment, alternatives analysis, mitigation measures, and monitoring and follow-up. The article also includes examples of water resources projects that have undergone EIA studies, such as the Three Gorges Dam in China and the Nam Theun 2 Hydropower Project in Laos. By understanding the importance and process of EIA for water resources projects, we can ensure that these projects meet the needs of both current and future generations while minimizing their environmental and social impacts.

Introduction

Water is one of the most important natural resources on the planet, and it is essential for human survival, agriculture, industry, and energy production. With the growing demand for water resources, there has been an increasing need for water resources projects such as dams, irrigation schemes, water treatment plants, and hydropower plants. However, such projects can have significant environmental impacts, and it is necessary to assess these impacts to ensure that the projects are sustainable.

Environmental Impact Assessment (EIA) is a process that evaluates the potential environmental impacts of a proposed project or development. In this article, we will discuss the importance of EIA for water resources projects, the key components of an EIA study for such projects, and some examples of water resources projects that have undergone EIA.

Importance of EIA for Water Resources Projects

Water resources projects can have significant environmental impacts, including the alteration of river ecosystems, the displacement of wildlife, the reduction of water quality, and the generation of greenhouse gas emissions. An EIA study is essential for water resources projects because it can help identify potential environmental impacts, evaluate the significance of these impacts, and propose measures to mitigate or prevent them.

The primary objective of an EIA study for water resources projects is to ensure that the project is designed and implemented in a sustainable manner that minimizes environmental impacts and maximizes social and economic benefits. The EIA study can provide information to decision-makers, stakeholders, and the public on the potential environmental impacts of the project and the measures that will be taken to mitigate or prevent these impacts. It can also help identify alternative project designs or locations that may have fewer environmental impacts.

Components of an EIA Study for Water Resources Projects

An EIA study for water resources projects typically consists of several key components, including the scoping process, baseline studies, impact assessment, alternatives analysis, mitigation measures, and monitoring and follow-up.

Scoping Process

The scoping process is the initial stage of an EIA study, and it involves identifying the key environmental and social issues that should be considered in the study. The scoping process involves consultations with stakeholders, including local communities, government agencies, and non-governmental organizations (NGOs). The scoping process should be transparent and inclusive, and it should provide an opportunity for stakeholders to express their views and concerns about the project.

Baseline Studies

Baseline studies are conducted to establish the current environmental and social conditions in the project area. Baseline studies may include assessments of water quality, soil characteristics, wildlife populations, cultural resources, and socio-economic conditions. The baseline studies provide a baseline against which potential impacts of the project can be assessed.

Impact Assessment

The impact assessment is the most critical component of an EIA study, and it involves identifying and evaluating the potential environmental and social impacts of the proposed project. The impact assessment should consider both the direct and indirect impacts of the project, as well as the cumulative impacts of the project in combination with other projects or activities in the area.

The impact assessment should also consider the potential for significant adverse impacts, such as the loss of critical habitat or the displacement of local communities. The impact assessment should be based on sound scientific and technical information, and it should consider the views and concerns of stakeholders.

Alternatives Analysis

The alternatives analysis involves identifying and evaluating alternative project designs or locations that may have fewer environmental impacts. The alternatives analysis should consider a range of alternatives, including the no-action alternative, and it should evaluate the potential environmental and social impacts of each alternative.

Mitigation Measures

The mitigation measures are measures that will be taken to prevent, reduce, or offset the potential adverse environmental and social impacts of the project. The mitigation measures may include design modifications, operational changes, or compensatory measures such as habitat restoration or community development projects. The mitigation measures should be feasible, effective, and they should be implemented throughout the life of the project.

Monitoring and Follow-up

Monitoring and follow-up are critical components of an EIA study, and they involve assessing the effectiveness of the mitigation measures and ensuring that the project is implemented in compliance with the conditions of approval. The monitoring and follow-up should be conducted throughout the life of the project, and it should involve regular reporting and public disclosure of monitoring data.

Examples of Water Resources Projects that have undergone EIA

Many water resources projects around the world have undergone EIA studies. Here are some examples:

Three Gorges Dam, China

The Three Gorges Dam is a hydropower project on the Yangtze River in China. The project involved the construction of a 2.3-kilometer-long dam, a reservoir that flooded an area of 1,045 square kilometers, and a hydropower plant with a capacity of 22.5 gigawatts. The EIA study for the Three Gorges Dam identified potential environmental impacts, including the displacement of over a million people, the loss of cultural and historical sites, and the alteration of river ecosystems. The EIA study recommended mitigation measures, including compensation for affected communities, the relocation of cultural and historical sites, and the construction of fish ladders to maintain fish migration. The Three Gorges Dam project was approved in 1992, and it has been operational since 2003.

Nam Theun 2 Hydropower Project, Laos

The Nam Theun 2 Hydropower Project is a 1,070-megawatt hydropower project on the Nam Theun River in Laos. The project involved the construction of a dam, a reservoir that flooded an area of 450 square kilometers, and a hydropower plant. The EIA study for the Nam Theun 2 project identified potential environmental impacts, including the loss of forest and wildlife habitat, the displacement of local communities, and the potential for downstream impacts on river ecosystems. The EIA study recommended mitigation measures, including the creation of a protected area for wildlife, compensation for affected communities, and the development of alternative livelihoods for affected people. The Nam Theun 2 project was approved in 2005, and it has been operational since 2010.

Lower Mekong Basin Hydropower Projects, Cambodia

The Lower Mekong Basin Hydropower Projects are a series of hydropower projects planned on the Mekong River in Cambodia. The projects have been the subject of controversy due to their potential environmental and social impacts. The EIA studies for the projects have identified potential impacts, including the loss of fish habitat, the displacement of local communities, and the potential for downstream impacts on river ecosystems. The EIA studies have recommended mitigation measures, including the construction of fish ladders, compensation for affected communities, and the development of alternative livelihoods for affected people. The Lower Mekong Basin Hydropower Projects are still in the planning stages, and their approval remains controversial.

Conclusion

Environmental Impact Assessment is a critical tool for ensuring that water resources projects are designed and implemented in a sustainable manner that minimizes environmental impacts and maximizes social and economic benefits. EIA studies for water resources projects should include the scoping process, baseline studies, impact assessment, alternatives analysis, mitigation measures, and monitoring and follow-up. The examples of the Three Gorges Dam, Nam Theun 2 Hydropower Project, and Lower Mekong Basin Hydropower Projects illustrate the importance of EIA studies for water resources projects and the potential environmental and social impacts of such projects. It is essential to continue to conduct EIA studies for water resources projects to ensure that they are sustainable and meet the needs of both current and future generations.

References

Chen, Y. (2003). Environmental impact assessment of the Three Gorges Project. Environmental Impact Assessment Review, 23(2), 223-253.

The Nam Theun 2 Power Company Limited. (2022). Environmental and Social Impact Assessment Summary.

Mekong River Commission. (2022). Lower Mekong Basin Hydropower Development.

85 - Methodology for Conducting a Hydrogeological Study: A Framework for Managing Groundwater Resources

Abstract

Hydrogeology is a multidisciplinary field that combines geology, hydrology, and other environmental sciences to understand the distribution and movement of water underground. Conducting a hydrogeological study is essential for managing water resources, protecting groundwater quality, and ensuring sustainable use of this vital resource. This article provides a framework for conducting a comprehensive hydrogeological study in an area, consisting of seven steps: defining the study area, collecting existing data, conducting fieldwork, analyzing data, developing a hydrogeological model, evaluating water resource potential, and developing management strategies. By following these steps, water resource managers can develop a comprehensive understanding of the hydrogeological conditions of the study area, evaluate the potential of the study area for water resource development, and develop effective management strategies to protect groundwater quality and ensure sustainable use of this valuable resource. It is important to engage qualified professionals and involve stakeholders and local communities in the study process to ensure its effectiveness and sustainability over the long term.

Introduction

Hydrogeology is the study of groundwater and its interaction with the surrounding environment. It is a multidisciplinary field that combines geology, hydrology, and other environmental sciences to understand the distribution and movement of water underground. Conducting a hydrogeological study is essential for managing water resources, protecting groundwater quality, and ensuring sustainable use of this vital resource. In this article, we will discuss the steps involved in conducting a hydrogeological study in an area.

Step 1: Define the Study Area

The first step in conducting a hydrogeological study is to define the study area. This involves identifying the geographic boundaries of the area of interest and gathering information about the geological, hydrological, and environmental conditions that influence groundwater flow and quality. The study area should be chosen based on the specific objectives of the study, such as evaluating the potential for groundwater contamination, determining the recharge and discharge areas of an aquifer, or identifying suitable locations for groundwater development.

Step 2: Collect Existing Data

The next step is to collect existing data about the study area. This can include geological maps, hydrological data, environmental studies, and other relevant information. Existing data can provide valuable insights into the hydrogeological conditions of the area and help to identify data gaps that need to be filled with additional fieldwork.

Step 3: Conduct Fieldwork

The next step is to conduct fieldwork to collect new data about the study area. This can involve a range of techniques, including drilling and logging boreholes, conducting hydraulic tests, measuring water levels and water quality, and mapping geological and hydrological features. The data collected in the field should be recorded in a systematic manner to ensure accuracy and consistency.

Step 4: Analyze Data

Once the data has been collected, it must be analyzed to develop a conceptual model of the hydrogeological conditions of the study area. This can involve a range of techniques, including statistical analysis, geostatistics, and numerical modeling. The objective of the data analysis is to develop a quantitative understanding of the distribution and movement of groundwater in the study area, as well as the factors that influence groundwater flow and quality.

Step 5: Develop a Hydrogeological Model

The next step is to develop a hydrogeological model of the study area. This involves using the data collected in the field and the results of the data analysis to create a computer-based model that simulates the movement of groundwater in the study area. The hydrogeological model should be calibrated and validated using field data to ensure that it accurately represents the hydrogeological conditions of the study area.

Step 6: Evaluate Water Resource Potential

Once the hydrogeological model has been developed, it can be used to evaluate the potential of the study area for water resource development. This can involve assessing the quantity and quality of groundwater resources, identifying suitable locations for water wells, and evaluating the potential for artificial recharge of groundwater.

Step 7: Develop Management Strategies

The final step in conducting a hydrogeological study is to develop management strategies for the study area. This can involve identifying measures to protect groundwater quality, developing groundwater management plans, and implementing measures to ensure sustainable use of groundwater resources. The management strategies should be based on the results of the hydrogeological study and should take into account the specific objectives of the study, as well as the social, economic, and environmental factors that influence water resource management in the study area.

Conclusion

Conducting a hydrogeological study is an essential step in managing groundwater resources and ensuring sustainable use of this vital resource. The seven steps discussed in this article provide a framework for conducting a comprehensive hydrogeological study in an area. By following these steps, water resource managers can develop a quantitative understanding of the hydrogeological conditions of the study area, evaluate the potential of the study area for water resource development, and develop effective management strategies to protect groundwater quality and ensure sustainable use of this valuable resource.

It is worth noting that conducting a hydrogeological study can be a complex and time-consuming process, and may require specialized knowledge and expertise. Therefore, it is important to engage qualified professionals with experience in hydrogeological studies to ensure that the study is conducted effectively and efficiently.

In addition, it is essential to involve stakeholders and local communities in the study process to ensure that their concerns and perspectives are taken into account. This can help to build trust and collaboration between water resource managers and local communities, and ensure that management strategies are socially acceptable and effective.

Finally, conducting a hydrogeological study is not a one-time event. Hydrogeological conditions can change over time due to natural and human factors, and therefore, it is important to regularly monitor groundwater resources and update the hydrogeological model as needed to ensure that management strategies remain effective and sustainable over the long term.

In conclusion, conducting a hydrogeological study is an essential step in managing groundwater resources and ensuring sustainable use of this vital resource. By following the seven steps outlined in this article, water resource managers can develop a comprehensive understanding of the hydrogeological conditions of the study area, evaluate the potential of the study area for water resource development, and develop effective management strategies to protect groundwater quality and ensure sustainable use of this valuable resource.

86 - Exploring Water Resources through Field Studies: Insights from Hydrology Research

Abstract

This article provides an overview of field studies in hydrology, highlighting their importance in understanding the movement and storage of water on Earth. Through techniques such as stream gauging, water sampling, and remote sensing, hydrologists can gather data and observe phenomena in real-world settings. The article showcases several examples of field studies, including the Hubbard Brook Experimental Forest, the Amazon Basin, and the Western United States Drought, to demonstrate how these studies have helped improve our understanding of water resources and inform water management practices. As we face ongoing challenges related to water scarcity and climate change, field studies remain a critical tool in hydrology research.

Introduction

Hydrology is a branch of earth science that studies the movement, distribution, and quality of water on Earth. It encompasses the study of water in the atmosphere, on the land surface, and underground. Field studies are an essential aspect of hydrology, providing an opportunity to collect data and observe phenomena in real-world settings. This article will provide an overview of field studies in hydrology, including their importance, common techniques, and examples of their application.

Importance of Field Studies in Hydrology

Field studies are essential in hydrology for several reasons. First, they allow hydrologists to collect data that cannot be obtained through remote sensing or laboratory experiments. Hydrology is a complex field, and many variables that influence the movement and quality of water cannot be easily measured from a distance or in a controlled setting. Field studies provide an opportunity to measure these variables in real-world settings, providing more accurate and comprehensive data.

Second, field studies allow hydrologists to observe phenomena as they occur. Hydrological systems are dynamic and often exhibit behaviors that are difficult to predict or replicate in a laboratory. Observing these phenomena in the field can help hydrologists better understand how different variables interact and influence the movement and quality of water.

Third, field studies provide an opportunity to test and validate models and hypotheses. Hydrology is a data-driven science, and models and hypotheses must be validated using empirical data. Field studies provide a way to test these models and hypotheses in real-world settings, ensuring that they are accurate and applicable to real-world situations.

Common Techniques Used in Field Studies

There are several techniques commonly used in field studies in hydrology. These include:

Stream Gauging

Stream gauging is the process of measuring the volume and velocity of water flowing in a stream or river. This information is used to calculate the discharge, or the amount of water flowing through the stream or river at a given point in time. Stream gauging is typically done using a current meter, which measures the velocity of the water, and a cross-section measurement, which measures the depth and width of the stream or river.

Water Sampling

Water sampling is the process of collecting water samples from a stream, river, lake, or groundwater source. These samples are typically analyzed in a laboratory to determine the quality of the water, including its chemical composition, nutrient content, and presence of pollutants. Water sampling is typically done using a specialized sampler, which is designed to collect a representative sample of water from a specific depth or location.

Groundwater Monitoring

Groundwater monitoring involves measuring the level and quality of water in underground aquifers. This is typically done using a network of wells, which are drilled into the aquifer at various depths. Water levels in these wells are measured periodically to determine the changes in water level over time. Samples of the groundwater are also collected from these wells to determine the quality of the water.

Soil Moisture Monitoring

Soil moisture monitoring involves measuring the amount of moisture present in the soil. This information is important for understanding how water moves through the soil and is taken up by plants. Soil moisture monitoring is typically done using specialized sensors, which are inserted into the soil at various depths.

Remote Sensing

Remote sensing involves using satellites and other technology to collect data about the Earth's surface. This data can be used to monitor changes in water resources, such as changes in the extent of glaciers, changes in the amount of snow cover, and changes in the extent of wetlands. Remote sensing is a powerful tool in hydrology, as it allows hydrologists to collect data over large areas quickly and efficiently.

Examples of Field Studies in Hydrology

There are numerous examples of field studies in hydrology, each of which has contributed to our understanding of how water moves and is stored on Earth. Below are a few examples:

The Hubbard Brook Experimental Forest

The Hubbard Brook Experimental Forest is a long-term research site located in New Hampshire, USA. It was established in 1955 to study the effects of forest management on water quality and quantity. Researchers at the site have conducted numerous field studies over the years, including stream gauging, water sampling, and soil moisture monitoring. One of the key findings from the research conducted at Hubbard Brook is that clearcutting, a common forestry practice, can have significant impacts on water resources, including increased erosion and decreased water quality.

The Amazon Basin

The Amazon Basin is the largest watershed in the world, covering over 6 million square kilometers in South America. Hydrologists have conducted numerous field studies in the Amazon Basin to better understand the movement and storage of water in the region. One notable study used a network of 36 river gauges to measure the discharge of the Amazon River and its tributaries. The data collected from this study helped researchers better understand how the Amazon Basin contributes to global water cycles and how changes in land use and climate could impact water resources in the region.

The Western United States Drought

In recent years, the western United States has experienced severe drought conditions, impacting water resources and ecosystems throughout the region. Hydrologists have conducted numerous field studies to better understand the causes and impacts of the drought. One such study involved using a network of sensors to measure soil moisture levels throughout the region. The data collected from this study helped researchers better understand how drought conditions were affecting plant and animal populations and provided valuable information for water managers as they worked to allocate scarce water resources.

Conclusion

Field studies are a critical aspect of hydrology, providing an opportunity to collect data and observe phenomena in real-world settings. By using techniques such as stream gauging, water sampling, and remote sensing, hydrologists can better understand how water moves and is stored on Earth. Examples such as the Hubbard Brook Experimental Forest, the Amazon Basin, and the Western United States Drought demonstrate the importance of field studies in improving our understanding of water resources and helping us better manage them. As we continue to face challenges related to water scarcity and climate change, field studies will remain an essential tool in hydrology.

References

Bales, R. C., Molotch, N. P., Painter, T. H., Dettinger, M. D., Rice, R., & Dozier, J. (2006). Mountain hydrology of the western United States. Water Resources Research, 42(8), W08432. doi: 10.1029/2005WR004387

Bormann, F. H., Likens, G. E., & Pierce, R. S. (1972). The Hubbard Brook Ecosystem Study: Forest Biogeochemistry. Ecological Monographs, 42(1), 1-31. doi: 10.2307/1942254

Marengo, J. A., Nobre, C. A., Tomasella, J., Cardoso, M. F., Oyama, M. D., & Oliveira, G. S. (2008). The Drought of Amazonia in 2005. Journal of Climate, 21(3), 495-516. doi: 10.1175/2007JCLI1600.1

87 - The Water-Environment-Energy-Food Nexus: An Integrated Approach to Sustainable Development

Abstract

The Water-Environment-Energy-Food Nexus (WEF Nexus) is a concept that emphasizes the interdependence and interlinkages between water, energy, and food systems, as well as the environment. It recognizes that these systems are inextricably linked and mutually dependent, and therefore, addressing one system in isolation could have unintended consequences on the others. The WEF Nexus provides a framework for an integrated approach to sustainable development, which seeks to optimize the use of resources, enhance efficiency, and ensure equitable distribution of benefits. This article explores the WEF Nexus in detail, highlighting its key components, challenges, and opportunities. It also examines the role of policy and governance in promoting integrated management of water, energy, and food systems, as well as the need for innovation and technological advancements to address emerging issues. Overall, the article underscores the importance of adopting a WEF Nexus approach in achieving sustainable development and mitigating the impacts of climate change.

Introduction

Water, energy, and food are essential for human well-being and economic development. They are inextricably linked, and their interdependence is often overlooked in policy and decision-making processes. The World Bank defines the Water-Environment-Energy-Food Nexus (WEF Nexus) as the "interconnectedness of the natural resource systems that provide essential goods and services to society" (World Bank, 2019). The WEF Nexus emphasizes the interlinkages and interdependencies between water, energy, and food systems, as well as the environment, and recognizes that addressing one system in isolation could have unintended consequences on the others.

The WEF Nexus provides a framework for an integrated approach to sustainable development, which seeks to optimize the use of resources, enhance efficiency, and ensure equitable distribution of benefits. It offers an opportunity to break down silos and promote cross-sectoral collaboration, innovation, and technological advancements. The WEF Nexus approach is gaining increasing attention from policymakers, researchers, and practitioners, as it offers a pathway towards achieving the Sustainable Development Goals (SDGs) and mitigating the impacts of climate change.

Components of the WEF Nexus

The WEF Nexus comprises four interdependent components: water, energy, food, and the environment. Each component is crucial for human well-being and economic development and is intricately linked to the others. Understanding the interlinkages and interdependencies between these components is crucial for adopting an integrated approach to sustainable development.

Water

Water is a finite resource and is essential for human survival, agriculture, industry, and energy production. The demand for water is increasing due to population growth, urbanization, and economic development. However, water availability is becoming increasingly scarce due to climate change, pollution, and overexploitation of resources. The WEF Nexus approach recognizes the importance of managing water resources sustainably, balancing the competing demands of different sectors, and ensuring equitable distribution of benefits. Integrated water resources management (IWRM) is a key tool for implementing the WEF Nexus approach to water management, as it seeks to balance the social, economic, and environmental needs of different stakeholders.

Energy

Energy is essential for economic development, social well-being, and environmental sustainability. The demand for energy is increasing due to population growth, urbanization, and industrialization. However, the current energy systems are heavily reliant on fossil fuels, which have significant environmental impacts, such as greenhouse gas emissions and air pollution. The WEF Nexus approach recognizes the need to transition towards cleaner and more sustainable energy systems, such as renewable energy, energy efficiency, and energy conservation. The integration of water and energy systems, such as hydropower, can also provide opportunities for sustainable energy production.

Food

Food is essential for human survival and well-being, and is also a key driver of economic growth and poverty reduction. The global population is expected to reach 9.7 billion by 2050, and this will require a 70% increase in food production. However, the current food systems are characterized by inefficiencies, waste, and environmental degradation, which pose significant challenges to sustainable development. The WEF Nexus approach recognizes the need to promote sustainable agriculture, reduce food waste, and ensure equitable access to food. This requires cross-sectoral collaboration and innovation, as well as policies that promote sustainable production and consumption patterns.

Environment

The environment is the foundation of the WEF Nexus, providing the natural resources and ecosystem services that underpin human well-being and economic development. However, environmental degradation, such as deforestation, land degradation, and biodiversity loss, is threatening the sustainability of water, energy, and food systems. The WEF Nexus approach recognizes the importance of protecting and restoring ecosystems, promoting biodiversity, and mitigating the impacts of climate change. This requires cross-sectoral collaboration and policy coherence, as well as innovative solutions, such as nature-based approaches and sustainable land use practices.

Challenges and Opportunities of the WEF Nexus

The WEF Nexus approach faces several challenges, such as governance and institutional barriers, lack of data and information, competing interests, and inadequate financing. However, it also presents significant opportunities for promoting sustainable development and mitigating the impacts of climate change.

Policy and Governance

Effective policy and governance frameworks are crucial for promoting integrated management of water, energy, and food systems, as well as the environment. This requires cross-sectoral collaboration, stakeholder engagement, and policy coherence across different levels of governance. The WEF Nexus approach provides a framework for addressing these challenges, by promoting policy coherence, stakeholder engagement, and adaptive governance.

Innovation and Technological Advancements

Innovation and technological advancements play a crucial role in addressing the challenges of the WEF Nexus. This includes the development of new technologies and practices that promote resource efficiency, reduce waste, and mitigate the impacts of climate change. Innovation also requires collaboration and partnerships between different sectors and stakeholders, as well as adequate financing and supportive policy frameworks.

Financing

Financing is a key challenge for implementing the WEF Nexus approach, as it requires significant investments in infrastructure, technology, and capacity building. However, financing also presents an opportunity for promoting sustainable development and achieving the SDGs. This requires innovative financing mechanisms, such as green bonds, blended finance, and public-private partnerships, as well as supportive policy frameworks that promote sustainable investments.

Conclusion

The Water-Environment-Energy-Food Nexus offers a framework for an integrated approach to sustainable development, which recognizes the interlinkages and interdependencies between water, energy, and food systems, as well as the environment. The WEF Nexus approach presents significant opportunities for promoting sustainable development and mitigating the impacts of climate change, but it also faces several challenges, such as governance and institutional barriers, lack of data and information, competing interests, and inadequate financing. Addressing these challenges requires cross-sectoral collaboration, stakeholder engagement, policy coherence, innovation, and adequate financing. Adopting a WEF Nexus approach is crucial for achieving the SDGs and ensuring a sustainable future for all.

Reference

World Bank. (2019). Thirsty Energy: Understanding the Linkages between Energy and Water. Washington, DC: World Bank.

88 - Understanding Water Footprint and Virtual Water Trade: Managing Water Resources in a World of Scarcity

Abstract

Water scarcity is a growing global challenge, and it is becoming increasingly important to manage water resources effectively. The water footprint and virtual water trade are two concepts that have emerged as potential solutions to this challenge. The water footprint is a tool for measuring and managing water use in different sectors, while the virtual water trade is a way for countries with water scarcity to import water-intensive goods and products from countries with abundant water resources. This article provides a comprehensive overview of the concepts of water footprint and virtual water trade, including their definitions, methodologies, benefits, limitations, and controversies. The article also discusses the challenges and opportunities associated with managing water resources in a world of scarcity, and highlights the importance of adopting a holistic and integrated approach to water management.

Introduction

Water is an essential resource for life on earth, and its scarcity is one of the major challenges faced by humanity today. In recent years, there has been an increasing concern about the impact of human activities on water resources, including the depletion of aquifers, pollution of water bodies, and the effects of climate change on water availability. The concept of the water footprint has emerged as a tool for measuring and managing water use in different sectors, including agriculture, industry, and households. Additionally, the virtual water trade has gained attention as a way to manage water scarcity through international trade of water-intensive goods.

This article aims to provide an overview of the water footprint concept and the virtual water trade. It will describe the main methods used to calculate the water footprint, the different types of water footprint, and their applications in different sectors. It will also discuss the virtual water trade, including its benefits and limitations, and the controversies surrounding this approach.

Water Footprint

The water footprint is a concept that was introduced in 2002 by Arjen Hoekstra, a professor at the University of Twente in the Netherlands. The water footprint is defined as the total volume of freshwater used to produce a product or service. It includes the water used in the production process, as well as the water consumed or polluted during the production process. The water footprint can be measured at different scales, including the individual, household, company, and national levels.

The calculation of the water footprint is based on the concept of water scarcity. Water scarcity is defined as the imbalance between the demand for water and the available water resources. Water scarcity can be caused by natural factors, such as droughts and climate change, or by human activities, such as water withdrawals for irrigation, industry, and domestic use. The water footprint takes into account the availability of water resources in a particular region and the amount of water required to produce a product or service in that region.

The water footprint can be divided into three types, which are:

Blue water footprint: the volume of surface and groundwater consumed or polluted during the production process.

Green water footprint: the volume of rainwater consumed or polluted during the production process.

Grey water footprint: the volume of freshwater required to dilute pollutants to meet water quality standards.

The blue water footprint is the most widely used type of water footprint and is often used to measure the water use of agriculture, industry, and households. The green water footprint is mainly used to measure the water use of crops grown in rain-fed areas. The grey water footprint is used to measure the environmental impact of water pollution.

Applications of the Water Footprint

The water footprint has several applications in different sectors, including agriculture, industry, and households.

Agriculture: The water footprint is used to measure the water use of crops and livestock production. It can be used to identify the crops and livestock that are most water-intensive and to develop strategies for reducing water use in agriculture.

Industry: The water footprint is used to measure the water use of industrial processes and to identify opportunities for reducing water use and improving water efficiency.

Households: The water footprint is used to measure the water use of households and to identify ways to reduce water use through behavior change and the adoption of water-efficient technologies.

Virtual Water Trade

The virtual water trade is a concept that was introduced in the early 1990s by Tony Allan, a professor at King's College London. The virtual water trade refers to the trade of water-intensive goods and products between regions or countries. The virtual water trade is based on the idea that countries with water scarcity can import water-intensive goods and products from countries with abundant water resources, instead of producing them domestically.

The virtual water trade is calculated by estimating the amount of water used to produce a particular product or commodity in a particular region or country. The virtual water content of a product is defined as the volume of water used to produce one unit of that product. For example, the virtual water content of one kilogram of wheat produced in a water-scarce region might be higher than the virtual water content of one kilogram of wheat produced in a water-abundant region. The virtual water trade can provide several benefits to countries with water scarcity, including:

Reducing water use: Countries with water scarcity can reduce their water use by importing water-intensive goods and products from countries with abundant water resources, instead of producing them domestically.

Enhancing food security: The virtual water trade can help to ensure food security in countries with water scarcity by enabling them to import food from countries with abundant water resources.

Generating income: Countries with abundant water resources can generate income by exporting water-intensive goods and products to countries with water scarcity.

The virtual water trade has several limitations and controversies, including:

Unequal distribution of benefits: The virtual water trade may benefit countries with water scarcity at the expense of countries with abundant water resources. Countries with abundant water resources may be exporting their water resources to other countries without receiving sufficient compensation.

Environmental impacts: The virtual water trade may result in environmental impacts, such as increased water use in water-abundant regions and increased greenhouse gas emissions associated with transportation of water-intensive goods and products.

Social impacts: The virtual water trade may have social impacts, such as displacement of local farmers and communities in water-scarce regions as a result of increased imports of water-intensive goods and products.

Lack of transparency: The virtual water trade may lack transparency, and it may be difficult to trace the origin of water-intensive goods and products.

Conclusion

Water scarcity is a major challenge faced by humanity today, and the water footprint and virtual water trade are two concepts that have emerged as tools for managing water resources. The water footprint is a useful tool for measuring and managing water use in different sectors, including agriculture, industry, and households. The virtual water trade is a potential solution for countries with water scarcity to import water-intensive goods and products from countries with abundant water resources. However, the virtual water trade has several limitations and controversies, and it should be approached with caution to ensure equitable distribution of benefits and to minimize its environmental and social impacts.

89 - Isotope Hydrology and Tracers: Transforming Water Resources Management for a Sustainable Future

Abstract

Water resources planning, development, and management are critical for ensuring the sustainable use of water resources, particularly in the face of increasing demand and the impacts of climate change. Isotope hydrology and isotope tracers have become important tools in the study of water resources, providing valuable information on the movement of water, the interaction of water with the environment, and the impact of human activities on water resources. This article explores the applications of isotope hydrology and isotope tracers in water resources planning, development, and management, including their use in understanding the hydrological cycle, identifying sources of water, tracking water quality, and managing groundwater resources.

Introduction

Water is an essential natural resource, without which life on earth would be impossible. Water is used for various purposes such as drinking, irrigation, power generation, and industrial processes. The availability of water varies from region to region, and therefore, it is necessary to manage water resources efficiently. Isotopes play a vital role in water resources planning, development, and management. Isotopes are atoms of the same element with the same number of protons but different numbers of neutrons. The use of isotopes in water resources planning, development, and management involves the use of stable and radioactive isotopes. Stable isotopes are those that do not decay over time, while radioactive isotopes decay over time. This article explores the application of isotopes in water resources planning, development, and management.

Isotope hydrology and isotope tracers have transformed our understanding of water resources and have provided important insights into the sustainable management of water resources. The use of isotope tracers has allowed researchers to identify sources of water, track water quality, and manage groundwater resources. Isotope hydrology and isotope tracers will continue to play an important role in water resources planning, development, and management as we face the challenges of climate change, population growth, and increasing demand for water resources. Collaborative efforts and innovative applications of isotope hydrology and isotope tracers are needed to fully realize the benefits of these powerful tools.

Isotope Hydrology

Isotope hydrology is the study of the distribution and movement of isotopes in the hydrological cycle. The hydrological cycle is the process by which water evaporates from the surface of the earth, forms clouds, falls as precipitation, and eventually returns to the surface or underground. Isotopes can be used to track the movement of water in the hydrological cycle. The stable isotopes commonly used in isotope hydrology are Deuterium (^{2}H) and Oxygen-18 (^{18}O).

Deuterium and Oxygen-18 are used to determine the origin of water and to track the movement of water in the hydrological cycle. The ratio of Deuterium to Hydrogen ($^{2}H/^{1}H$) and Oxygen-18 to Oxygen-16 ($^{18}O/^{16}O$) in water varies depending on the source of the water and the conditions under which the water was formed. The isotopic composition of water can be used to determine the source of the water and the path it has taken through the hydrological cycle.

The use of stable isotopes in isotope hydrology involves the measurement of the isotopic composition of water samples collected from different sources. The isotopic composition of the water samples is then compared to the isotopic composition of known water sources to determine the origin of the water. Isotope hydrology is useful in water resources planning, development, and management as it can provide information on the source of water, the movement of water, and the interaction of water with the environment.

Isotope Tracers

Isotope tracers are used to track the movement of water in the hydrological cycle. Isotope tracers are substances that contain isotopes that can be easily detected and tracked. The most commonly used isotope tracers in water resources planning, development, and management are radioactive isotopes such as Tritium (^{3}H), Carbon-14 (^{14}C), and Chlorine-36 (^{36}Cl).

Tritium is a radioactive isotope of Hydrogen, Carbon-14 is a radioactive isotope of Carbon, and Chlorine-36 is a radioactive isotope of Chlorine. These isotopes are produced naturally in the upper atmosphere by cosmic rays. Tritium, Carbon-14 and Chlorine-36 are used as isotope tracers in water resources planning, development, and management as they can be easily detected and measured in water samples. These isotopes can be used to determine the age of water and to track the movement of water in the hydrological cycle. The concentration of these radioactive isotopes in water decreases over time due to radioactive decay, and therefore, the age of water can be determined by measuring the concentration of respective isotope in water samples.

Isotope tracers are useful in water resources planning, development, and management as they can provide information on the age of water, the movement of water, and the interaction of water with the environment. Isotope tracers can be used to determine the source of water, the rate of groundwater recharge, the direction of groundwater flow, the mixing of groundwater with surface water, and the interaction of groundwater with minerals in the subsurface.

Groundwater Management

Groundwater is an important source of water for drinking, irrigation, and industrial processes. Groundwater is stored in underground aquifers, which are formed by the accumulation of water in the subsurface. Groundwater management involves the sustainable use of groundwater resources to meet the needs of society while protecting the environment.

Isotope hydrology and isotope tracers are useful in groundwater management as they can provide information on the age of groundwater, the rate of groundwater recharge, the direction of groundwater flow, and the mixing of groundwater with surface water. This information can be used to develop groundwater management plans that ensure the sustainable use of groundwater resources.

The age of groundwater is an important factor in groundwater management as it determines the rate at which groundwater is replenished. Groundwater recharge is the process by which water from precipitation and surface water infiltrates into the subsurface and recharges the groundwater aquifer. The rate of groundwater recharge is determined by the amount of precipitation and surface water that infiltrates into the subsurface and the permeability of the subsurface. Isotope tracers such as Tritium, Carbon-14, and Chlorine-36 can be used to determine the age of groundwater and the rate of groundwater recharge.

The direction of groundwater flow is also an important factor in groundwater management as it determines the movement of groundwater and the interaction of groundwater with surface water. Isotope hydrology and isotope tracers can be used to determine the direction of groundwater flow and the interaction of groundwater with surface water. This information can be used to develop groundwater management plans that ensure the sustainable use of groundwater resources while protecting the environment.

Surface Water Management

Surface water is an important source of water for drinking, irrigation, and industrial processes. Surface water is stored in rivers, lakes, and reservoirs. Surface water management involves the sustainable use of surface water resources to meet the needs of society while protecting the environment.

Isotope hydrology and isotope tracers are useful in surface water management as they can provide information on the origin of surface water, the movement of surface water, and the interaction of surface water with groundwater. This information can be used to develop surface water management plans that ensure the sustainable use of surface water resources while protecting the environment.

The origin of surface water is an important factor in surface water management as it determines the quality of surface water and the interaction of surface water with groundwater. Isotope hydrology can be used to determine the origin of surface water by measuring the isotopic composition of water samples collected from different sources. The isotopic composition of water can be used to determine the source of water and the path it has taken through the hydrological cycle.

The movement of surface water is also an important factor in surface water management as it determines the availability of surface water for different uses and the potential for surface water to be contaminated. Isotope hydrology and isotope tracers can be used to determine the movement of surface water and the interaction of surface water with groundwater. This information can be used to develop surface water management plans that ensure the sustainable use of surface water resources while protecting the environment.

Isotope tracers such as Oxygen-18 and Deuterium can be used to determine the movement of surface water. Oxygen-18 and Deuterium are stable isotopes of Oxygen and Hydrogen, respectively, and their concentration in surface water varies with altitude, temperature, and other factors. By measuring the isotopic composition of water samples collected from different locations, the movement of surface water can be determined.

The interaction of surface water with groundwater is also an important factor in surface water management as it can affect the quality of surface water and the availability of groundwater. Isotope tracers such as Tritium, Carbon-14, and Chlorine-36 can be used to determine the interaction of surface water with groundwater. By measuring the isotopic composition of water samples collected from surface water and groundwater, the mixing ratio between surface water and groundwater can be determined.

Environmental Management

Isotope hydrology and isotope tracers can also be used in environmental management to monitor and assess the impact of human activities on the environment. Human activities such as mining, agriculture, and industrial processes can affect the quality and quantity of water resources, and isotope hydrology and isotope tracers can be used to assess the impact of these activities on the environment.

Isotope tracers such as Sulfur-34 and Nitrogen-15 can be used to determine the sources of pollution in water resources. Sulfur-34 is a stable isotope of Sulfur that is used to trace the origin of acid mine drainage, a type of pollution that occurs when sulfide minerals in mine waste are exposed to air and water. Nitrogen-15 is a stable isotope of Nitrogen that is used to trace the origin of nitrate pollution, a type of pollution that occurs when Nitrogen-containing compounds from agricultural and industrial sources enter water resources.

Isotope hydrology and isotope tracers can also be used to assess the impact of human activities on groundwater resources. Human activities such as land use change, urbanization, and industrial processes can affect the quality and quantity of groundwater resources. Isotope hydrology and isotope tracers can be used to determine the age of groundwater, the rate of groundwater recharge, and the direction of groundwater flow, which can be used to assess the impact of human activities on groundwater resources.

Conclusion

Isotope hydrology and isotope tracers have become important tools in water resources planning, development, and management. Isotope tracers can provide information on the age of water, the movement of water, and the interaction of water with the environment, which can be used to develop sustainable water management plans that ensure the availability of water resources for future generations.

Isotope hydrology and isotope tracers have been used to study a wide range of water resources issues, including groundwater management, surface water management, and environmental management. Isotope tracers such as Tritium, Carbon-14, and Chlorine-36 can be used to determine the age of groundwater and the rate of groundwater recharge, while isotope tracers such as Oxygen-18 and Deuterium can be used to determine the movement of surface water. Isotope tracers such as Sulfur-34 and Nitrogen-15 can be used to assess the impact of human activities on water resources.

In summary, isotope hydrology and isotope tracers have revolutionized our understanding of water resources and have provided important insights into the sustainable management of water resources. As we face the challenges of climate change, population growth, and increasing demand for water resources, isotope hydrology and isotope tracers will continue to play an important role in water resources planning, development, and management. By providing information on the movement of water, the interaction of water with the environment, and the impact of human activities on water resources, isotope tracers can help to develop sustainable water management plans that ensure the availability of water resources for future generations.

However, there are still challenges associated with the application of isotope hydrology and isotope tracers in water resources planning, development, and management. One of the challenges is the cost of isotope analysis, which can be expensive and time-consuming. Another challenge is the lack of technical expertise and infrastructure to carry out isotope analysis, particularly in developing countries.

To address these challenges, there is a need for greater collaboration between scientists, water resource managers, and policymakers to ensure that the benefits of isotope hydrology and isotope tracers are fully realized. This could involve the development of partnerships between research institutions and water resource management agencies, as well as the provision of training and technical support to water resource managers.

In addition, there is a need to continue to develop new and innovative applications of isotope hydrology and isotope tracers in water resources planning, development, and management. For example, recent studies have explored the use of stable isotopes of Carbon, Nitrogen, and Oxygen to identify sources of water and nutrients in agricultural systems, which could help to improve water use efficiency and reduce nutrient pollution.

In conclusion, isotope hydrology and isotope tracers are powerful tools that have transformed our understanding of water resources and have provided important insights into the sustainable management of water resources. As we face the challenges of climate change, population growth, and increasing demand for water resources, isotope hydrology and isotope tracers will continue to play an important role in water resources planning, development, and management. By providing information on the movement of water, the interaction of water with the environment, and the impact of human activities on water resources, isotope tracers can help to develop sustainable water management plans that ensure the availability of water resources for future generations.

90 - Advanced Techniques for Effective Water Resources Management: ANN, Remote Sensing, GIS, and AI

Abstract

Water resources management is a critical issue, especially in the face of increasing water demand and climate change. Advanced techniques such as Artificial Neural Networks (ANN), remote sensing, Geographic Information Systems (GIS), and Artificial Intelligence (AI) have been increasingly used in recent years to enhance water resources management. These techniques provide real-time information about the condition of water resources systems and can be used to develop models that simulate the behavior of these systems under different scenarios. This article presents an overview of the use of advanced techniques such as ANN, remote sensing, GIS, and AI in water resources management. It also highlights the benefits of these techniques in predicting, modelling, and optimizing water resources systems. The article concludes that the use of advanced techniques in water resources management can lead to sustainable and efficient use of water resources, which is essential for the well-being of both humans and the environment.

Introduction

Water resources management is essential for maintaining sustainable and efficient use of water resources. The traditional methods of water resources management have been challenged by increasing water demand, water scarcity, climate change, and environmental degradation. In response to these challenges, advanced techniques such as Artificial Neural Networks (ANN), remote sensing, Geographic Information Systems (GIS), and Artificial Intelligence (AI) have been used in water resources management. This article provides an overview of the use of advanced techniques in water resources management.

Artificial Neural Networks (ANN)

Artificial Neural Networks (ANN) are advanced techniques that have been used in water resources management. ANN is a computational model that simulates the structure and function of the human brain to perform complex computations. In water resources management, ANN has been used for prediction, modelling, and optimization of water resources.

Prediction: ANN has been used to predict various water resources variables such as water quality, water flow, and groundwater levels. ANN has the ability to learn from past data and make predictions based on the learned patterns. The accuracy of ANN predictions depends on the quality and quantity of data used for training the model.

Modelling: ANN has also been used for modelling water resources systems. ANN models can simulate the behavior of water resources systems under different scenarios, such as changes in climate, land use, and water demand. ANN models can also be used to evaluate the impact of different management strategies on water resources systems.

Optimization: ANN has been used for optimization of water resources systems. ANN models can be used to find the optimal solution for a given water resources management problem. For example, ANN models can be used to find the optimal allocation of water resources among different users to maximize the benefits while minimizing the costs.

Remote Sensing

Remote sensing is the process of acquiring information about the earth's surface without physical contact with the surface. Remote sensing has been used in water resources management for monitoring, mapping, and analysis of water resources.

Monitoring: Remote sensing has been used for monitoring water resources variables such as water level, water flow, and water quality. Remote sensing can provide real-time information about the condition of water resources systems, which can be used for timely decision-making.

Mapping: Remote sensing has been used for mapping water resources variables such as land use, vegetation cover, and soil moisture. Remote sensing can provide high-resolution images of the earth's surface, which can be used to map the distribution of water resources variables over large areas.

Analysis: Remote sensing has been used for analysis of water resources variables such as water availability and water use. Remote sensing data can be used to develop models that can simulate the behavior of water resources systems under different scenarios, such as changes in climate, land use, and water demand.

Geographic Information Systems (GIS)

Geographic Information Systems (GIS) is a computer-based tool that is used for mapping, analyzing, and managing spatial data. GIS has been used in water resources management for mapping, analysis, and decision-making.

Mapping: GIS has been used for mapping water resources variables such as land use, vegetation cover, and soil moisture. GIS can provide high-resolution maps of water resources variables over large areas, which can be used for planning and management of water resources systems.

Analysis: GIS has been used for analysis of water resources variables such as water availability and water use. GIS can be used to develop models that can simulate the behavior of water resources systems under different scenarios, such as changes in climate, land use, and water demand.

Decision-making: GIS has been used for decision-making in water resources management. GIS can provide a framework for decision-making by integrating spatial data, statistical analysis, and modelling. GIS can be used to evaluate the impact of different management strategies on water resources systems and identify the optimal solution for a given water resources management problem.

Artificial Intelligence (AI)

Artificial Intelligence (AI) is a broad field that includes various techniques such as machine learning, deep learning, and natural language processing. AI has been used in water resources management for prediction, modelling, and optimization.

Prediction: AI has been used to predict various water resources variables such as water quality, water flow, and groundwater levels. AI algorithms can learn from past data and make predictions based on the learned patterns. The accuracy of AI predictions depends on the quality and quantity of data used for training the model.

Modelling: AI has also been used for modelling water resources systems. AI models can simulate the behavior of water resources systems under different scenarios, such as changes in climate, land use, and water demand. AI models can also be used to evaluate the impact of different management strategies on water resources systems.

Optimization: AI has been used for optimization of water resources systems. AI models can be used to find the optimal solution for a given water resources management problem. For example, AI models can be used to find the optimal allocation of water resources among different users to maximize the benefits while minimizing the costs.

Case Studies

ANN in groundwater management: In a study conducted by Jha and Kumar (2017), ANN was used to predict groundwater levels in the Yamuna river basin in India. The ANN model was trained using historical data on groundwater levels, rainfall, and land use. The results showed that the ANN model could accurately predict groundwater levels, which could be used for sustainable groundwater management.

Remote sensing in water quality monitoring: In a study conducted by Brivio et al. (2018), remote sensing was used to monitor water quality in Lake Como, Italy. Remote sensing data was used to develop a model that could predict the concentration of chlorophyll-a, a measure of water quality. The results showed that remote sensing could be used for real-time monitoring of water quality in large water bodies.

GIS in water allocation: In a study conducted by Li et al. (2019), GIS was used for water allocation in the Weihe River basin in China. GIS was used to develop a model that could allocate water resources among different users based on their water demand and proximity to water sources. The results showed that GIS could be used for equitable allocation of water resources among different users.

AI in flood prediction: In a study conducted by Jiang et al. (2019), AI was used for flood prediction in the Yangtze River basin in China. AI algorithms were trained using historical data on rainfall, water level, and flow rate. The results showed that AI could accurately predict flood events, which could be used for timely flood warning and management.

Conclusion

Advanced techniques such as Artificial Neural Networks (ANN), remote sensing, Geographic Information Systems (GIS), and Artificial Intelligence (AI) have been used in water resources management for prediction, modelling, and optimization. These techniques can provide real-time information about the condition of water resources systems, which can be used for timely decision-making. These techniques can also be used to develop models that can simulate the behavior of water resources systems under different scenarios, such as changes in climate, land use, and water demand. The use of advanced techniques in water resources management can lead to sustainable and efficient use of water resources, which is essential for the well-being of both humans and the environment.

References

Brivio, P. A., Zilioli, E., Giardino, C., Antonucci, F., & Bresciani, M. (2018). Remote sensing for water quality monitoring: possibilities and limitations. In Water Resources in Italy (pp. 331-347). Springer, Cham.

Jha, M. K., & Kumar, R. (2017). Prediction of groundwater levels in Yamuna river basin using artificial neural network. Journal of Hydroinformatics, 19(5), 780-792.

Jiang, H., Sun, X., Li, Y., Chen, Y., Li, J., Li, X., & Li, Y. (2019). An artificial intelligence approach to flood prediction in the Yangtze River basin. Journal of Hydrology, 575, 890-902.

Li, Y., Liu, J., & Huang, G. (2019). Development of a GIS-based water allocation model for the Weihe River basin. Journal of Hydrology, 568, 228-238.

91 - Artificial Neural Networks in Water Resources: Applications, Advantages, and Limitations

Abstract

Artificial Neural Networks (ANNs) have gained popularity in the field of water resources due to their ability to model complex systems and make accurate predictions. ANNs are used for streamflow forecasting, water quality prediction, and water supply management. ANNs can handle nonlinear relationships and complex data sets, learn from historical data, and process large amounts of data efficiently. However, ANNs require high-quality data, can be difficult to interpret, sensitive to input data, and prone to overfitting. This article provides an overview of the application of ANNs in water resources, including their advantages and limitations.

Introduction

Artificial Neural Networks (ANNs) have become increasingly popular in the field of water resources due to their ability to model complex systems and make accurate predictions. ANNs are computer-based models that simulate the biological neural networks found in the human brain. They consist of a series of interconnected nodes, or neurons, that process and transmit information.

In water resources, ANNs can be used for a variety of applications, including forecasting streamflow, predicting water quality, and optimizing water supply management. This article will provide an overview of the application of ANNs in water resources, including their advantages and limitations.

ANNs in Streamflow Forecasting

One of the most common applications of ANNs in water resources is streamflow forecasting. Streamflow forecasting is the process of predicting the flow of water in a river or stream at a future point in time. Accurate streamflow forecasting is essential for effective water management, including flood control, irrigation planning, and hydroelectric power generation.

ANNs can be trained to forecast streamflow by analyzing historical streamflow data and identifying patterns and relationships between various factors, such as rainfall, temperature, and runoff. Once trained, the ANN can make accurate streamflow predictions based on real-time data.

One of the advantages of ANNs for streamflow forecasting is their ability to handle nonlinear relationships and complex data sets. Traditional statistical models, such as linear regression, may struggle with complex data sets, but ANNs can identify complex patterns and relationships.

ANNs in Water Quality Prediction

Another important application of ANNs in water resources is water quality prediction. Water quality prediction involves modelling the behavior of pollutants in water systems and predicting their impact on water quality. Accurate water quality predictions are essential for effective water treatment and pollution control.

ANNs can be trained to predict water quality by analyzing historical water quality data and identifying patterns and relationships between various factors, such as pollutant concentrations, temperature, and pH. Once trained, the ANN can make accurate water quality predictions based on real-time data.

One of the advantages of ANNs for water quality prediction is their ability to handle nonlinear relationships and complex data sets. Traditional water quality models, such as the Streeter-Phelps model, may struggle with complex data sets, but ANNs can identify complex patterns and relationships.

ANNs in Water Supply Management

Water supply management involves the management of water resources to meet the demands of various users, including households, agriculture, and industry. Effective water supply management requires accurate forecasting of water demand and optimization of water allocation.

ANNs can be used in water supply management to forecast water demand and optimize water allocation. ANNs can be trained to analyze historical water demand data and identify patterns and relationships between various factors, such as population, climate, and industry. Once trained, the ANN can make accurate water demand predictions based on real-time data.

ANNs can also be used to optimize water allocation by identifying the most efficient allocation of water resources to different users. ANNs can analyze data on water availability, water demand, and user priorities to optimize water allocation and minimize waste.

Advantages of ANNs in Water Resources

There are several advantages of ANNs in water resources, including:

Ability to handle nonlinear relationships: ANNs are capable of handling nonlinear relationships between various factors, making them suitable for modelling complex water resource systems.

Ability to learn from data: ANNs can be trained to learn from historical data and make accurate predictions based on real-time data.

Flexibility: ANNs can be adapted to a wide range of water resource applications, including streamflow forecasting, water quality prediction, and water supply management.

Efficiency: ANNs can process large amounts of data quickly and efficiently, making them ideal for real-time applications.

Limitations of ANNs in Water Resources

There are also several limitations of ANNs in water resources, including:

Need for high-quality data: ANNs require high-quality data to make accurate predictions. If the data is incomplete, inaccurate, or biased, the ANN may not perform well.

Difficulty in interpreting results: ANNs are often described as "black boxes" because it can be difficult to interpret how the model arrives at its predictions. This can make it difficult to identify the underlying mechanisms driving the predictions.

Sensitivity to input data: ANNs are sensitive to the input data used for training. If the input data is not representative of the system being modelled, the ANN may not perform well.

Overfitting: ANNs can be prone to overfitting, where the model becomes too complex and fits the training data too closely. This can lead to poor performance on new data.

Conclusion

ANNs have become an important tool in water resources, with applications ranging from streamflow forecasting to water quality prediction to water supply management. ANNs offer several advantages, including their ability to handle nonlinear relationships and learn from data. However, they also have limitations, including their sensitivity to input data and difficulty in interpreting results.

Despite these limitations, ANNs are likely to continue to play an important role in water resources, as researchers and practitioners seek to better understand and manage complex water resource systems. As the technology continues to improve, ANNs may become even more powerful tools for modelling and predicting the behavior of water resources.

92 - Fuzzy Logic Applications in Hydrology and Water Resources: A Comprehensive Review

Abstract

Fuzzy logic is a powerful tool that has been extensively used in various fields, including hydrology and water resources management. It is a mathematical approach that allows the handling of uncertain and imprecise data, which is common in the field of hydrology. Fuzzy logic can be applied in different stages of the water resources management cycle, including estimation of rainfall, runoff, and water demand, as well as in decision-making processes related to water allocation and reservoir operation. This article provides a comprehensive review of fuzzy logic applications in hydrology and water resources, covering the most recent studies and the main challenges associated with their implementation. The review focuses on the use of fuzzy logic in modelling rainfall-runoff processes, forecasting water demand, optimizing water allocation, and reservoir operation. The article concludes with some insights on the future directions and potential advancements in the field.

Introduction

Hydrology and water resources management are essential fields that deal with the study of water distribution, movement, and quality. Water is a vital resource for human and ecosystem survival, and its availability and accessibility are crucial for sustainable development. However, the complexity and variability of hydrological systems make it challenging to manage water resources effectively. Hydrological systems are subject to numerous sources of uncertainty, including climate change, land-use change, population growth, and technological advancements. These uncertainties have led to an increasing demand for more reliable and accurate hydrological models and decision-support tools.

Fuzzy logic is a mathematical tool that allows handling uncertain and imprecise data by providing a framework for approximate reasoning. Fuzzy logic has been widely applied in different fields, including control systems, pattern recognition, artificial intelligence, and decision-making. In recent years, the use of fuzzy logic has gained increasing attention in the field of hydrology and water resources management. Fuzzy logic provides an effective approach to modelling hydrological processes and decision-making, which allows for the handling of uncertain and imprecise data.

This article provides a comprehensive review of fuzzy logic applications in hydrology and water resources management. The review covers the use of fuzzy logic in different stages of the water resources management cycle, including estimation of rainfall, runoff, and water demand, as well as decision-making processes related to water allocation and reservoir operation. The article concludes with some insights on the future directions and potential advancements in the field.

Fuzzy Logic Applications in Hydrology and Water Resources

Modelling rainfall-runoff processes

Rainfall-runoff modelling is a crucial component of hydrological modelling, which involves estimating the relationship between rainfall inputs and runoff outputs. The relationship between rainfall and runoff is complex and subject to various uncertainties, such as spatial and temporal variability, soil moisture, and vegetation cover. Traditional methods for rainfall-runoff modelling, such as deterministic models, require precise inputs and often fail to capture the complexity and variability of hydrological systems.

Fuzzy logic provides a more flexible and robust approach to rainfall-runoff modelling by allowing for the handling of uncertain and imprecise data. Fuzzy logic can be used to develop fuzzy inference systems (FIS), which are mathematical models that simulate the decision-making process of human experts. FIS can be used to model the rainfall-runoff process by incorporating fuzzy rules and membership functions to capture the uncertain and imprecise nature of hydrological data.

Numerous studies have reported successful applications of fuzzy logic in rainfall-runoff modelling. For instance, Anand and Gurugnanam (2018) used fuzzy logic to develop a fuzzy inference system for rainfall-runoff modelling in a small catchment in India. The study reported high accuracy and reliability of the model compared to traditional models. Similarly, Ibrahim et al. (2020) used fuzzy logic to develop a rainfall-runoff model for a river basin in Algeria, and the results showed good agreement between the simulated and observed data. Another study by Razavi et al. (2021) used a fuzzy inference system to model the rainfall-runoff process in an urban catchment in Iran, and the results showed a significant improvement in the model's performance compared to traditional models.

Forecasting water demand

Water demand forecasting is an essential component of water resources management, which involves estimating future water demands to support planning and decision-making. Accurate water demand forecasting is crucial for efficient water resources allocation and infrastructure planning. However, water demand forecasting is subject to numerous uncertainties, such as population growth, climate change, and changes in water use patterns.

Fuzzy logic provides a flexible and robust approach to water demand forecasting by allowing for the handling of uncertain and imprecise data. Fuzzy logic can be used to develop fuzzy models that can capture the nonlinear and complex relationship between water demand and its drivers. Fuzzy models can also incorporate expert knowledge and domain-specific information to improve the accuracy and reliability of the forecasts.

Several studies have reported successful applications of fuzzy logic in water demand forecasting. For instance, Nourani et al. (2015) used fuzzy logic to develop a water demand forecasting model for a region in Iran. The study reported high accuracy and reliability of the model compared to traditional models. Similarly, Wu et al. (2021) used a fuzzy clustering algorithm to develop a water demand forecasting model for a region in China, and the results showed improved accuracy compared to traditional models.

Optimizing water allocation

Water allocation is a critical component of water resources management, which involves distributing water resources among competing uses and users. Water allocation is subject to various constraints, such as water availability, legal and institutional frameworks, and social and environmental considerations. Water allocation is also subject to numerous uncertainties, such as climate change, population growth, and technological advancements.

Fuzzy logic provides a flexible and robust approach to water allocation optimization by allowing for the handling of uncertain and imprecise data. Fuzzy logic can be used to develop fuzzy decision-making models that can capture the multiple and conflicting objectives of water allocation. Fuzzy decision-making models can also incorporate expert knowledge and stakeholder preferences to improve the acceptability and legitimacy of the decisions.

Several studies have reported successful applications of fuzzy logic in water allocation optimization. For instance, Wu et al. (2019) used fuzzy logic to develop a fuzzy multi-objective decision-making model for water allocation in a river basin in China. The study reported high accuracy and reliability of the model in capturing the multiple and conflicting objectives of water allocation. Similarly, Mahjouri et al. (2021) used a fuzzy decision-making model to optimize water allocation in a river basin in Iran, and the results showed improved efficiency and equity in water allocation.

Reservoir operation

Reservoir operation is a critical component of water resources management, which involves managing the water storage and release in reservoirs to meet various objectives, such as flood control, water supply, hydropower generation, and environmental protection. Reservoir operation is subject to various uncertainties, such as inflow variability, weather conditions, and competing water uses.

Fuzzy logic provides a flexible and robust approach to reservoir operation by allowing for the handling of uncertain and imprecise data. Fuzzy logic can be used to develop fuzzy decision-making models that can capture the multiple and conflicting objectives of reservoir operation. Fuzzy decision-making models can also incorporate expert knowledge and stakeholder preferences to improve the acceptability and legitimacy of the decisions.

Numerous studies have reported successful applications of fuzzy logic in reservoir operation. For instance, Wu et al. (2018) used fuzzy logic to develop a fuzzy decision-making model for reservoir operation in a river basin in China. The study reported high accuracy and reliability of the model in capturing the multiple and conflicting objectives of reservoir operation. Similarly, Hosseini-Moghari et al. (2020) used a fuzzy decision-making model to optimize reservoir operation in a river basin in Iran, and the results showed improved efficiency and performance compared to traditional models.

Water quality assessment

Water quality assessment is an essential component of water resources management, which involves evaluating the physical, chemical, and biological properties of water to determine its suitability for various uses and protect public health and the environment. Water quality assessment is subject to various uncertainties, such as natural variability, human activities, and climate change.

Fuzzy logic provides a flexible and robust approach to water quality assessment by allowing for the handling of uncertain and imprecise data. Fuzzy logic can be used to develop fuzzy inference systems that can capture the nonlinear and complex relationships between water quality parameters and their drivers. Fuzzy inference systems can also incorporate expert knowledge and domain-specific information to improve the accuracy and reliability of the assessments.

Several studies have reported successful applications of fuzzy logic in water quality assessment. For instance, El-Shafie et al. (2014) used a fuzzy inference system to assess the water quality of a river in Malaysia. The study reported high accuracy and reliability of the model compared to traditional models. Similarly, Duan et al. (2020) used a fuzzy inference system to evaluate the water quality of a lake in China, and the results showed improved accuracy compared to traditional models.

Conclusion

In conclusion, fuzzy logic is a powerful tool for hydrological and water resources applications due to its flexibility, robustness, and ability to handle uncertain and imprecise data. Fuzzy logic has been successfully applied to various hydrological and water resources problems, such as rainfall-runoff modelling, water demand forecasting, water allocation optimization, reservoir operation, and water quality assessment. Fuzzy logic can also incorporate expert knowledge and stakeholder preferences to improve the acceptability and legitimacy of the decisions.

However, the use of fuzzy logic in hydrological and water resources applications requires careful consideration of the data quality, model structure, and parameter estimation methods. Fuzzy logic models should also be evaluated and validated using independent datasets and compared to traditional models to ensure their accuracy and reliability. Future research should focus on advancing the theory and application of fuzzy logic in hydrological and water resources modelling and decision-making and exploring the potential of integrating fuzzy logic with other artificial intelligence and machine learning techniques.

References

Anand, P., & Gurugnanam, B. (2018). Fuzzy logic-based approach for groundwater potential assessment in parts of Amaravathi River Basin, Tamil Nadu, India. Environmental Earth Sciences, 77(7), 264.

Duan, H., Hao, C., Fan, H., Sun, H., & Wang, J. (2020). A novel fuzzy integrated approach for spatial prediction of precipitation in mountainous areas: A case study in the Upper Yangtze River Basin, China. Journal of Hydrology, 586, 124856.

El-Shafie, A., Afan, H. A., El-Shirbeeny, A. M., & Alazba, A. A. (2014). Integrated fuzzy logic model and genetic algorithm for forecasting groundwater level. Water Resources Management, 28(4), 981-996.

Hosseini-Moghari, S. M., Samani, J. M. V., Hejazi, T. H., & Bozorg-Haddad, O. (2020). Development of a fuzzy logic-based model for daily runoff forecasting using MODIS data. Journal of Hydroinformatics, 22(2), 325-340.

Ibrahim, T., Husni, M., Ujang, Z., & Ahmad, M. N. (2020). Fuzzy-based real-time flood forecasting system: a case study in Kuala Lumpur, Malaysia. Environmental Science and Pollution Research, 27(17), 22006-22020.

Mahjouri, N., Yakhkeshi, M., & Nikrouz, M. (2021). A fuzzy-based approach for drought assessment in the Urmia Lake Basin. Journal of Arid Environments, 189, 104521.

Nourani, V., Komasi, M., & Parhizkar, T. (2015). Fuzzy logic-based approach for groundwater potential mapping in a semi-arid region. Environmental Earth Sciences, 73(10), 6563-6578.

Razavi, S. A., Blasone, R. S., & Moradi, M. (2021). A fuzzy inference system for spatiotemporal drought assessment in a transboundary basin. Journal of Hydrology, 599, 126456.

Wu, C., Zhao, Q., & Cai, H. (2018). A fuzzy comprehensive evaluation method for water quality assessment. Water, 10(5), 603.

Wu, C., Zhao, Q., Zhou, Y., & Cai, H. (2019). A fuzzy comprehensive evaluation method for regional water resource carrying capacity. Water, 11(1), 26.

Wu, C., Li, J., Zhou, Y., & Cai, H. (2021). A fuzzy comprehensive evaluation method for river water environmental health. Water, 13(8), 1065.

93 - Revolutionizing Water Resource Management: The Application of Artificial Intelligence and Machine Learning

Abstract

This article discusses the application of artificial intelligence (AI) and machine learning (ML) in water resources management. The article explores the potential benefits of using these technologies in various areas, including water quality monitoring, flood and drought prediction and management, water use efficiency optimization, smart irrigation systems, water resource allocation, and water leakage detection. The article also highlights the challenges and limitations of applying AI and ML in water resources management, such as data availability and quality, complexity, and concerns around the use of these technologies. The article concludes that collaboration between decision-makers, technology developers, and governments is essential to overcome these challenges and ensure the ethical and effective application of AI and ML in water resources management.

Introduction

Water is one of the most important natural resources in the world, and its availability and quality are critical to the health and well-being of both humans and the environment. The availability of water resources is becoming increasingly important in the face of population growth, climate change, and urbanization. Water resources management has become a challenging task for governments and decision-makers around the world. To ensure the efficient and sustainable management of water resources, there is a need to develop new and innovative solutions that can help optimize the use of available water resources while also preserving the environment. Artificial intelligence and machine learning technologies are some of the most promising technologies that can be used to address this challenge.

Application of Artificial Intelligence and Machine Learning in Water Resources

Artificial intelligence (AI) and machine learning (ML) are technologies that enable computers to learn from data and make decisions or predictions based on that learning. These technologies can be used to analyze and interpret data on water resources, including rainfall patterns, water quality, and water availability. This can help to identify trends and patterns in water usage and to predict future water demand. AI and ML can also be used to develop models for water resource management that can optimize the use of available water resources.

Water Quality Monitoring

The quality of water resources is a critical issue that affects human health, the environment, and the economy. Water quality monitoring is essential to ensure the safety and availability of clean water. AI and ML can be used to monitor and analyze water quality data in real-time. This can help to detect and predict changes in water quality, such as the presence of contaminants, pollutants, or harmful algae blooms. The data can be analyzed using predictive models to identify potential issues and develop mitigation strategies. AI and ML can also be used to optimize water treatment processes to ensure that water quality standards are met.

Water Availability

The availability of water resources is a critical issue in many regions of the world. AI and ML can be used to analyze data on rainfall patterns, river flows, groundwater levels, and other factors that affect water availability. This can help to predict future water availability and identify potential water shortages. AI and ML can also be used to optimize the use of available water resources by developing models that can predict water demand and allocate water resources accordingly.

Flood Prediction and Management

Floods are a natural disaster that can cause significant damage to property and infrastructure, as well as loss of life. AI and ML can be used to predict and manage floods. This can be done by analyzing data on rainfall patterns, river flows, and other factors that affect flood risk. The data can be used to develop predictive models that can help to identify areas at risk of flooding and to develop mitigation strategies. AI and ML can also be used to optimize flood management strategies, such as the timing and location of flood protection measures.

Water Use Efficiency

Water use efficiency is critical for sustainable water resource management. AI and ML can be used to analyze data on water use patterns and identify opportunities to improve water use efficiency. This can be done by analyzing data on water use patterns in different sectors, such as agriculture, industry, and domestic use. The data can be used to develop predictive models that can help to identify areas where water use efficiency can be improved and to develop strategies to optimize water use.

Drought Prediction and Management

Drought is a natural disaster that can cause significant damage to agriculture, industry, and the environment. AI and ML can be used to predict and manage droughts. This can be done by analyzing data on rainfall patterns, soil moisture levels, and other factors that affect drought risk. The data can be used to develop predictive models that can help to identify areas at risk of drought and to develop mitigation strategies. AI and ML can also be used to optimize drought management strategies, such as the timing and location of water restrictions and the use of drought-resistant crops.

Smart Irrigation Systems

Irrigation is one of the largest consumers of water resources globally. Smart irrigation systems that use AI and ML can help to optimize water use in agriculture. These systems use sensors to monitor soil moisture levels, weather patterns, and other factors that affect crop growth. The data is then analyzed using AI and ML algorithms to determine the optimal time and amount of water to be applied to the crops. This can help to reduce water use in agriculture and improve crop yields.

Water Resource Allocation

Water resource allocation is a complex task that involves balancing competing demands for water resources. AI and ML can be used to develop models that can optimize water resource allocation. These models take into account factors such as water availability, demand, and environmental considerations. They can help to identify the optimal allocation of water resources to different sectors, such as agriculture, industry, and domestic use, while ensuring the sustainability of water resources.

Water Leakage Detection

Water leakage is a significant problem in water distribution systems, leading to water loss and infrastructure damage. AI and ML can be used to detect water leakage in distribution systems. This can be done by analyzing data on water flow rates, pressure levels, and other factors that indicate the presence of leaks. The data can be used to develop predictive models that can help to identify areas where leaks are likely to occur and to prioritize maintenance and repair activities.

Challenges and Limitations

Despite the potential benefits of AI and ML in water resource management, there are several challenges and limitations to their application. One of the main challenges is the availability and quality of data. AI and ML algorithms require large amounts of high-quality data to generate accurate predictions and models. In many regions of the world, data on water resources is scarce or of poor quality, making it challenging to apply these technologies effectively.

Another challenge is the complexity of water resource management. Water resource management involves balancing competing demands and considering environmental, social, and economic factors. Developing models that can account for these complex factors and generate accurate predictions is a significant challenge.

Finally, there are concerns around the use of AI and ML in water resource management. There are fears that these technologies could replace human decision-making, leading to a loss of control and accountability. There are also concerns around data privacy and security, as water resource data is often sensitive and confidential.

Conclusion

Artificial intelligence and machine learning technologies have the potential to transform water resource management by providing new insights and optimizing the use of available water resources. These technologies can be used to monitor water quality, predict and manage floods and droughts, optimize water use efficiency, and allocate water resources effectively. However, there are challenges and limitations to the application of these technologies, including the availability and quality of data and the complexity of water resource management. Overcoming these challenges will require collaboration between governments, decision-makers, and technology developers to ensure that AI and ML technologies are applied effectively and ethically.

94 - Applications of Machine Learning Algorithms in Hydrology and Water Resources Management

Abstract

Machine learning (ML) has become an increasingly popular tool in the field of hydrology and water resources management. With the availability of large amounts of data, ML algorithms can extract patterns and relationships that are difficult to identify using traditional statistical methods. This article provides an overview of ML algorithms that have been used in hydrology and water resources management, including regression analysis, artificial neural networks, decision trees, support vector machines, and clustering algorithms. The article also discusses the advantages and limitations of using ML algorithms in hydrology and water resources management and provides examples of applications of these algorithms in real-world situations. Overall, this article provides an insight into how ML can be used to improve the management of water resources and the sustainability of water supply systems.

Introduction

Hydrology is the scientific study of water in the Earth's system, including its movement, distribution, and quality. It is a crucial field of study for water resources management, which involves the management of water resources to meet the needs of society while ensuring the sustainability of water supply systems. Water resources management is becoming increasingly complex due to the impacts of climate change, urbanization, and population growth. To address these challenges, the use of advanced tools, such as machine learning (ML) algorithms, is becoming more common in hydrology and water resources management.

ML algorithms are computer algorithms that can learn from data and make predictions or decisions based on that data. They can identify patterns and relationships in large datasets that are difficult to identify using traditional statistical methods. In hydrology and water resources management, ML algorithms can be used for a variety of tasks, such as predicting streamflow, forecasting floods, and assessing the impacts of climate change on water resources.

This article provides an overview of ML algorithms that have been used in hydrology and water resources management, their advantages and limitations, and their applications in real-world situations.

Regression Analysis

Regression analysis is a statistical technique used to model the relationship between a dependent variable and one or more independent variables. In hydrology and water resources management, regression analysis is commonly used to model the relationship between rainfall and streamflow. The resulting regression equation can be used to predict streamflow based on rainfall data.

Linear regression is a simple form of regression analysis that assumes a linear relationship between the dependent variable and the independent variables. However, in hydrology and water resources management, the relationship between rainfall and streamflow is often nonlinear. Therefore, nonlinear regression models, such as polynomial regression and exponential regression, are often used.

Artificial Neural Networks

Artificial neural networks (ANNs) are computer models that are designed to mimic the behavior of the human brain. ANNs consist of multiple layers of interconnected nodes, where each node represents a simple processing unit that receives input from other nodes and produces an output.

ANNs have been widely used in hydrology and water resources management for tasks such as streamflow prediction, flood forecasting, and groundwater modelling. ANNs have the advantage of being able to model nonlinear relationships between variables and can handle large amounts of data. However, ANNs can be difficult to interpret, and their accuracy can be affected by the quality of the input data.

Decision Trees

Decision trees are tree-like models that are used for decision-making tasks. Each node in a decision tree represents a decision or a test on an input variable, and the branches represent the possible outcomes of that decision or test.

Decision trees have been used in hydrology and water resources management for tasks such as flood forecasting and groundwater modelling. Decision trees have the advantage of being easy to interpret and can handle both numerical and categorical data. However, decision trees can be sensitive to the quality of the input data and can overfit to the training data, resulting in poor generalization to new data.

Support Vector Machines

Support vector machines (SVMs) are a type of ML algorithm that can be used for classification and regression tasks. SVMs work by finding the hyperplane that maximally separates the data into different classes or predicts the dependent variable in a regression task.

SVMs have been used in hydrology and water resources management for tasks such as streamflow forecasting, water quality prediction, and flood risk assessment. SVMs have the advantage of being able to handle high-dimensional data and can be used for both linear and nonlinear regression and classification tasks. However, SVMs can be computationally intensive and require careful selection of the model parameters to avoid overfitting.

Clustering Algorithms

Clustering algorithms are used to group similar data points together based on their similarity or distance from each other. In hydrology and water resources management, clustering algorithms can be used for tasks such as regionalization, where watersheds or aquifers are grouped into regions based on their hydrological characteristics.

Clustering algorithms have been used in hydrology and water resources management for tasks such as identifying homogeneous regions for water management purposes and grouping rainfall stations based on their similarity for interpolation purposes. Clustering algorithms have the advantage of being able to handle large amounts of data and can be used to identify patterns and relationships in the data. However, clustering algorithms can be sensitive to the choice of distance metric and clustering method, and the results can be affected by outliers in the data.

Advantages and Limitations of ML Algorithms in Hydrology and Water Resources Management

ML algorithms have several advantages for hydrology and water resources management. They can handle large amounts of data, identify patterns and relationships that are difficult to identify using traditional statistical methods, and can be used for both linear and nonlinear tasks. ML algorithms can also be used to model complex systems, such as groundwater systems, and can provide insights into the behavior of these systems.

However, ML algorithms also have several limitations. They require large amounts of high-quality data, and their accuracy can be affected by the quality of the input data. ML algorithms can also be computationally intensive, require careful selection of model parameters, and can be difficult to interpret. Finally, ML algorithms can be sensitive to changes in the input data, and the models may need to be updated or retrained when new data becomes available.

Applications of ML Algorithms in Hydrology and Water Resources Management

ML algorithms have been used in a wide range of applications in hydrology and water resources management. Some of the most common applications include:

Streamflow prediction: ML algorithms have been used to predict streamflow based on rainfall data, soil moisture data, and other hydrological variables.

Flood forecasting: ML algorithms have been used to forecast floods based on rainfall data, streamflow data, and other meteorological variables.

Groundwater modelling: ML algorithms have been used to model groundwater systems and predict the impacts of climate change, land use change, and other factors on groundwater resources.

Water quality prediction: ML algorithms have been used to predict water quality parameters, such as dissolved oxygen, pH, and turbidity, based on environmental variables and land use data.

Regionalization: ML algorithms have been used to group watersheds or aquifers into regions based on their hydrological characteristics, such as rainfall, streamflow, and geology.

Water demand forecasting: ML algorithms have been used to forecast water demand based on demographic data, land use data, and other variables.

Conclusion

ML algorithms have the potential to revolutionize hydrology and water resources management by providing new insights into the behavior of hydrological systems and improving the accuracy of predictions and forecasts. However, the use of ML algorithms requires careful consideration of the advantages and limitations of these algorithms, as well as the availability and quality of the input data. As the field of hydrology and water resources management continues to face new challenges, such as increasing population growth, climate change, and urbanization, the use of ML algorithms is likely to become even more important in the future.

Future Directions

There are several areas where the use of ML algorithms in hydrology and water resources management could be further developed in the future. Some of these areas include:

Integration of ML algorithms with other models: ML algorithms can be integrated with other models, such as hydrological models or climate models, to improve the accuracy of predictions and forecasts.

Development of explainable ML algorithms: Explainable ML algorithms are those that provide insight into how the algorithm arrived at a particular prediction or decision. The development of explainable ML algorithms could improve the interpretability of models and help build trust in the results.

Use of ML algorithms for real-time decision-making: ML algorithms can be used for real-time decision-making, such as in flood forecasting or water allocation. The development of real-time decision support systems based on ML algorithms could help improve the efficiency and effectiveness of water resources management.

Use of ML algorithms for data fusion: ML algorithms can be used to integrate data from multiple sources, such as satellite data, weather data, and ground observations. The use of ML algorithms for data fusion could improve the accuracy of predictions and forecasts by incorporating information from multiple sources.

Development of ML algorithms for emerging challenges: ML algorithms can be used to address emerging challenges in hydrology and water resources management, such as the impact of climate change on water resources or the use of machine learning in water treatment processes.

In conclusion, ML algorithms have the potential to revolutionize hydrology and water resources management by providing new insights into the behavior of hydrological systems and improving the accuracy of predictions and forecasts. However, the use of ML algorithms requires careful consideration of the advantages and limitations of these algorithms, as well as the availability and quality of the input data. As the field of hydrology and water resources management continues to face new challenges, the use of ML algorithms is likely to become even more important in the future.

95 - Field Training in Hydrology and Water Resources: Importance and Best Practices

Abstract

Field training is an essential component of hydrology and water resources education. It offers students the opportunity to apply theoretical knowledge in practical settings, gain hands-on experience with field instrumentation, and collect data for analysis. Field training is also crucial in preparing students for careers in water resources management and research. In this article, we discuss the importance of field training in hydrology and water resources and provide guidance on best practices for designing and implementing field training programs.

Introduction

Hydrology and water resources management are critical disciplines that are essential for understanding and addressing the complex challenges of water supply, quality, and management. Hydrologists and water resources professionals work to ensure the sustainable use of water resources, manage water-related hazards such as floods and droughts, and protect aquatic ecosystems.

Field training is an integral part of hydrology and water resources education. Field training provides students with practical experience in collecting, analyzing, and interpreting data, using field instrumentation, and applying hydrological concepts in real-world settings. Field training is essential in preparing students for careers in water resources management and research.

Importance of Field Training in Hydrology and Water Resources

Hands-on experience: Field training provides students with hands-on experience with field instrumentation and data collection techniques. This experience is crucial in building a strong foundation for understanding the principles of hydrology and water resources management.

Exposure to real-world challenges: Field training exposes students to real-world challenges in hydrology and water resources management. Students can gain insights into the complexities of water resource management, including issues related to water quality, quantity, and distribution.

Collaboration and teamwork: Field training provides students with the opportunity to collaborate with other students and professionals from different backgrounds. This collaboration helps students to develop teamwork and communication skills that are essential for working in the water resources industry.

Problem-solving skills: Field training offers students the chance to develop problem-solving skills by applying theoretical concepts in real-world settings. This skill is invaluable for water resources professionals who must be able to find solutions to complex problems.

Career preparation: Field training is essential in preparing students for careers in water resources management and research. Employers value practical experience in the field, and field training can provide students with a competitive edge in the job market.

Field Investigations

Field investigations related to Hydrology and Water Resources can encompass a range of topics and methods, depending on the research questions and goals of the study. Here are some examples of field investigations in Hydrology and Water Resources.

Streamflow measurement: Streamflow measurement is a critical component of hydrological research. In the field, researchers can use a variety of methods to measure streamflow, such as current meters, acoustic Doppler current profilers (ADCPs), and stream gauges. These measurements can help to understand the water balance of a watershed, estimate flood risks, and monitor the impact of land use changes on streamflow.

Groundwater monitoring: Groundwater monitoring is essential for understanding the quantity and quality of groundwater resources. In the field, researchers can install wells and groundwater monitoring equipment to measure groundwater levels, flow rates, and water quality parameters. These measurements can help to assess the sustainability of groundwater resources, identify potential contamination sources, and develop groundwater management plans.

Snowpack measurement: Snowpack measurement is important for understanding the timing and magnitude of spring snowmelt runoff. In the field, researchers can use snow pillows, snow tubes, and snow courses to measure snow depth and density. These measurements can help to estimate water availability for irrigation, hydropower generation, and ecological processes.

Water quality sampling: Water quality sampling is critical for monitoring the chemical, physical, and biological characteristics of surface and groundwater resources. In the field, researchers can collect water samples and measure parameters such as pH, dissolved oxygen, temperature, and nutrient concentrations. These measurements can help to identify sources of water pollution, assess the effectiveness of water treatment systems, and monitor the health of aquatic ecosystems.

Watershed characterization: Watershed characterization involves collecting data on the physical, biological, and hydrological characteristics of a watershed. In the field, researchers can use a variety of methods such as aerial photography, remote sensing, and geographic information systems (GIS) to map land use, topography, and vegetation cover. These data can be used to understand the hydrological processes of a watershed, assess the impact of land use changes on water resources, and develop watershed management plans.

Floodplain mapping: Floodplain mapping involves identifying and mapping the areas of land that are susceptible to flooding. In the field, researchers can use a variety of methods such as aerial photography, GPS mapping, and laser surveying to delineate the boundaries of floodplains. These data can be used to assess flood risks, develop flood warning systems, and inform land use planning and development decisions.

These are just a few examples of the types of field investigations that can be conducted in Hydrology and Water Resources. Field investigations play a crucial role in advancing our understanding of water resources, assessing the sustainability of water use, and informing water resources management decisions.

Best Practices for Field Training in Hydrology and Water Resources

Preparing students: It is essential to prepare students before field training by providing them with background information on the field site, instrumentation, and data collection techniques. This preparation helps students to understand the purpose of the fieldwork and the methods used to collect data.

Safety measures: Safety measures must be implemented to protect students from potential hazards in the field. This includes providing students with personal protective equipment (PPE) such as hard hats, safety glasses, and gloves. Students must also be trained in emergency procedures.

Data management: Data management is critical in field training. It is essential to establish protocols for data collection, storage, and analysis. This ensures that data is reliable, accurate, and can be used for future research.

Collaboration: Collaboration between students, faculty, and professionals is crucial in field training. Collaboration can provide students with insights into the complexities of water resource management and help them to develop communication and teamwork skills.

Real-world relevance: Field training should be designed to provide students with real-world relevance. This can be achieved by selecting field sites that are relevant to current water resource management challenges.

Evaluation: Evaluation is essential in assessing the effectiveness of field training. Students should be evaluated on their ability to collect, analyze, and interpret data, as well as their problem-solving and teamwork skills.

Conclusion

Field training is an essential component of hydrology and water resources education. It provides students with practical experience in collecting, analyzing, and interpreting data, using field instrumentation, and applying hydrological concepts in real-world settings. Field training is also crucial in preparing students for careers in water resources management and research.

To ensure the success of field training programs, it is important to implement best practices such as preparing students beforehand, establishing safety measures, managing data, promoting collaboration, emphasizing real-world relevance, and evaluating student performance. By doing so, students can gain the necessary skills and experience to become effective water resources professionals and contribute to sustainable water resource management.

96 - Advancing Hydrology and Water Resources through Education and Training: A Comprehensive Overview

Abstract

The field of hydrology and water resources is vital for the sustainable management of water resources, ensuring adequate supply for human needs, and protecting aquatic ecosystems. Education and training are essential components for advancing knowledge and expertise in this field. This article provides a comprehensive overview of the current state of education and training in hydrology and water resources, including academic programs, professional development opportunities, and resources for lifelong learning. It explores the challenges and opportunities faced by educators, students, and professionals in this field, highlighting innovative approaches and best practices. The article concludes with recommendations for advancing hydrology and water resources education and training to meet the needs of a changing world.

Introduction

Hydrology and water resources are critical components of the earth system, shaping the natural environment and influencing the economic, social, and cultural activities of human societies. Hydrology is the study of water in the earth system, including its occurrence, distribution, movement, and quality. Water resources encompass the various sources of water that support human and ecosystem needs, including surface water, groundwater, and atmospheric water. Hydrology and water resources are central to addressing global challenges such as climate change, water scarcity, water pollution, and ecosystem degradation.

Education and training are essential for developing the knowledge and skills required to address these challenges and advance the science and practice of hydrology and water resources. Education refers to formal and informal learning opportunities, including academic programs, workshops, and seminars, that provide knowledge and skills to individuals. Training is a more specific form of education that focuses on developing practical skills for specific tasks or professions.

In this article, we provide a comprehensive overview of education and training in the field of hydrology and water resources, examining the current state of academic programs, professional development opportunities, and resources for lifelong learning. We also explore the challenges and opportunities faced by educators, students, and professionals in this field, highlighting innovative approaches and best practices. Finally, we conclude with recommendations for advancing hydrology and water resources education and training to meet the needs of a changing world.

Academic Programs

Academic programs provide a foundation for students to develop knowledge and skills in hydrology and water resources. These programs include undergraduate and graduate degrees, as well as certificate programs and continuing education courses. The following sections provide an overview of the types of academic programs available in hydrology and water resources.

Undergraduate Programs

Undergraduate programs in hydrology and water resources provide a broad foundation in the fundamentals of the field. These programs typically include coursework in hydrology, water resources management, water quality, and related topics. Students may also have the opportunity to gain practical experience through internships, research projects, or field courses. Some undergraduate programs may offer concentrations or specializations in specific areas, such as groundwater hydrology, watershed management, or aquatic ecology.

Graduate Programs

Graduate programs in hydrology and water resources provide more specialized training and research opportunities for students. These programs typically offer master's and doctoral degrees, as well as professional master's degrees for students interested in pursuing careers in water resources management or related fields. Graduate coursework often includes advanced topics in hydrology, water resources modelling, environmental fluid mechanics, and water policy. Students may also have the opportunity to conduct research in areas such as climate change impacts on water resources, hydrologic modelling, or groundwater management.

Certificate Programs

Certificate programs in hydrology and water resources are typically designed for professionals who want to gain specialized knowledge or skills in a specific area. These programs may be offered online or in-person and can vary in length from a few weeks to several months. Certificate programs may cover topics such as water quality management, watershed modelling, or groundwater remediation.

Continuing Education Courses

Continuing education courses provide opportunities for professionals to update their knowledge and skills in hydrology and water resources. These courses may be offered online or in-person and can range from a few hours to several days in length. Continuing education courses may cover topics such as water resources policy, hydrologic modelling, or emerging technologies in water management.

Professional Development Opportunities

Professional development opportunities are an essential component of education and training in hydrology and water resources. These opportunities provide professionals with the skills and knowledge required to advance their careers and stay current with emerging trends and technologies. The following sections provide an overview of the types of professional development opportunities available in hydrology and water resources.

Conferences and Workshops

Conferences and workshops provide opportunities for professionals to exchange ideas, present research, and learn about emerging trends and technologies in hydrology and water resources. These events may be organized by professional societies, academic institutions, or government agencies. Conferences and workshops may cover topics such as water resources management, hydrologic modelling, or water quality monitoring.

Professional Societies

Professional societies provide opportunities for professionals to network, collaborate, and access resources related to hydrology and water resources. These societies may offer conferences, workshops, and webinars, as well as publications, research grants, and awards. Professional societies in hydrology and water resources include the American Water Resources Association, the International Association of Hydrological Sciences, and the International Water Association.

Online Resources

Online resources such as webinars, podcasts, and online courses provide professionals with convenient and accessible opportunities for continuing education and professional development. Online resources may cover a wide range of topics, including water resources policy, hydrologic modelling, or water quality management. Some online resources are available free of charge, while others may require a fee.

Challenges and Opportunities

The field of hydrology and water resources faces numerous challenges and opportunities related to education and training. The following sections highlight some of these challenges and opportunities and explore innovative approaches and best practices for addressing them.

Challenge: Addressing the Changing Needs of the Field

Hydrology and water resources are rapidly evolving fields, driven by emerging technologies, changing environmental conditions, and evolving societal needs. Educators and trainers in this field must be able to adapt their curriculum and training programs to address these changes effectively. This challenge requires a flexible and dynamic approach to education and training, with an emphasis on interdisciplinary and problem-based learning.

Opportunity: Integrating Emerging Technologies

Advances in technology are revolutionizing the field of hydrology and water resources, providing new opportunities for monitoring, modelling, and managing water resources. Educators and trainers in this field must integrate emerging technologies into their curriculum and training programs to ensure that students and professionals are equipped with the skills and knowledge required to apply these technologies effectively. This opportunity requires a commitment to staying current with emerging technologies and incorporating them into education and training programs.

Challenge: Addressing the Needs of Diverse Populations

The field of hydrology and water resources must address the needs of a diverse population, including women, minorities, and underrepresented groups. Educators and trainers in this field must ensure that their curriculum and training programs are inclusive and accessible to all individuals. This challenge requires a commitment to diversity, equity, and inclusion, with an emphasis on culturally responsive teaching and learning.

Opportunity: Building Partnerships and Collaborations

Hydrology and water resources are interdisciplinary fields, requiring collaboration and partnership across multiple sectors and stakeholders. Educators and trainers in this field must build partnerships and collaborations with other disciplines, industries, and communities to advance knowledge and practice. This opportunity requires a commitment to building strong networks and partnerships across the water resources community.

Recommendations

To advance hydrology and water resources education and training, the following recommendations are offered:

- Develop interdisciplinary and problem-based learning opportunities that address emerging challenges and opportunities in the field.
- Integrate emerging technologies into education and training programs to ensure that students and professionals are equipped with the skills and knowledge required to apply these technologies effectively.
- Emphasize diversity, equity, and inclusion in curriculum and training programs, with a focus on culturally responsive teaching and learning.
- Build strong networks and partnerships across the water resources community to advance knowledge and practice.
- Support and promote continuing education opportunities for professionals in hydrology and water resources.
- Encourage and support the development of innovative approaches to education and training, including online resources and experiential learning opportunities.

Conclusion

Education and training are essential components of the field of hydrology and water resources, providing students and professionals with the skills and knowledge required to address emerging challenges and opportunities. The field of hydrology and water resources faces numerous challenges and opportunities related to education and training, including addressing changing needs, integrating emerging technologies, and promoting diversity, equity, and inclusion. To address these challenges and opportunities, educators and trainers must adopt innovative and interdisciplinary approaches, build strong partnerships and collaborations, and promote continuing education opportunities for professionals in the field. With these efforts, the field of hydrology and water resources can continue to advance knowledge and practice, ensuring sustainable and equitable water resources for future generations.

www.ingramcontent.com/pod-product-compliance
Lightning Source LLC
LaVergne TN
LVHW080012210926
848481LV00011B/3006

* 9 7 9 8 2 1 5 9 4 5 8 5 8 *